SOLUTIONS MANUAL

to accompany

PHYSICAL CHEMISTRY

SEVENTH EDITION

ROBERT A. ALBERTY
Professor of Chemistry
Massachusetts Institute of Technology

JOHN WILEY & SONS
New York Chichester Brisbane Sydney Toronto Singapore

ISBN 0 471 82900 5
Printed in the United States of America

10 9 8 7 6 5 4

PREFACE

This manual gives solutions of the problems in the first set of the text R. A. Alberty, PHYSICAL CHEMISTRY, 7th ed., Wiley, New York, 1987, and it gives answers for the second set.

Working problems is an important part of learning physical chemistry. Not all knowledge of physical chemistry is quantitative, but much of it is. Since physical chemistry utilizes physics and mathematics to predict and interpret chemical phenomena, there are many opportunities to use quantitative methods.

The availability of hand-held electronic calculators has made it much easier to work physical chemistry problems and has made it possible to include more difficult problems.

The units of physical quantities are usually shown in solving problems in this manual. It is important to develop the habit of using units and cancelling them to obtain the units for the answer because this helps prevent errors.

In this book, as in your handwritten lecture notes, there is no distinction between italic (sloping) and roman (upright) type, but it is important to note that in the printed literature italic type is used for symbols for physical quantities and roman type is used for units.

I am indebted to many physical chemists who have recommended problems and who have suggested improvements in this SOLUTIONS MANUAL.

Robert A. Alberty

Cambridge, Massachusetts
February 1986

C O N T E N T S

Part I
THERMODYNAMICS

Part II
QUANTUM CHEMISTRY

Part III
CHEMICAL DYNAMICS

Part IV
STRUCTURES

PART ONE

THERMODYNAMICS

CHAPTER 1: Zeroth Law of Thermodynamics and Equations of State

1.1 The perfect gas law also represents the behavior of mixtures of gases at low pressures. The volume is then interpreted as the molar volume of the mixture, that is the volume of a mole of the mixture. The partial pressure of gas i in a mixture is defined as y_iP, where y_i is its mole fraction, and P is the total pressure. Ten grams of N_2 is mixed with 5 g of O_2 and held at 25 °C at 0.750 bar. (a) What are the mole fractions of N_2 and O_2? (b) What are the partial pressures of N_2 and O_2? (c) What is the molar volume assuming the molecules do not interact? (d) What is the actual volume assuming the molecules do not interact?

SOLUTION

$$m_{N_2} = \frac{10\ \text{g}}{28.013\ \text{g mol}^{-1}} = 0.357\ \text{mol}$$

$$m_{O_2} = \frac{5\ \text{g}}{32{,}000\ \text{g mol}^{-1}} = 0.156\ \text{mol}$$

$$P_{N_2} = (0.357/0.513)(0.750\ \text{bar}) = 0.522\ \text{bar}$$

$$P_{O_2} = (0.156/0.513)(0.750\ \text{bar}) = 0.228\ \text{bar}$$

$$V = \frac{RT}{P} = \frac{(8.3144\ \text{J K}^{-1}\ \text{mol}^{-1})(298.15\ \text{K})}{0.750 \times 10^5\ \text{Pa}}$$

$$= 0.0331\ \text{m}^3\ \text{mol}^{-1}$$

The gas contains 0.357 + 0.156 = 0.513 mol and so the actual volume is (0.513 mol)(0.0331 m^3 mol^{-1}).

$$= 0.0170\ \text{m}^3 \quad \text{or} \quad 17.0\ \text{L}$$

1.2 Calculate the second virial coefficient of hydrogen at 0 °C from the fact that the molar volumes at 50.7, 101.3, 202.6, and 303.9 bar are 0.4634, 0.2386, 0.1271, and 0.09004 L mol^{-1}, respectively.

SOLUTION

$$\frac{PV}{RT} = 1 + \frac{B}{V} + \frac{C}{V^2} + \cdots$$

P/bar	50.7	101.3	202.6	303.9
V/L mol^{-1}	0.4634	0.2386	0.1271	0.09004
PV/RT	1.035	1.064	1.134	1.205
(1/V)/mol L^{-1}	2.158	4.191	7.868	11.106

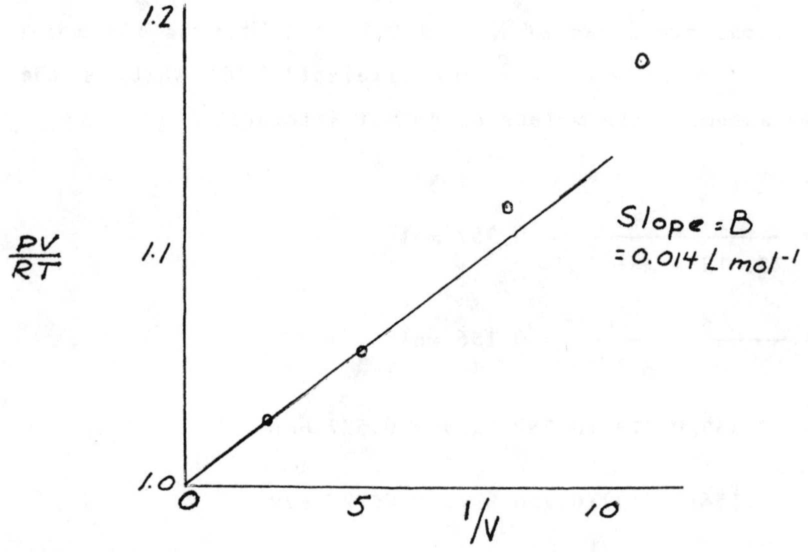

1.3 The second virial coefficient B of methyl isobutyl ketone is −1580 cm^3 mol^{-1} at 120 °C. Compare its compressibility factor at this temperature with that of a perfect gas at 1 bar.

SOLUTION

$$Z = \frac{PV}{RT} = 1 + \frac{B}{V} = 1 + \frac{BP}{RT}$$

In the B/V term, V may be replaced by RT/P if $Z \approx 1$, since the approximation is made in a small correction term. At 1 bar

$$Z = 1 + \frac{(-1.58\ \text{L mol}^{-1})(1\ \text{bar})}{(0.08314\ \text{L bar K}^{-1}\ \text{mol}^{-1})(393.15\ \text{K})}$$

$$= 0.952$$

The compressibility factor for a perfect gas is of course unity.

1.4 Using Fig. 1.4 calculate the compressibility factor Z for $NH_3(g)$ at 400 K and 50 bar.

SOLUTION

$$B = -110\ \text{cm}^3\ \text{mol}^{-1}$$

$$B' = \frac{B}{RT} = \frac{-(110\ \text{cm}^3\ \text{mol}^{-1})(10^{-3}\ \text{L cm}^{-3})}{(0.08314\ \text{L bar K}^{-1}\ \text{mol}^{-1})(400\ \text{K})}$$

$$= -3.31 \times 10^{-3}\ \text{bar}^{-1}$$

$$Z = 1 + B'P = 1 - (3.31 \times 10^{-3}\ \text{bar}^{-1})(50\ \text{bar})$$

$$= 0.835$$

1.5 Derive the expressions for van der Waals constants a and b in terms of the critical temperature and pressure; that is, derive equations 1.35 and 1.36 from 1.32 - 1.34.

SOLUTION

Equations 1.32 and 1.33 may be written

$$\frac{RT_c}{(V_c - b)^2} = \frac{2a}{V_c^{\,3}} \qquad (1)$$

$$\frac{2RT_c}{(V_c - b)^3} = \frac{6a}{V_c^{\,4}} \qquad (2)$$

Division of the first equation by the second yields

$$V_c = 3b \qquad (3)$$

Substitution of this expression in equation 1 yields

$$T_c = 8a/27Rb \qquad (4)$$

Substitution of equations 3 and 4 in equation 1.34 yields

$$P_c = a/27b^2 \qquad (5)$$

Since there are three relations (equations 3 - 5) between the van der Waals constants and the critical constants, a and b may be expressed in terms of T_c and P_c or T_c and V_c. Critical pressures are generally known more accurately than critical volumes, and so a and b are generally calculated using equations 1.35 and 1.36.

1.6 Calculate the second virial coefficient of methane at 400 K and 300 K from its van der Waals constants, and compare these results with Fig. 1.4.

SOLUTION

For methane a = 2.283 L^2 bar mol^{-2} and b = 0.04278 L mol^{-1}. From equation 1.28, the virial coefficient at 400 K is

$$B = b - a/RT = 0.04278 - 2.283/(0.083144)(400)$$
$$= -0.026 \text{ L mol}^{-1}$$

compared with -0.020 L mol^{-1} from Fig. 1.4. The calculated virial coefficient at 300 K is -0.048 L mol, compared with 0.050 L mol^{-1} in Fig. 1.4.

1.7 The critical temperature of carbon tetrachloride is 283.1 °C. The densities in grams per cubic centimeter of the liquid ρ_1 and vapor ρ_v at different temperatures are as follows:

t/°C	100	150	200	250	270	280
ρ_1	1.4343	1.3215	1.1888	0.9980	0.8666	0.7634
ρ_v	0.0103	0.0304	0.0742	0.1754	0.2710	0.3597

What is the critical molar volume of CCl_4? It is found that the mean of the densities of the liquid and vapor does not vary rapidly with temperature and can be represented by

$$\frac{\rho_1 + \rho_v}{2} = AT + B$$ where A and B are constants.

The extrapolated value of the average density at the critical temperature is the critical density. The molar volume V_c at the critical point is equal to the molar mass divided by the critical density.

SOLUTION

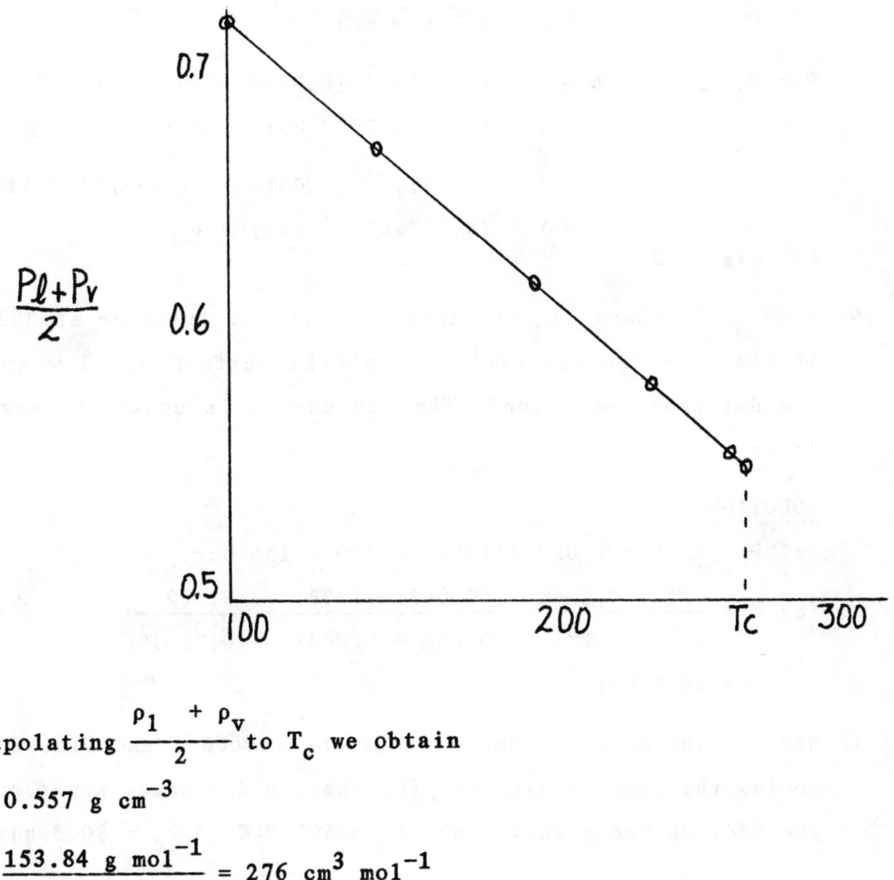

Extrapolating $\frac{\rho_1 + \rho_v}{2}$ to T_c we obtain

$$\rho_c = 0.557 \text{ g cm}^{-3}$$

$$V_c = \frac{153.84 \text{ g mol}^{-1}}{0.557 \text{ g cm}^{-3}} = 276 \text{ cm}^3 \text{ mol}^{-1}$$

1.8 Use the van der Waals constants for H_2 and O_2 in Table 1.3 to calculate the initial slopes of the plots of the compressibility factor Z versus P.

SOLUTION

$$Z = 1 + \frac{1}{RT}\left(b - \frac{a}{RT}\right) P + \ldots$$

At 273.15 K, $RT = (0.083144 \text{ L bar K}^{-1} \text{ mol}^{-1})(273.15 \text{ K})$
$= 22.71 \text{ L bar mol}^{-1}$

For H_2, $Z = 1 + (22.71)^{-1} [0.0266 - (0.2476/22.71)] P + \ldots$
$= 1 + (6.92 \times 10^{-4} \text{ bar}^{-1}) P + \ldots$

For O_2, $Z = 1 + (22.71)^{-1} [0.0318 - (1.378/22.71)] P + \ldots$
$= 1 - (4.07 \times 10^{-3} \text{ bar}) P + \ldots$

See Fig. 1.2

1.9 A mole of ethane is contained in a 200 ml-cylinder at 373 K. What is the pressure according to (a) the perfect gas law and (b) the van der Waals equation? The van der Waals constants are given in Table 1.3.

SOLUTION

(a) $P = RT/V = (.08314)(373)/0.200 = 155 \text{ bar}$

(b) $$P = \frac{RT}{V - b} - \frac{a}{V^2} = \frac{(0.08314)(373)}{0.200 - 0.0638} - \frac{5.562}{(0.200)^2}$$
$= 88.6 \text{ bar}$

1.10 What is the molar volume of n-hexane at 660 K and 91 bar according to (a) the perfect gas law, (b) the van der Waals equation, and (c) the Redlich-Kwong equation? $T_c = 507.7$ K $P_c = 30.3$ bar

SOLUTION

(a) $$V = \frac{RT}{P} = \frac{(0.08314 \text{ L bar K}^{-1} \text{ mol}^{-1})(660 \text{ K})}{91 \text{ bar}} = 0.603 \text{ L mol}^{-1}$$

$$(b)\ a = \frac{27\ R^2\ T_c^{\ 2}}{64\ P_c} = \frac{(27)(0.08314)^2(507.7)^2}{(64)(30.3)}$$

$$= 24.81\ L^2\ \text{bar mol}^{-2}$$

$$b = \frac{RT_c}{8\ P_c} = \frac{(0.08314)(507.7)}{8(30.3)}$$

$$= 0.174\ L\ \text{mol}^{-1}$$

$$P = \frac{RT}{V - b} - \frac{a}{V^2}$$

$$91\ \text{bar} = \frac{(0.08314)(660\ K)}{V - 0.174} - \frac{24.81}{V^2}$$

Rather than solving a cubic equation, substituting successive values of V shows that

$$V = 0.39\ L\ \text{mol}^{-1}.$$

$$(c)\ a = 0.4275\ R^2\ T_c^{\ 2.5}/P_c = 0.4275(.08314)^2(507.7)^{2.5}/30.3$$

$$= 566.4\ L^2\ \text{bar}\ K^{1/2}\ \text{mol}^{-2}$$

$$b = 0.0866\ RT_c/P_c = (0.0866)(.08314)(507.7)/30.3$$

$$= 0.121\ L\ \text{mol}^{-1}$$

$$P = \frac{RT}{V - b} - \frac{a}{T^{1/2}\ V(V + b)}$$

$$= \frac{(.08314)(660)}{V - 0.121} - \frac{566.4}{(660)^{1/2}\ V(V + 0.121)}$$

Trying successive values of V shows that $V = 0.40\ L\ \text{mol}^{-1}$.

1.11 Derive equation 1.25 that gives the boundary of the physically unrealizable region for a van der Waals gas and show that the maximum pressure satisfying this equation is the critical pressure.

SOLUTION

$$P = \frac{RT}{V - b} - \frac{a}{V^2}$$

$$\left(\frac{\partial P}{\partial V}\right)_T = 0 = -\frac{RT}{(V-b)^2} + \frac{2a}{V^3}$$

Eliminating RT between these equations yields

$$P = \frac{a}{V^2}\left(1 - \frac{2b}{V}\right)$$

The maximum pressure according to this equation is obtained by differentiating with respect to V.

$$\frac{dP}{dV} = 0 = -\frac{2a}{V^3} + \frac{6ab}{V^4} \quad \text{so that} \quad b = \frac{V_c}{3} = \frac{RT_c}{8P_c}$$ where equation 1.37 has been used.

1.12 When pressure is applied to a liquid, its volume decreases. Assuming that the isothermal compressibility $\kappa = -\frac{1}{V}\left(\frac{\partial V}{\partial P}\right)_T$ is independent of pressure, derive an expression for the volume as a function of pressure.

SOLUTION

$$\int_{V_1}^{V_2} \frac{dV}{V} = -\kappa \int_{P_1}^{P_2} dP$$

$$\ln \frac{V_2}{V_1} = -\kappa (P_2 - P_1)$$

$$V_2 = V_1 e^{-\kappa(P_2 - P_1)}$$

1.13 The coefficient of cubic expansion coefficient α is defined by $\alpha = \left(\frac{1}{V}\frac{\partial V}{\partial T}\right)_P$ and the isothermal compressibility is defined by

$\kappa = \left(\frac{1}{V}\frac{\partial V}{\partial P}\right)_T$ Calculate these quantities for a perfect gas.

SOLUTION

$$V = RT/P$$

$$\left(\frac{\partial V}{\partial T}\right)_P = \frac{R}{P}$$

$$\alpha = 1/T$$

$$\left(\frac{\partial V}{\partial P}\right)_T = \frac{RT}{P^2}$$

$$\kappa = \frac{RT}{VP^2} = 1/P$$

1.14 Assuming that the atmosphere is isothermal at 0 °C and that the average molar mass of air is 29 g mol^{-1}, calculate the atmospheric pressure at 20,000 ft above sea level.

SOLUTION

$$h = (2.0 \times 10^4 \text{ ft})(12 \text{ in ft}^{-1})(2.54 \text{ cm in}^{-1})(10^{-2} \text{ m cm}^{-1}) = 6096 \text{ m}$$

$$P = P_o e^{-gMh/RT}$$

$$P = (1.013 \text{ bar}) \exp\left[\frac{-(9.8 \text{ m s}^{-2})(29 \times 10^{-3} \text{ kg mol}^{-1})(6096 \text{ m})}{(8.314 \text{ J K}^{-1} \text{ mol}^{-1})(273 \text{ K})}\right]$$

$$= 0.472 \text{ bar}$$

1.15 Calculate the pressure and composition of air on the top of Mt. Everest assuming the atmosphere has a temperature of 0 °C independent of altitude (h = 29,141 ft).

SOLUTION

$$h = (29{,}141 \text{ ft})(12 \text{ in ft}^{-1})(2.54 \text{ cm in}^{-1})(0.01 \text{ m cm}^{-1}) = 8{,}882 \text{ m}$$

For O_2, $P = (0.2 \text{ bar}) \exp(-9.8 \times 32 \times 10^{-3} \times 8882/8.314 \times 273)$
$= 0.059 \text{ bar}$

For N_2, $P = (0.8 \text{ bar}) \exp(-9.8 \times 28 \times 10^{-3} \times 8882/8.314 \times 273)$
$= 0.274 \text{ bar}$

The total pressure is 0.333 bar, and $y_{O_2} = 0.177$ and $y_{N_2} = 0.823$.

1.16 $B = -K_c$

1.17 21.7 cm^3 mol^{-1}

1.18 $275.9\ cm^3\ mol^{-1}$

1.19 $\kappa = 0.28$

1.20 $T_c \approx 435$ K (actual 425 K)

$V_c = 0.261$ L mol^{-1} (actual 0.255 L mol^{-1})

$P_c = 39$ bar (actual 38.0 bar)

1.22 0.324 nm

1.23 -1.96×10^{-3} and -0.60×10^{-4} bar^{-1}

1.24 (a) 19.8 (b) 1621 (c) 2739 bar

1.25 (a) 100 (b) 152 bar

1.26 $\alpha = 1/T(1 + bP/RT)$ $\kappa = 1/P(1 + bP/RT)$

1.27 $V = V_o \exp[\alpha(T - T_o)]$

$V = V_o [1 + \alpha(T - T_o)]$

1.28 0.843 bar

1.29 3.27×10^{-9} bar $y_{O_2} = 0.015$ $y_{N_2} = 0.985$

CHAPTER 2: First Law of Thermodynamics

2.1 How much work is done when a person weighing 75 kg (165 lbs) climbs the Washington monument, 555 ft high? How many kilojoules must be supplied to do this muscular work, assuming that 25% of the energy produced by the oxidation of food in the body can be converted into muscular mechanical work?

SOLUTION

$$w = mgh$$

$$\text{work} = (\text{mass})(\text{acceleration of gravity})(\text{height})$$

$$= (75\ \text{kg})(9.806\ \text{m s}^{-2})(555\ \text{ft})(12\ \text{in ft}^{-1})(2.54 \times 10^{-2}\ \text{m in}^{-1})$$

$$= 124.4\ \text{kJ}$$

The energy needed is four times greater than the work done.

$$E = 4(124.4\ \text{kJ})$$

$$= 497.6\ \text{kJ}$$

2.2 The surface tension of water is 71.97×10^{-3} N m^{-1} or 71.97×10^{-3} J m^{-2} at 25 °C. Calculate the surface energy in joules of 1 mol of water dispersed as a mist containing droplets 1 μm (10^{-4} cm) in radius. The density of water may be taken as 1.00 g cm^{-3}.

SOLUTION

The ratio of area to volume for a droplet is the same as the ratio of area to volume for the whole 18 cm^3.

$$V = \frac{4}{3}\pi r^3 = \frac{4}{3}\pi(10^{-6}\ \text{m})^3$$

$$A = 4\pi r^2 = 4\pi(10^{-6}\ \text{m})^2$$

$$\frac{\text{Total area}}{18 \times 10^{-6}\ \text{m}^3} = \frac{4\pi(10^{-6}\ \text{m})^2}{\frac{4}{3}\pi(10^{-6}\ \text{m})^3} \qquad \text{Total area} = 54\ \text{m}^2$$

Surface energy = $(54\ m^2)(71.97 \times 10^{-3}\ J\ m^{-2})$

$= 3.89\ J$

2.3 A mole of sodium metal is added to water. How much work is done on the atmosphere by the subsequent reaction if the temperature is 25 °C?

SOLUTION

$Na(s) + H_2O(l) = NaOH(soln) + 1/2\ H_2(g)$

$V = RT/P = (8.314\ J\ K^{-1}\ mol^{-1})(298\ K)/(101325\ Pa)$

$= 0.0245\ m^3\ mol^{-1}$

$w = -P\Delta V = -(101325\ Pa)(0.0123\ m^3\ mol^{-1})$

$= -1.24\ kJ\ mol^{-1}$

2.4 One mole of nitrogen at 25 °C and 1 bar is expanded reversibly and isothermally to a pressure of 0.132 bar. (a) What is the value of w? (b) What is the value of w if the nitrogen is expanded against a constant pressure of 0.132 bar?

SOLUTION

(a) $w = RT\ \ln(P_2/P_1)$

$= (8.314\ J\ K^{-1}\ mol^{-1})(298.15\ K)\ \ln 0.132$

$= -5.03\ kJ\ mol^{-1}$

(b) $V_1 = RT/P_1 = (8.314)(298.15)/101325 = 0.0245\ m^3$

$V_2 = (8.314)(298.15)/(0.132)(101325) = 0.185\ m^3$

$w = -P\Delta V = -\ (0.132 \times 10^5)(0.185 - 0.0245)$

$= -2.12\ kJ\ mol^{-1}$

2.5 Calculate the work done on the gas in the reversible isothermal (298.15 K) expansion of a mole of N_2 from 1 L to 10 L assuming it is (a) an ideal gas and (b) a van der Waals gas (See Table 1.3).

SOLUTION

(a) $w = -RT \ln \frac{V_2}{V_1} = -(8.314)(298.15) \ln 10$

$= -5.71 \text{ kJ mol}^{-1}$

(b) $w = -RT \ln \frac{V_2 - b}{V_1 - b} + a \left(\frac{1}{V_1} - \frac{1}{V_2}\right)$

We can leave V in L mol^{-1} in the first term because this unit cancels, but we must convert a to SI base units and express V_1 and V_2 in SI units in the second term.

$a = (1.408 \text{ L}^2 \text{ bar mol}^{-2})(10^5 \text{ Pa bar}^{-1})(10^{-3} \text{ m}^3 \text{ L}^{-1})^2$

$= 0.1408 \text{ m}^3 \text{ Pa mol}^{-2}$

$w = -(8.314)(298.15) \ln \frac{10 - 0.0391}{1 - 0.0391} + 0.1408 \left(\frac{1}{10^{-3}} - \frac{1}{10^{-2}}\right)$

$= -5.67 \text{ kJ mol}^{-1}$

2.6 Are the following expressions exact differentials?

(a) $xy^2 dx - x^2 y\, dy$ (b) $\frac{dx}{y} - \frac{x}{y^2} dy$

SOLUTION

(a) Taking the cross derivatives

$\frac{\partial xy^2}{\partial y} = 2xy$ $\frac{\partial(-x^2 y)}{\partial x} = -2xy$

Since the cross derivatives are unequal, the expression is not an exact differential.

(b) $\frac{\partial(1/y)}{\partial y} = -\frac{1}{y^2}$ $\frac{\partial(-x/y^2)}{\partial x} = -\frac{1}{y^2}$

Since the cross derivatives are equal, the expression is an exact differential.

2.7 Over narrow ranges of temperature and pressure, the differential expression for the volume of a fluid as a function of temperature and pressure can be integrated to obtain

$$V = Ke^{\alpha T} e^{-\kappa P}$$

(α and κ are defined in Section 1.10). Show that V is a state function.

SOLUTION

The test of a state function is that it forms exact differentials. For an exact differential the mixed second derivatives are equal.

$$\frac{\partial V}{\partial T} = Ke^{-\kappa P} e^{\alpha T} \alpha$$

$$\frac{\partial}{\partial P}\left(\frac{\partial V}{\partial T}\right) = Ke^{\alpha T} e^{-\kappa P} (-\kappa\alpha)$$

$$\frac{\partial V}{\partial P} = Ke^{\alpha T} e^{-\kappa P} (-\kappa)$$

$$\frac{\partial}{\partial T}\left(\frac{\partial V}{\partial P}\right) = Ke^{\alpha T} e^{-\kappa P} (-\kappa\alpha)$$

Thus V for a substance that can be described in this way is a state function.

2.8 A mole of liquid water is vaporized at 100 °C and 1.013 bar. The heat of vaporization is 40.69 kJ mol^{-1}. What are the values of (a) w_{rev}, (b) q, (c) ΔU, and (d) ΔH?

SOLUTION

(a) Assuming that water vapor is a perfect gas and that the volume of liquid water is negligible,

$$w = -P\Delta V = -RT$$
$$= -(8.314 \times 10^{-3} \text{ kJ K}^{-1} \text{ mol}^{-1})(373.15 \text{ K})$$
$$= -3.10 \text{ kJ mol}^{-1}$$

(b) The heat of vaporization is 40.69 kJ mol^{-1}, and, since heat is absorbed, q has a positive sign.

$$q = 40.69 \text{ kJ mol}^{-1}$$

(c) $\Delta U = q + w$

$$= (40.69 - 3.10) \text{ kJ mol}^{-1}$$

$$= 37.59 \text{ kJ mol}^{-1}$$

(d) $\Delta H = \Delta U + \Delta(PV) = \Delta U + P\Delta V$

$$= \Delta U + RT$$

$$= 37.59 \text{ kJ mol}^{-1} + (8.314 \times 10^{-3} \text{ kJ K}^{-1} \text{ mol}^{-1})(373.15 \text{ K})$$

$$= 40.69 \text{ kJ mol}^{-1}$$

2.9 Calculate $H^\circ(2000\text{ K}) - H^\circ(0\text{ K})$ for H(g).

SOLUTION

Equation 2.50 may be integrated to obtain

$$H^\circ(T_2) - H^\circ(T_1) = \int_{T_1}^{T_2} C_P^\circ \, dT$$

$$H^\circ(2000\text{ K}) - H^\circ(0\text{ K}) = \int_0^{2000} \frac{5}{2} R \, dT = \frac{5}{2} R(2000) = 41.572 \text{ kJ mol}^{-1}$$

Table A.2 yields $6.197 + 35.376 = 41.573 \text{ kJ mol}^{-1}$

(Note that for O(g) a slightly higher value is obtained because there is some absorption of heat by excitation to higher electronic levels.)

2.10 Considering H_2O to be a rigid nonlinear molecule, what value of C_P for the gas would be expected classically? If vibration is taken into account, what value is expected? Compare these values of C_P with the actual values of 298 and 3000 K in Table A.1.

SOLUTION

A rigid molecule has translational and rotational energy. The translational contribution to C_V is

$\frac{3}{2}$ R = 12.471 J K^{-1} mol^{-1}. Since H_2O is a nonlinear molecule, it has three rotational degrees of freedom, and so the rotational contribution to C_V is $\frac{3}{2}$ R = 12.471 J K^{-1} mol^{-1}. Thus C_P for the rigid molecule is $C_P = C_V + R = 33.258$ J K^{-1} mol^{-1}

Since H_2O is a nonlinear molecule, the number of vibrational degrees of freedom is 3N − 6 = 3. Since each vibrational degree of freedom contributes R to the heat capacity, classical theory predicts

$$C_P = 33.258 \text{ J K}^{-1} \text{ mol}^{-1} + 3R$$
$$= 58.201 \text{ J K}^{-1} \text{ mol}^{-1}$$

The experimental value of C_P at 298 K is 33.577 J K^{-1} mol^{-1}, which is only slightly higher than the value expected for a rigid molecule. The experimental value of C_P at 3000 K is 55.664 J K^{-1} mol^{-1}, which is only slightly less than the classical expectation for a vibrating water molecule.

2.11 The heat capacities of a gas may be represented by $C_P = \alpha + \beta T + \gamma T^2$

For N_2 $\alpha = 26.984$, $\beta = 5.910 \times 10^{-3}$, and $\gamma = -3.377 \times 10^{-7}$, when C_P is expressed in J K^{-1} mol^{-1}. How much heat is required to heat a mole of N_2 from 300 K to 1000 K?

SOLUTION

$$q = \int_{300}^{1000} (26.984 + 5.910 \times 10^{-3}\, T - 3.377 \times 10^{-7}\, T^2)dT$$

$$= 26.984(1000 - 300) + \frac{1}{2}(5.910 \times 10^{-3})(1000^2 - 300^2) - \frac{1}{3}(3.377 \times 10^{-7})(1000^3 - 300^3)$$

$$= 21.468 \text{ kJ mol}^{-1}$$

2.12 One mole of methane (considered to be a perfect gas) initially at 25 °C and 1 bar pressure is heated at constant pressure until the volume has doubled. The variation of the molar heat capacity with absolute temperature is given by $C_P = 22.34 + 48.1 \times 10^{-3}\ T$ where C_P is in J K^{-1} mol^{-1}. Calculate (a) ΔH, and (b) ΔU.

SOLUTION

$$\Delta H = \int_{298.1\ K}^{596.2\ K} (22.34 + 48.1 \times 10^{-3}\ T)dT$$

$$= (22.34)(298.1) + [(48.1 \times 10^{-3})/2](596.2^2 - 298.1^2)$$

$$= 13.07\ kJ\ mol^{-1}$$

$$\Delta U = \Delta H - P\Delta V = 13.07 - P(RT/P)$$

$$= 13.07 - (8.314 \times 10^{-3}\ kJ\ K^{-1}\ mol^{-1})(298.1\ K)$$

$$= 10.59\ kJ\ mol^{-1}$$

2.13 One mole of perfect gas at 25 °C and 100 bar is allowed to expand reversibly and isothermally to 5 bar. Calculate (a) the work done on the gas in joules, (b) the heat absorbed in joules, (c) ΔU, and (d) ΔH.

SOLUTION

(a) $w = RT \ln (P_2/P_1)$

$= (8.314)(298) \ln (5/100)$

$= -7.42\ kJ\ mol^{-1}$

(b) $q = -w = 7.42\ kJ\ mol^{-1}$

(c) $\Delta U = 0$ since the gas is perfect

(d) $\Delta H = \Delta U + \Delta(PV) = 0$

2.14 Calculate the temperature increase and final pressure of helium if a mole is compressed adiabatically and reversibly from 44.8 L at 0 °C to 22.4 L.

SOLUTION

$$\gamma = C_P/C_V = (\tfrac{5}{2}\ R)/(\tfrac{3}{2}\ R) = \tfrac{5}{3}$$

$$\frac{T_1}{T_2} = \left(\frac{V_2}{V_1}\right)^{\gamma - 1}$$

$$T_2 = (273.25\ K)(44.8\ L/22.4\ L)^{2/3}$$

$$= 433.6\ K \quad \text{or} \quad 160.4\ ^\circ C$$

Thus the temperature increase is 160.4 °C. The final pressure is given by

$$P = \frac{RT}{V} = \frac{(0.08314\ L\ bar\ K^{-1}\ mol^{-1})\ (433.6\ K)}{22.4\ L\ mol^{-1}}$$

$$= 1.609\ bar$$

2.15 A mole of argon is allowed to expand adiabatically from a pressure of 10 bar and 298.15 K to 1 bar. What is the final temperature and how much work can be done?

SOLUTION

$$\gamma = C_P/C_V = (\tfrac{5}{2}\ R)/(\tfrac{3}{2}\ R) = \tfrac{5}{3} \qquad (\gamma - 1)/\gamma = 2/5$$

$$\frac{T_1}{T_2} = \left(\frac{P_1}{P_2}\right)^{(\gamma - 1)/\gamma}$$

$$T_2 = (298.15\ K)(1/10)^{2/5} = 118.70\ K$$

$$w = \int_{T_1}^{T_2} C_V\ dT = \tfrac{3}{2}\ R(T_2 - T_1)$$

$$= \tfrac{3}{2}\ (8.314\ J\ K^{-1}\ mol^{-1})(118.70\ K - 298.15\ K) = 2238\ J\ mol^{-1}$$

Thus the maximum work that can be done on the surroundings is 2238 J mol^{-1}.

2.16 A tank contains 20 liters of compressed nitrogen at 10 bar and 25 °C. Calculate the maximum work (in joules) which can be obtained when gas is allowed to expand to 1 bar pressure (a) isothermally and (b) adiabatically.

SOLUTION

(a) For the isothermal expansion of 1 mol

$$w_{rev} = RT \ln \frac{P_2}{P_1}$$

$$= (8.314 \text{ J K}^{-1} \text{ mol}^{-1})(298.15 \text{ K}) \ln (1/10)$$

$$= -5708 \text{ J mol}^{-1}$$

There are

$(10 \text{ bar})(20 \text{ L})/(0.08314 \text{ L bar K}^{-1} \text{ mol}^{-1})(298 \text{ K}) = 8.07 \text{ mol}$.

Therefore the maximum work done on the surroundings is 46.1 kJ.

(b) For the adiabatic expansion we will assume that $\gamma = C_P/C_V$ has the value it has at room temperature. From Table A.2

$$\gamma = 29.125/(29.125 - 8.314) = 1.399$$

$$\frac{T_1}{T_2} = \left(\frac{P_1}{P_2}\right)^{(\gamma - 1)/\gamma} \qquad T_2 = (298.15 \text{ K})(1/10)^{0.285} = 154.7 \text{ K}$$

$$w = \int_{T_1}^{T_2} C_V \, dT = C_V(T_2 - T_1)$$

$$= (20.811 \text{ J K}^{-1} \text{ mol}^{-1})(154.7 - 298.15 \text{ K})$$

$$= -2.99 \text{ kJ mol}^{-1}$$

For 8.07 moles the maximum work done on the surroundings is 24.1 kJ.

2.17 In an adiabatic calorimeter, oxidation of 0.4362 gram of naphthalene caused a temperature rise of 1.707 °C. The heat capacity of the calorimeter and water was 10,290 J K^{-1}. If corrections for oxidation of the wire and residual nitrogen are neglected, what is the enthalpy of combustion of naphthalene per mole?

SOLUTION

$$\Delta H = \frac{(10{,}290\ \text{J K}^{-1})(1.707\ \text{K})(128.19\ \text{g mol}^{-1})}{(0.4362\ \text{g})(1000\ \text{J kJ}^{-1})}$$

$$= -5163\ \text{kJ mol}^{-1}$$

2.18 The following reactions might be used to power rockets.

(1) $H_2(g) + \frac{1}{2}\,O_2(g) = H_2O(g)$

(2) $CH_3OH(1) + 1\frac{1}{2}\,O_2(g) = CO_2(g) + 2H_2O(g)$

(3) $H_2(g) + F_2(g) = 2HF(g)$

(a) Calculate the enthalpy changes at 25 °C for each of these reactions per kilogram of reactants.

(b) Since the thrust is greater when the molar mass of the exhaust gas is lower, divide the heat per kilogram by the molar mass of the product (or the average molar mass in the case or reaction 2) and arrange the above reactions in order of effectiveness on the basis of thrust.

SOLUTION

(a) (1) $\Delta H = -241.818\ \text{kJ mol}^{-1}$

$= (-241.818\ \text{kJ mol}^{-1})(1000\ \text{g kg}^{-1})/(18\ \text{g mol}^{-1})$

$= -13.4\ \text{MJ kg}^{-1}$

(2) $\Delta H = -393.509 + 2(-241.818) + 238.66$

$= -638.49\ \text{kJ mol}^{-1}$

$= (-638.49\ \text{kJ mol}^{-1})(1000\ \text{g kg}^{-1})/(80\ \text{g mol}^{-1})$

$= -7.98\ \text{MJ kg}^{-1}$

(3) $\Delta H = 2(-271.1) = -542.2 \text{ kJ mol}^{-1}$

$= (-542.2 \text{ kJ mol}^{-1})(1000 \text{ g kg}^{-1})/(40 \text{ g mol}^{-1})$

$= -13.6 \text{ MJ kg}^{-1}$

(b) (1) $-13.4/18 = -0.744$

(2) $$\frac{-7.98}{\frac{1}{3}(44 + 2 \times 18)} = -0.299$$

(3) $-13.6/20 = -0.680$

(1) > (3) > (2)

2.19 Calculate the enthalpy of formation of PCl_5(cr), given the heats of the following reactions at 25 °C.

$2P(cr) + 3Cl_2(g) = 2PCl_3(l) \quad \Delta H^\circ = -635.13 \text{ kJ mol}^{-1}$

$PCl_3(l) + Cl_2(g) = PCl_5(cr) \quad \Delta H^\circ = -137.28 \text{ kJ mol}^{-1}$

SOLUTION

Multiplying the second reaction by 2 and adding the two reactions yields

$2P(cr) + 5Cl_2(g) = 2PCl_5(cr) \quad \Delta H^\circ = -909.69 \text{ kJ mol}^{-1}$

$\Delta H_f^\circ[PCl_5(cr)] = (-909.69 \text{ kJ mol}^{-1})/2$

$= -454.85 \text{ kJ mol}^{-1}$

2.20 Calculate ΔH° for the dissociation $O_2(g) = 2O(g)$ at 0, 298, and 3000 K. In Section 14.2 the enthalpy change for dissociation at 0 K will be found to be equal to the spectroscopic dissociation energy D_0.

SOLUTION

$\Delta H^\circ \text{ (0 K)} = 2(246.785) = 493.570 \text{ kJ mol}^{-1}$

$\Delta H^\circ \text{ (298 K)} = 2(249.170) = 498.340 \text{ kJ mol}^{-1}$

ΔH° (3000 K) = 2(256.722) = 513.444 kJ mol^{-1}

The spectroscopic dissociation energy of O_2 is given as 5.115 eV in Table 15.4. This can be converted to kJ mol^{-1} by multiplying by 96.485 kJ V^{-1} mol^{-1} to obtain 493.521 kJ mol^{-1}.

2.21 Compare the enthalpies of combustion of $CH_4(g)$ to $CO_2(g)$ and $H_2O(g)$ at 298 and 2000 K.

$$CH_4(g) + 2O_2(g) = CO_2(g) + 2H_2O(g)$$

<u>SOLUTION</u>

$\Delta H^\circ(298\ K) = -393.522 + 2(-241.827) - (-74.873)$

$= -802.303$ kJ mol^{-1}

$\Delta H^\circ(2000\ K) = -396.639 + 2(-251.668) - (-92.462)$

$= -807.513$ kJ mol^{-1}

2.22 Calculate ΔH°_{298} for

$H_2(g) + F_2(g) = 2HF(g)$

$H_2(g) + Cl_2(g) = 2HCl(g)$

$H_2(g) + Br_2(g) = 2HBr(g)$

$H_2(g) + I_2(g) = 2HI(g)$

<u>SOLUTION</u>

(a) 2(-271.1) = -542.2 kJ mol^{-1}

(b) 2(-92.31) = -184.62 kJ mol^{-1}

(c) 2(-36.40) - 30.91 = -103.71 kJ mol^{-1}

(d) 2(26.48) - 62.44 = -9.48 kJ mol^{-1}

2.23 Methane may be produced from coal in a process represented by the following steps, where coal is approximated by graphite:

$$2C(cr) + 2H_2O(g) = 2CO(g) + 2H_2(g)$$

$$CO(g) + H_2O(g) = CO_2(g) + H_2(g)$$

$$CO(g) + 3H_2(g) = CH_4(g) + H_2O(g)$$

The sum of these three reactions is

$$2C(cr) + 2H_2O(g) = CH_4(g) + CO_2(g)$$

What is ΔH° at 500 K for each of these reactions? Check that the sum of the ΔH°'s of the first three reactions is equal to ΔH° for the fourth reaction. From the standpoint of heat balance would it be better to develop a process to carry out the overall reactions in three separate reactors, or in a single reactor?

SOLUTION

$$\Delta H^\circ_{500} = 2(-110.02) - 2(-243.83) = 267.62 \text{ kJ mol}^{-1}$$

$$\Delta H^\circ_{500} = -393.68 - (-110.02) - (-243.83) = -39.83 \text{ kJ mol}^{-1}$$

$$\Delta H^\circ_{500} = -80.82 - 243.83 - (-110.02) = -214.63 \text{ kJ mol}^{-1}$$

$$\Delta H^\circ_{500} = -80.82 - 393.68 - 2(-243.83) = 13.16 \text{ kJ mol}^{-1}$$

Since the first reaction is very endothermic, there is an advantage in carrying the subsequent reactions out in the same reactor so that they can provide heat.

2.24 Compare the enthalpy of combustion of $CH_4(g)$ to $CO_2(g)$ and $H_2O(l)$ at 298 K with the sum of the enthalpies of combustion of graphite and $2H_2(g)$, from which $CH_4(g)$ can in principle be produced.

SOLUTION

$$CH_4(g) + 2O_2(g) = CO_2(g) + 2H_2O(l)$$

$$\Delta H^\circ(298 \text{ K}) = -393.51 + 2(-285.83) - (-74.81) = -890.36 \text{ kJ mol}^{-1}$$

$$2H_2(g) + O_2(g) = 2H_2O(l) \qquad \Delta H^\circ = -571.66 \text{ kJ mol}^{-1}$$

$$C(\text{graphite}) + O_2(g) = CO_2(g) \qquad \Delta H^\circ = -393.51 \text{ kJ mol}^{-1}$$

The sum of the enthalpy changes for the last two reactions (-965.17 kJ mol^{-1}) is more negative than the enthalpy change for the first reaction by the enthalpy of formation of $CH_4(g)$.

2.25 Calculate the heat of hydration of $Na_2SO_4(s)$ from the integral heats of solution of $Na_2SO_4(s)$ and $Na_2SO_4 \cdot 10\ H_2O(s)$ in infinite amounts of H_2O, which are -2.34 kJ mol^{-1} and 78.87 kJ mol^{-1}, respectively. Enthalpies of hydration cannot be measured directly because of the slowness of the phase transition.

SOLUTION

$$Na_2SO_4(s) = Na_2SO_4(ai) \qquad \Delta H^\circ = -2.43 \text{ kJ mol}^{-1}$$

$$Na_2SO_4(ai) = Na_2SO_4 \cdot 10\ H_2O(s) \qquad \Delta H^\circ = -78.87 \text{ kJ mol}^{-1}$$

$$Na_2SO_4(s) + 10\ H_2O(l) = Na_2SO_4 \cdot 10\ H_2O(s) \qquad \Delta H^\circ = -81.21 \text{ kJ mol}^{-1}$$

2.26 Calculate the integral heat of solution of one mole of HCl(g) in 200 $H_2O(l)$.

$$HCL(g) + 200\ H_2O(l) = HCl \text{ in } 200\ H_2O$$

SOLUTION

$$\Delta H^\circ_{298} = \Delta_f H^\circ\ [HCl \text{ in } 200\ H_2O] - \Delta_f H^\circ [HCl(g)]$$

$$= -166.272 - (-92.307) = -73.965 \text{ kJ mol}^{-1}$$

2.27 Calculate the enthalpies of reaction at 25 °C for the following reactions in dilute aqueous solutions:

(a) $HCl(ai) + NaBr(ai) = HBr(ai) + NaCl(ai)$

(b) $CaCl_2(ai) + Na_2CO_3(ai) = CaCO_3(s) + 2\ NaCl(ai)$

SOLUTION

(a) $\Delta H^\circ = 0$ because all of the reactants and products are completely ionized.

$$H^+ + Cl^- + Na^+ + Br^- = H^+ + Br^- + Na^+ + Cl^-$$

(b) $Ca^{2+}(ai) + CO_3^{2-}(ai) = CaCO_3(s)$

$$\Delta H^\circ = -1206.92 - (-542.83) - (-677.14) = 13.05 \text{ kJ mol}^{-1}$$

2.28 Calculate $\Delta H^\circ(298.15 \text{ K})$ for the solution process

$$CaCl_2 \cdot 6\ H_2O(cr) = CaCl_2 \text{ in } 400\ H_2O + 6\ H_2O(l)$$

For $CaCl_2 \cdot H_2O(cr)$, $\Delta_f H^\circ = -2607.9 \text{ kJ mol}^{-1}$ and for $CaCl_2$ in 400 H_2O, $\Delta_f H^\circ = -874.982 \text{ kJ mol}^{-1}$.

SOLUTION

$$\Delta H^\circ = -874.982 + 6(-285.830) - (-2607.9) = 17.9 \text{ kJ mol}^{-1}$$

Notice that the water of hydration is included explicitly in the calculation, but the water of dilution is not.

2.29 What is the heat evolved in freezing water at −10 °C given that

$H_2O(l) = H_2O(cr)$ $\quad \Delta H^\circ(273 \text{ K}) = -6004 \text{ J mol}^{-1}$ and

$C_p(H_2O,l) = 75.3 \text{ J K}^{-1} \text{ mol}^{-1}$ and $C_p(H_2O,s) = 36.8 \text{ J K}^{-1} \text{ mol}^{-1}$

SOLUTION

$$\Delta H^\circ(263 \text{ K}) = H^\circ(273 \text{ K}) + [C_{P,H_2O(cr)} - C_{P,H_2O(l)}] \times (263 \text{ K} - 273 \text{ K})$$

$$= -6004 \text{ J mol}^{-1} + (-38.5 \text{ J K}^{-1} \text{ mol}^{-1})(-10 \text{ K})$$

$$= -5619 \text{ J mol}^{-1}$$

2.30 What is the enthalpy change for the vaporization of water at 0 °C? This value may be estimated from Table A.1 by assuming that the heat capacities of $H_2O(l)$ and $H_2O(g)$ are independent of temperature from 0 to 25 °C.

SOLUTION

$$H_2O(l) = H_2O(g)$$

$$\Delta H^\circ(298\ K) = -241.818 - (-285.830) = 44.011\ kJ\ mol^{-1}$$

$$\Delta H_2^\circ = \Delta H_1^\circ + \Delta C_P(T_2 - T_1)$$

$$\Delta H^\circ(273\ K) = \Delta H^\circ(298\ K) + [C_P(H_2O,g) - C_P(H_2O,l)](273 - 298)$$

$$= 44{,}011 + (33.577 - 75.291)(-25) = 45{,}054\ J\ mol^{-1}$$

2.31 Calculate the enthalpy of dissociation of $H_2(g)$ at 3000 K using $\Delta_f H^\circ(H,\ 298.15\ K) = 217.999\ kJ\ mol^{-1}$ and $H^\circ - H^\circ_{298}$ values in Table A.2.

SOLUTION

$$\begin{array}{ccccc} & H_2(g) & \overset{2(217.999)}{\underset{298\ K}{=}} & 2H(g) & \\ 88.743 & \downarrow & & \downarrow & 2(56.162) \\ & H_2(g) & \underset{3000\ K}{=} & 2H(g) & \end{array}$$

$$\Delta H^\circ(3000) = -88.743 + 2(217.999) + 2(56.162)$$

$$= 459.579\ kJ\ mol^{-1} \quad (4.74551\ eV)$$

2.32 Calculate the standard enthalpy of formation of methane at 1000 K from the value at 298.15 K using the $H^\circ - H^\circ_{298}$ data in Table A.2.

SOLUTION

$$\begin{array}{lccccc} 1000\ K & C(graphite) & + & 2H_2(g) & = & CH_4(g) \\ & \uparrow (H^\circ_{1000} - H^\circ_{298})_C & & \uparrow 2(H^\circ_{1000} - H^\circ_{298})_{H_2} & & \uparrow (H^\circ_{1000} - H^\circ_{298})_{CH_4} \\ 298\ K & C(graphite) & + & 2H_2(g) & = & CH_4(g) \end{array}$$

The change in state represented by the first reaction can be accomplished by cooling one mole of graphite and two moles of hydrogen to 298 K, converting them to methane, and then heating the methane to 1000 K.

$$\Delta_f H^\circ_{1000} = -(H^\circ_{1000} - H^\circ_{298})_C - 2(H^\circ_{1000} - H^\circ_{298})_{H_2} + \Delta_f H^\circ(CH_4, 298\ K) + (H^\circ_{1000} - H^\circ_{298})_{CH_4}$$

$$= -11.816 - 2(20.686) - 74.873 + 38.179$$

$$= -89.882\ kJ\ mol^{-1}$$

2.33 Estimate the heat capacity at constant pressure at 298 K for ethane using the Benson method. Compare this result with the value calculated using values in Table 2.2.

SOLUTION

$$C_P = 2(25.90) = 51.80\ J\ K^{-1}\ mol^{-1}$$

$$C_P = 9.404 + (159.836 \times 10^{-3})(298.15) - (462.28 \times 10^{-7})(298.15)^2$$

$$= 52.95\ J\ K^{-1}\ mol^{-1}$$

2.34 Estimate the enthalpies of formation of n-butane and iso-butane at 298.15 K using the Benson method.

SOLUTION

For n-butane, $\Delta_f H^\circ = 2(-42.68) + 2(-20.63) = -126.62$ kJ mol^{-1}

For iso-butane, $\Delta_f H^\circ = 3(-42.68) - 7.95 = -135.99$ kJ mol^{-1}

The values in R. A. Alberty and C. Gehrig, J. Phys. Chem. Ref. Data, 13, 1173(1984) are -125.7 and -134.6 kJ mol^{-1}, respectively.

2.35 914.3 m

2.36 2.91 kWh 0.172 4.4 kg

2.37 (a) -1.72 (b) kJ mol^{-1}

2.38 $\partial C_V/\partial \ln V \neq \partial(RT)/\partial T = R$

$\partial C_V/\partial \ln V = \partial R/\partial \ln T = 0$

2.40 (a) 1.99 (b) −23.30 (c) −23.30 (d) −21.31 kJ mol^{-1}

2.41 106.780 kJ mol^{-1}

2.42 At 298 K HI is a rigid diatomic molecule. At 200 K its vibration is almost completely excited. At 298 K I_2 is almost completely vibrationally excited. At 2000 K it absorbs more energy per degree because of electronic excitation.

2.43

	CO	CO_2	NH_3	CH_4
C_P(classical)	37.413	62.354	83.136	108.077
C_P(3000 K)	37.217	62.229	79.496	101.391

The values are in J K^{-1} mol^{-1}.

2.44 40.874 kJ mol^{-1}

2.45 (a) −5.70 (b) 5.70 (c) 0 (d) 0 kJ mol^{-1}

2.46 (a) 567 K (b) 9.42 bar (c) 5527 J mol^{-1}

2.47 (a) 0.495 (b) 0.307 bar

2.48 (a) 41.84 (b) 25.56 kJ

2.49

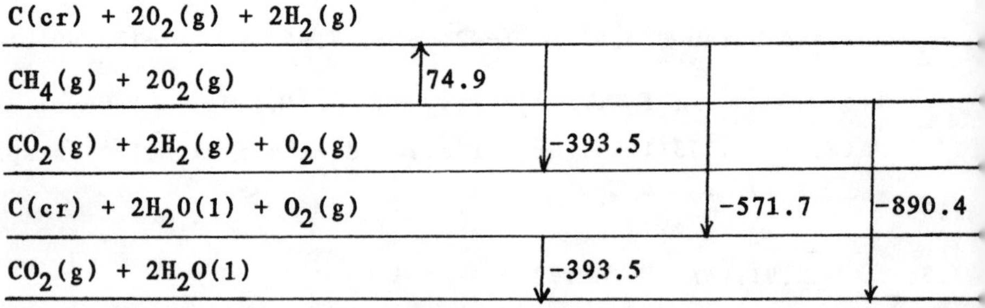

2.50 (a) −253.42 (b) −249.70 kJ mol^{-1}

2.51 -41.51 kJ mol^{-1}

2.52 12.01 kJ mol^{-1}

2.53 -585 ± 8 kJ mol^{-1} and -589 ± 2 kJ mol^{-1}

2.54 (a) -120.9 (b) -50.2 (c) -19.92 (d) -45.2 kJ mol^{-1}

2.55 225.756 kJ mol^{-1} is absorbed.

2.56 -214.627, -800.521, -1015.148 kJ mol^{-1}

2.57 432.074, 435.998, 459.578 kJ mol^{-1}

2.58 214.627, -97.92, 116.71 kJ mol^{-1}. Single reactor.

2.59 (a) -94.1 (b) -359.0 kJ mol^{-1}

2.60 (a) 0.038 (b) 10.03 kJ mol^{-1}

2.61 -44.037 kJ mol^{-1}

2.62 2432.6 J g^{-1}

2.63 1640.558 kJ mol^{-1}

2.64 -89.881 kJ mol^{-1}

2.65 2600 K

CHAPTER 3: Second and Third Laws of Thermodynamics

3.1 Theoretically, how high could a gallon of gasoline lift an automobile weighing 2800 lb against the force of gravity, if it is assumed that the cylinder temperature is 2200 K and the exit temperature 1200 K? (Density of gasoline = 0.80 g cm^{-3}; 1 lb = 453.6 g; 1 ft = 30.48 cm; 1 L = 0.2642 gal. Heat of combustion of gasoline = 46.9 kJ g^{-1}.)

SOLUTION

$$q = \frac{(46.9 \times 10^3\ J\ g^{-1})(1\ gal)(10^3\ cm^3\ L^{-1})(0.80\ g\ cm^{-3})}{0.2642\ gal\ L^{-1}}$$

$$= 14.2 \times 10^7\ J$$

$$w = q\,\frac{T_2 - T_1}{T_2} = \frac{(14.2 \times 10^7\ J)(2200\ K - 1200\ K)}{(2200\ K)} = 6.45 \times 10^7\ J$$

$$= mgh = (2800\ lb)(0.4536\ kg\ lb^{-1})(9.8\ m\ s^{-2})(0.3048\ m\ ft^{-1})h$$

$$h = 17{,}000\ ft$$

3.2 (a) What is the maximum work that can be obtained from 1000 J of heat supplied to a steam engine with a high-temperature reservoir at 100 °C if the condenser is at 20 °C? (b) If the boiler temperature is raised to 150 °C by the use of superheated steam under pressure, how much more work can be obtained?

SOLUTION

$$(a)\ w = q\,\frac{T_2 - T_1}{T_2} = (1000\ J)\,\frac{80\ K}{373.1\ K} = 214\ J$$

$$(b)\ w = (1000\ J)\,\frac{130\ K}{423.1\ K} = 307\ J \quad \text{or} \quad 93\ J\ \text{more than (a)}$$

3.3 A heat pump is to be used to maintain the temperature of a building at 18 °C when the outside temperature is -5 °C. For a frictionless heat pump how much work must be expended to obtain a joule of heat?

SOLUTION

$$|w| = |q_1|(T_1 - T_2)/T_1 = (1\ J)(23\ K)/(291\ K)$$

$$= 0.079\ J$$

3.4 What is the entropy change for the freezing of one mole of water at 0 °C? The heat of fusion is 333.5 J g^{-1}.

SOLUTION

$$\Delta S = \frac{\Delta H}{T} = \frac{-(333.5\ J\ g^{-1})(18.015\ g\ mol^{-1})}{273.15\ K} = -22.00\ J\ K^{-1}\ mol^{-1}$$

3.5 In the reversible isothermal expansion of a perfect gas at 300 K from 1 to 10 liters, where the gas has an initial pressure of 20.27 bar, calculate (a) ΔS for the gas and (b) ΔS for all systems involved in the expansion.

SOLUTION

$$\text{(a)}\ n = \frac{PV}{RT} = \frac{(20.27\ bar)(1\ L)}{(0.08314\ L\ bar\ K^{-1}\ mol^{-1})(300\ K)} = 0.812\ mol$$

$$\Delta S = nR\ \ln\frac{V_2}{V_1} = (0.812\ mol)(8.314\ J\ K^{-1}\ mol^{-1})\ \ln\frac{10}{1}$$

$$= 15.56\ J\ K^{-1}$$

(b) ΔS = 0 since the process is carried out reversibly. The heat gained by the gas is equal to the heat lost by the heat reservoir, and both bodies are at the same temperature.

3.6 Assuming the heat capacity of water is independent of temperature, calculate the net change in entropy when 1 mol of water at 0 °C is mixed with 1 mol of water at 100 °C. Assume the heat capacity is

$(4.184\ J\ K^{-1}\ g^{-1})(18\ g\ mol^{-1}) = 75.3\ J\ K^{-1}\ mol^{-1}$, and the heat capacity of the calorimeter is negligible.

SOLUTION

The entropy change for the cold water is

$$\Delta S_C = (75.3\ J\ K^{-1}\ mol^{-1})\ \ln \frac{323.15}{273.15} = 12.65\ J\ K^{-1}\ mol^{-1}$$

The entropy change for the hot water is

$$\Delta S_H = (75.3\ J\ K^{-1}\ mol^{-1})\ \ln \frac{323.15}{373.15} = -10.83\ J\ K^{-1}\ mol^{-1}$$

The net entropy change, $1.82\ J\ K^{-1}\ mol^{-1}$, is positive, as expected for an irreversible process.

3.7 A mole of propane gas is allowed to expand from 2 L to 10 L. What is the change in entropy of the propane, assuming (a) it is a perfect gas and (b) it is a van der Waals gas?

SOLUTION

$$\text{(a)}\ \Delta S = R\ \ln \frac{V_2}{V_1} = 8.3144\ \ln \frac{10\ L\ mol^{-1}}{2\ L\ mol^{-1}} = 13.38\ J\ K^{-1}\ mol^{-1}$$

$$\text{(b)}\ \Delta S = R\ \ln \frac{V_2 - b}{V_1 - b} = 8.3144\ \ln \frac{10 - .084}{2 - .084} = 13.67\ J\ K^{-1}\ mol^{-1}$$

3.8 Calculate the increase in entropy of a mole of silver that is heated at constant pressure from 0 to 30 °C if the value of C_P in this temperature range is considered to be constant at $25.48\ J\ K^{-1}\ mol^{-1}$.

SOLUTION

$$\Delta S = C_P\ \ln \frac{T_2}{T_1} = (25.48\ J\ K^{-1}\ mol^{-1})\ \ln \frac{303}{273}$$

$$= 2.657\ J\ K^{-1}\ mol^{-1}$$

3.9 Calculate the change in entropy of a mole of aluminum which is heated from 600 °C to 700 °C. The melting point of aluminum is

660 °C, the heat of fusion is 393 J g^{-1}, and the heat capacities of the solid and liquid may be taken as 31.8 and 34.3 J K^{-1} mol^{-1}, respectively.

SOLUTION

$$\Delta S = \int_{T_1}^{T_f} \frac{C_{P,s}}{T}\, dT + \frac{\Delta H_f}{T_f} + \int_{T_f}^{T_2} \frac{C_{P,l}}{T}\, dT$$

$$= C_{P,s} \ln \frac{T_f}{T_1} + \frac{\Delta H_f}{T_f} + C_{P,l} \ln \frac{T_2}{T_f}$$

$$= (31.8 \text{ J K}^{-1} \text{ mol}^{-1}) \ln \frac{933 \text{ K}}{873 \text{ K}} + \frac{(27 \text{ g mol}^{-1})(393 \text{ J g}^{-1})}{933 \text{ K}} +$$

$$(34.3 \text{ J K}^{-1} \text{ mol}^{-1}) \ln \frac{973 \text{ K}}{933 \text{ K}} = 14.92 \text{ J K}^{-1} \text{ mol}^{-1}$$

3.10 A mole of steam is condensed at 100 °C and the water is cooled to 0 °C and frozen to ice. What is the entropy change of the water? Consider that the average specific heat of liquid water is 4.2 J K^{-1} g^{-1}. The heat of vaporization at the boiling point and the heat of fusion at the freezing point are 2258.1 and 333.5 J g^{-1}, respectively.

SOLUTION

$$\Delta S = -\frac{\Delta H_{vap}}{T_{vap}} + \int_{373 \text{ K}}^{273 \text{ K}} \frac{C_P}{T}\, dT - \frac{\Delta H_{fus}}{T_{fus}}$$

$$= -\frac{(2258.1 \text{ J g}^{-1})(18.016 \text{ g mol}^{-1})}{373.15 \text{ K}} +$$

$$(75.379 \text{ J K}^{-1} \text{ mol}^{-1}) \int_{373 \text{ K}}^{273 \text{ K}} d \ln T - \frac{(333.5 \text{ J g}^{-1})(18.016 \text{ g mol}^{-1})}{273.15 \text{ K}}$$

$$= -154.4 \text{ J K}^{-1} \text{ mol}^{-1}$$

3.11 Calculate the increase in entropy of nitrogen when it is heated from 25 to 1000 °C (a) at constant pressure, and (b) at constant volume. Given: $C_P = 26.9835 + 5.9622 \times 10^{-3}\ T - 3.377 \times 10^{-7}\ T^2$ in J K^{-1} mol^{-1}.

SOLUTION

$$\text{(a)}\ \Delta S = \int_{298}^{1273} \frac{C_P}{T} = \int_{298}^{1273} \left(\frac{26.9835}{T} + 5.9622 \times 10^{-3} - 3.377 \times 10^{-7}\ T\right)dT$$

$$= 26.9835 \ln \frac{1273}{298} + 5.9622 \times 10^{-3}(1273 - 298) -$$

$$\left(\frac{1}{2}\right)3.377 \times 10^{-7}(1273^2 - 298^2) = 45.25\ \text{J K}^{-1}\ \text{mol}^{-1}$$

$$\text{(b)}\ \Delta S = \int_{298}^{1273} \frac{C_P - 8.314}{T}\, dT = \int_{298}^{1273} \frac{C_P}{T}\, dT - 8.314 \ln \frac{1273}{298}$$

$$= 45.25 - 8.314 \ln \frac{1273}{298} = 33.18\ \text{J K}^{-1}\ \text{mol}^{-1}$$

3.12 One mole of ammonia (considered to be a perfect gas) initially at 25 °C and 1 bar pressure is heated at constant pressure until the volume has trebled. Calculate (a) q, (b) w, (c) ΔH, (d) ΔU, and (e) ΔS. Given: $C_P = 25.895 + 32.999 \times 10^{-3}\ T - 30.46 \times 10^{-7}\ T^2$ in J K^{-1} mol^{-1}.

SOLUTION

$$\text{(a)}\ q = \int_{T_1}^{T_2} C_P\ dT$$

$$= \int_{298}^{894} [25.895 + 32.999 \times 10^{-3}\ T - 30.46 \times 10^{-7}\ T^2]dT$$

$$= (25.895)(596) + \frac{32.999 \times 10^{-3}}{2}(894^2 - 298^2) - \frac{30.46 \times 10^{-3}}{3}(894^3 - 298^3)$$

$$= 26.4 \text{ kJ mol}^{-1}$$

(b) $w = -P\Delta V = -R(T_2 - T_1) = -(8.314 \text{ J K}^{-1} \text{ mol}^{-1})(596 \text{ K})$

$$= -4.96 \text{ kJ mol}^{-1}$$

(c) $\Delta H = q_p = 26.4 \text{ kJ mol}^{-1}$

(d) $\Delta U = q + w = 26.4 - 5.0 = 21.4 \text{ kJ mol}^{-1}$

(e) $$S = \int_{298}^{894} \frac{C_P}{T} dT$$

$$= \int_{298}^{894} [\frac{25.895}{T} + 32.999 \times 10^{-3} - 30.46 \times 10^{-7} T]dT$$

$$= 25.895 \ln \frac{894}{298} + 32.999 \times 10^{-3}(894 - 298) - \frac{30.46 \times 10^{-7}}{2}(894^2 - 298^2)$$

$$= 46.99 \text{ J K}^{-1} \text{ mol}^{-1}$$

3.13 A mole of perfect gas is expanded isothermally and reversibly from 30 L to 100 L at 300 K. (a) What are the values of ΔU, ΔS, w and q? (b) If the expansion is carried out irreversibly by allowing the gas to expand into an evacuated container, what are the values of ΔU, ΔS, w, and q?

SOLUTION

(a) $\Delta U = 0$

$$\Delta S = R \ln \frac{V_2}{V_1} = (8.314 \text{ J K}^{-1} \text{ mol}^{-1}) \ln \frac{100 \text{ L}}{30 \text{ L}}$$

$$= 10.01 \text{ J K}^{-1} \text{ mol}^{-1}$$

$$w = -RT \ln \frac{V_2}{V_1} = -(8.314\ \text{J K}^{-1}\ \text{mol}^{-1})(300\ \text{K}) \ln \frac{100\ \text{L}}{30\ \text{L}}$$

$$= -3.003\ \text{kJ mol}^{-1}$$

$$q = 3.003\ \text{kJ mol}^{-1}$$

(b) $\Delta U = 0$ $\qquad$ $\Delta S = 10.010\ \text{J K}^{-1}\ \text{mol}^{-1}$

$w = 0$ $\qquad$ $q = 0$

3.14 The temperature of a mole of perfect monatomic gas is increased from 300 K to 500 K. What is the change in entropy (a) if the volume is held constant and (b) if the pressure is held constant?

SOLUTION

(a) $$\Delta S = C_V \ln \frac{T_2}{T_1} = \frac{3}{2}(8.314\ \text{J K}^{-1}\ \text{mol}^{-1}) \ln \frac{500\ \text{K}}{300\ \text{K}}$$

$$= 6.371\ \text{J K}^{-1}\ \text{mol}^{-1}$$

(b) $$\Delta S = C_P \ln \frac{T_2}{T_1} = \frac{5}{2}(8.314\ \text{J K}^{-1}\ \text{mol}^{-1}) \ln \frac{500\ \text{K}}{300\ \text{K}}$$

$$= 10.618\ \text{J K}^{-1}\ \text{mol}^{-1}$$

3.15 A mole of argon undergoes the following change in state

$$P_1 = 20\ \text{bar},\ T_1 = 300\ \text{K} \rightarrow P_2 = 1\ \text{bar},\ T_2 = 200\ \text{K}$$

What is the change in entropy, assuming it is a perfect gas?

SOLUTION

Since argon is monatomic, equation 3.55 yields

$$\Delta S = \frac{5}{2}(8.314) \ln \frac{200}{300} - 8.314 \ln \frac{1}{20} = 16.48\ \text{J K}^{-1}\ \text{mol}^{-1}$$

3.16 What is the change in entropy of a mole of liquid benzene at 25 °C when the pressure is raised to 1000 bar? The coefficient of thermal expansion α is $1.237 \times 10^{-3}\ \text{K}^{-1}$, the density is $0.879\ \text{g cm}^{-3}$, and the molar mass is $78.11\ \text{g mol}^{-1}$.

SOLUTION

$$\Delta S = -V\alpha\Delta P$$

$$= -\left(\frac{78.11 \text{ g mol}^{-1}}{0.879 \text{ g cm}^{-3}}\right)(10^{-2} \text{ m cm}^{-1})^3(1.237 \times 10^{-3} \text{ K}^{-1})(1000 \text{ bar})$$

$$= -10.99 \text{ J K}^{-1} \text{ mol}^{-1}$$

3.17 Ten moles of H_2 and two moles of D_2 are mixed at 25 °C. What is the value of ΔS°?

SOLUTION

$$\Delta S = -R(n_1 \ln y_1 + n_2 \ln y_2)$$

$$= -(8.314 \text{ J K}^{-1} \text{ mol}^{-1})\left[(10 \text{ mol}) \ln \frac{10}{12} + (2 \text{ mol}) \ln \frac{2}{12}\right]$$

$$= 44.95 \text{ J K}^{-1}$$

3.18 Section 3.13 showed that the entropy of mixing of 4 atoms of A and 4 atoms of B in an idealized crystal is k ln 70. When Stirling's approximation is used, equation 3.68 is obtained. Equation 3.68 will give only an approximate answer for the 8 atom system, but calculate the result and see how good it is.

SOLUTION

$$\Delta S = k \ln 70 = 4.25 \text{ k}$$

$$\Delta S = -k[4 \ln \tfrac{1}{2} + 4 \ln \tfrac{1}{2}] = 8k \ln 2 = 5.55 \text{ k}$$

3.19 Calculate the molar entropy of liquid chlorine at its melting point, 172.12 K, from the following data obtained by W. F. Giauque and T. M. Powell.

T/K	15	20	25	30	35	40	50	60
$C_P/J\ K^{-1}\ mol^{-1}$	3.72	7.74	12.09	16.69	20.79	23.97	29.25	33.47

T/K	70	90	110	130	150	170	172.12
$C_P/J\ K^{-1}\ mol^{-1}$	36.32	40.63	43.81	47.24	51.04	55.10	M.P.

The heat of fusion is 6406 J mol^{-1}. Below 15 K it may be assumed that C_P is proportional to T^3.

SOLUTION

For $T < 15$ K, $C_P = CT^3$, $C = \dfrac{3.72\ J\ K^{-1}\ mol^{-1}}{(15\ K)^3}$

$$S_{15\ K} = \int_0^{15\ K} CT^3\,\frac{dT}{T} = [CT^3/3]_0^{15} = \frac{3.72}{15^3} \times \frac{15^3}{3}$$

$$= 1.24\ J\ K^{-1}\ mol^{-1}$$

The contributions from 15 K to the melting point are calculated using $C_P\Delta T/T$, where C_P is the average for the range and T is the average for the range.

K	T_{avg}	$C_{P,avg}$	ΔT
15–20	17.5	5.73	5
20–25	22.5	9.92	5
25–30	27.5	14.39	5
30–35	32.5	18.74	5
35–40	37.5	22.38	5
40–50	45	26.61	10
50–60	55	31.36	10
60–70	65	34.89	10
70–90	80	38.47	20
90–110	100	42.22	20
110–130	120	45.52	20
130–150	140	49.14	20
150–170	160	53.07	20
170–172.12	171.06	55.10	2.12

$$S_{\text{liquid},T_m} = 1.24 + \sum \frac{C_P \Delta T}{T} + \frac{\Delta H_{fus}}{T_m} = 1.24 + 69.29 + \frac{6406}{172.12}$$

$$= 107.75 \text{ J K}^{-1} \text{ mol}^{-1}$$

3.20 Calculate the molar entropy of carbon disulfide at 25 °C from the following heat-capacity data and the heat of fusion, 4389 J mol^{-1}, at the melting point (161.11 K).

T/K	15.05	20.15	29.76	42.22	57.52	75.54	89.37
C_P/J K^{-1} mol^{-1}	6.90	12.01	20.75	29.16	35.56	40.04	43.14

T/K	99.00	108.93	119.91	131.54	156.83	161-298
C_P/J K^{-1} mol^{-1}	45.94	48.49	50.50	52.63	56.62	75.48

SOLUTION

$$S^\circ(298.15 \text{ K}) = \frac{C_P(15.05 \text{ K})}{3} + \int_{15.05 \text{ K}}^{161.11 \text{ K}} \frac{C_P}{T} dT + \frac{\Delta H_{fus}}{161.11 \text{ K}} + \int_{161.11}^{298.15} \frac{C_P}{T} dT$$

The first integral may be approximated by multiplying the average value of C_P/T for each temperature interval by the width of the interval. Thus the first contribution is

$$\frac{1}{2}\left(\frac{6.90}{15.05} + \frac{12.01}{20.15}\right)(20.15 - 15.05) = 2.69 \text{ J K}^{-1} \text{ mol}^{-1}$$

The first integral has the value 74.69 J K^{-1} mol^{-1}. Thus

$$S^\circ(298.15 \text{ K}) = \frac{6.90}{3} + 74.69 + \frac{4389}{161.11} + 75.45 \ln \frac{298.15}{161.11}$$

$$= 150.67 \text{ J K}^{-1} \text{ mol}^{-1}$$

3.21 Using molar entropies from Table A.1, calculate ΔS° for the following reactions at 25 °C.

(a) $H_2(g) + \frac{1}{2} O_2(g) = H_2O(l)$

(b) $H_2(g) + Cl_2(g) = 2HCl(g)$

(c) Methane(g) + $\frac{1}{2} O_2(g)$ = Methanol(l)

SOLUTION

(a) $\Delta S° = 69.91 - 130.68 - \frac{1}{2}(205.13) = -163.34 \text{ J K}^{-1} \text{ mol}^{-1}$

(b) $\Delta S° = 2(186.908) - 130.684 - 223.066 = 20.066 \text{ J K}^{-1} \text{ mol}^{-1}$

(c) $\Delta S° = 126.8 - 186.264 - \frac{1}{2}(205.138) = -162.0 \text{ J K}^{-1} \text{ mol}^{-1}$

3.22 What is $\Delta S°$ for $H_2(g) = 2H(g)$ at 298, 1000, and 3000 K?

SOLUTION

$\Delta S°(298 \text{ K}) = 2(114.604) - 130.574 = 98.634 \text{ J K}^{-1} \text{ mol}^{-1}$

$\Delta S°(1000 \text{ K}) = 2(139.758) - 166.113 = 113.403 \text{ J K}^{-1} \text{ mol}^{-1}$

$\Delta S°(3000 \text{ K}) = 2(162.594) - 202.778 = 122.410 \text{ J K}^{-1} \text{ mol}^{-1}$

3.23 What is $\Delta S°$ (298 K) for

$$H_2O(l) = H^+(ao) + OH^-(ao)$$

Why is this change negative and not positive?

SOLUTION

$\Delta S = -10.75 - 69.92 = -80.67 \text{ J K}^{-1} \text{ mol}^{-1}$

The ions polarize neighboring water molecules and attract them. For this reason the product state is more ordered than the reactant state.

3.24 From electromotive force measurements it has been found that $\Delta S°$ for the reaction $\frac{1}{2} H_2(g) + AgCl(cr) = HCl(aq) + Ag(cr)$ is $-62.4 \text{ J K}^{-1} \text{ mol}^{-1}$ at 298.15 K. What is the value of $S°[Cl^-(aq)]$?

SOLUTION

Since HCl is completely dissociated in aqueous solution and $S^\circ[H^+(ao)] = 0$,

$$\Delta S^\circ = S^\circ[Cl^-(ao)] + S^\circ[Ag(cr)] - \frac{1}{2} S^\circ[H_2(g)] - S^\circ[AgCl(cr)]$$

$$-62.4 = S^\circ[Cl^-(ao)] + 42.55 - \frac{1}{2}(130.684) - 96.2$$

$$S^\circ[Cl^-(ao)] = 56.6 \text{ J K}^{-1} \text{ mol}^{-1}$$

3.25 Estimate the standard entropy of n-butane and iso-butane at 298.15 K using the Benson method.

SOLUTION

For n-butane, $S^\circ = 2(127.24) + 2(39.41) - 8.314 \ln (3^2 \times 2)$
$= 309.3 \text{ J K}^{-1} \text{ mol}^{-1}$

For iso-butane, $S^\circ = 3(127.24) - 50.50 - 8.314 \ln 3^4$
$= 303.8 \text{ J K}^{-1} \text{ mol}^{-1}$

3.26 The gas is not returned to its initial state.

3.27 (a) 16.1% (b) 34.8% (c) 37.1%
(d) 37.3% (e) 47.1% (f) 62.0%

3.28 $9.96 \text{ J K}^{-1} \text{ mol}^{-1}$

3.29 Since this quantity is always positive, the change is spontaneous.

3.30 (a) 8.50 (b) 76.61 (c) 5.00 $\text{J K}^{-1} \text{ mol}^{-1}$

3.31 $125.39 \text{ J K}^{-1} \text{ mol}^{-1}$

3.32 $19.262 \text{ J K}^{-1} \text{ mol}^{-1}$

3.33 $\Delta U = -2.252 \text{ kJ mol}^{-1}$ $\Delta S = 0$

$q = 0$ $w = -2.252 \text{ kJ mol}^{-1}$

3.34 $-0.0253\ J\ K^{-1}\ mol^{-1}$

3.35 $9.13\ J\ K^{-1}\ mol^{-1}$

3.36 For the system $\Delta S = 0.021\ J\ K^{-1}\ mol^{-1}$. Therefore the change is spontaneous.

3.37 $115.5\ J\ K^{-1}\ mol^{-1}$

3.38 $288.7\ J\ K^{-1}\ mol^{-1}$

3.39 (a) 80.479 (b) -32.93 (c) $-38.9\ J\ K^{-1}\ mol^{-1}$

3.40 -5.06, $45.426\ J\ K^{-1}\ mol^{-1}$

3.41 $316.6\ J\ K^{-1}\ mol^{-1}$

3.42 393.8, $365.0\ J\ K^{-1}\ mol^{-1}$

CHAPTER 4: Gibbs Energy and Helmholtz Energy

4.1 Derive the expression for $C_P - C_V$ for a gas with the following equation of state.

$$\left(P + \frac{a}{V^2}\right)V = RT$$

SOLUTION

Combining equations 4.41 and 4.47

$$C_P - C_V = T\left(\frac{\partial P}{\partial T}\right)_V \left(\frac{\partial V}{\partial V}\right)_P$$

$$P = \frac{RT}{V} - \frac{a}{V^2}$$

$$\left(\frac{\partial P}{\partial T}\right)_V = \frac{R}{V}$$

$$PV^2 = RTV - a$$

$$PV^2 - RTV + a = 0$$

$$2\,VP\left(\frac{\partial V}{\partial T}\right)_P - RV - RT\left(\frac{\partial V}{\partial T}\right)_P = 0$$

$$\left(\frac{\partial V}{\partial T}\right)_P = \frac{RV}{2\,PV - RT}$$

$$C_P - C_V = R/\left(\frac{2\,PV}{RT} - 1\right) = R\left(1 - \frac{2\,a}{VRT}\right)^{-1}$$

4.2 Show that C_P and C_V for a perfect gas are independent of volume and pressure.

SOLUTION

$$C_V = \left(\frac{\partial V}{\partial T}\right)_V$$

$$\left(\frac{\partial C_V}{\partial V}\right)_T = \frac{\partial^2 V}{\partial T \partial V} = \frac{\partial^2 V}{\partial V \partial T} = \frac{\partial}{\partial T}\left(\frac{\partial V}{\partial V}\right)_T = 0$$

since $\left(\frac{\partial V}{\partial V}\right)_T = 0$. The value of $(\partial V/\partial V)_T$ can be obtained from Equation 4.42.

$$\left(\frac{\partial C_V}{\partial P}\right)_T = \frac{\partial^2 V}{\partial P \partial T} = \frac{\partial^2 V}{\partial T \partial P} = \frac{\partial}{\partial T}\left(\frac{\partial V}{\partial P}\right)_T = \frac{\partial}{\partial T}\left(\frac{\partial V}{\partial V}\right)_T\left(\frac{\partial V}{\partial P}\right)_T = 0$$

since $(\partial V/\partial V)_T = 0$ for a perfect gas. Since $C_P = C_V + R$, $(\partial C_P/\partial V)_T = 0$ and $(\partial C_P/\partial P)_T = 0$.

4.3 (a) Integrate the Gibbs-Helmholtz equation to obtain an expression for ΔG_2 at temperature T_2 in terms of ΔG_1 at T_1, assuming ΔH is independent of temperature. (b) Obtain an expression for ΔG_2 using the more accurate approximation that $\Delta H = \Delta H_1 + (T - T_1)\Delta C_P$ where T_1 is an arbitrary reference temperature.

SOLUTION

(a) Using equation 4.51

$$\int d(\Delta G/T) = -\int (\Delta H/T^2)dT$$

$$\frac{\Delta G_2}{T_2} - \frac{\Delta G_1}{T_1} = -\Delta H\left(\frac{1}{T_1} - \frac{1}{T_2}\right)$$

$$\Delta G_2 = \Delta G_1 T_2/T_1 + \Delta H[1 - (T_2/T_1)]$$

(b)
$$\int d(\Delta G/T) = \int \frac{\Delta H_1}{T^2} dT + \Delta C_P \int \frac{(T - T_1)}{T^2} dT$$

$$\frac{\Delta G_2}{T_2} - \frac{\Delta G_1}{T_1} = -\Delta H_1\left(\frac{1}{T_1} - \frac{1}{T_2}\right) + \Delta C_P \ln\frac{T_2}{T_1} + T_1\Delta C_P\left(\frac{1}{T_1} - \frac{1}{T_2}\right)$$

$$\Delta G_2 = \Delta G_1 T_2/T_1 + [\Delta H_1 - T_1\Delta C_P]\left(1 - \frac{T_2}{T_1}\right) + T_2\Delta C_P \ln\frac{T_2}{T_1}$$

4.4 One mole of a perfect gas is allowed to expand reversibly and isothermally (25 °C) from a pressure of 1 bar to a pressure of 0.1

bar: (a) What is the change in Gibbs energy? (b) What would be the change in Gibbs energy if the process occurred irreversibly?

SOLUTION

(a) $\left(\frac{\partial G}{\partial P}\right)_T = V = \frac{RT}{P}$

$$\Delta G = RT \ln \frac{P_2}{P_1} = (8.314 \text{ J K}^{-1} \text{ mol}^{-1})(298.15 \text{ K}) \ln 0.1$$

$$= -5708 \text{ J mol}^{-1}$$

(b) $\Delta G = -5708 \text{ J mol}^{-1}$ because G is a state function and depends only on the initial state and the final state.

4.5 Calculate the change in Gibbs energy for the process

$H_2O(1,-10\ ^{\circ}C) = H_2O(cr,-10\ ^{\circ}C)$

The vapor pressure of water at -10 °C is 286.5 Pa, and the vapor pressure of ice at -10 °C is 260.0 Pa. The process may be carried out by the following reversible steps:

1. A mole of water is transferred at -10 °C from liquid to saturated vapor (P = 286.5 Pa). $\Delta G = 0$, since the two phases are in equilibrium.
2. The water vapor is allowed to expand from 286.5 to 260.0 Pa at -10 °C.
3. A mole of water is transferred at -10 °C from vapor at P = 260.0 Pa to ice at -10 °C.

SOLUTION

Since the first and third steps are transfers at equilibrium, there are no Gibbs energy changes associated with them.

$$\Delta G_2 = -RT \ln \frac{P_1}{P_2} = -(8.314 \text{ J K}^{-1} \text{ mol}^{-1})(263.15 \text{ K}) \ln \frac{286.5 \text{ Pa}}{260.0 \text{ Pa}}$$

$$= -212.5 \text{ J mol}^{-1}$$

4.6 A mole of perfect gas is compressed isothermally from 1 to 5 bar at 100 °C. (a) What is the Gibbs energy change? (b) What would have been the Gibbs energy change if the compression had been carried out at 0 °C?

SOLUTION

(a) $\Delta G^{\circ} = RT \ln \dfrac{P_2}{P_1} = (8.314 \text{ J K}^{-1} \text{ mol}^{-1})(373.15 \text{ K}) \ln 5$

$= 4993 \text{ J mol}^{-1}$

(b) $\Delta G^{\circ} = (8.314 \text{ J K}^{-1} \text{ mol}^{-1})(273.15 \text{ K}) \ln 5 = 3655 \text{ J mol}^{-1}$

4.7 What is the difference between the molar heat capacity of iron at constant pressure and constant volume at 25 °C? Given: $\alpha = 35.1 \times 10^{-6} \text{ K}^{-1}$, $\kappa = 0.52 \times 10^{-6} \text{ bar}^{-1}$, and the density is 7.86 g cm^{-3}.

SOLUTION

$$V = \frac{55.847 \text{ g mol}^{-1}}{7.86 \text{ g cm}^{-3}} = \frac{7.11 \text{ cm}^3 \text{ mol}^{-1}}{(10^2 \text{ cm m}^{-1})^3} = 7.11 \times 10^{-6} \text{ m}^3 \text{ mol}^{-1}$$

$$C_P - C_V = \alpha^2 TV/\kappa$$

$$= \frac{(35.1 \times 10^{-6} \text{ K}^{-1})^2 (298 \text{ K})(7.11 \times 10^{-6} \text{ m}^3 \text{ mol}^{-1})(10^5 \text{ Pa bar}^{-1})}{(0.52 \times 10^{-6} \text{ bar}^{-1})}$$

$$= 0.51 \text{ J K}^{-1} \text{ mol}^{-1}$$

4.8 When a liquid is compressed its Gibbs energy is increased. To a first approximation the increase in Gibbs energy can be calculated using $(\partial G/\partial P)_T = V$, assuming a constant molar volume. What is the change in Gibbs energy for liquid water when it is compressed to 1000 bar?

SOLUTION

$$\int_{G_1}^{G_2} dG = \int_{P_1}^{P_2} VdP$$

$$\Delta G = V\Delta P = (18 \times 10^{-6}\ m^3\ mol^{-1})(999\ bar)(10^5\ Pa\ bar^{-1})$$

$$= 1.8\ kJ\ mol^{-1}$$

4.9 Show that when a liquid is compressed, the change in Gibbs energy is given by

$\Delta G = V\Delta P - \frac{1}{2}\kappa V(\Delta P)^2$ if $\kappa\,\Delta P << 1$. The isothermal compressibility is represented by κ.

SOLUTION

$$\int_{G_1}^{G_2} dG = \int_{P_1}^{P_2} VdP = \int_{P_1}^{P_2} V\,e^{-\kappa(P_2 - P_1)}\,dP$$

$$\Delta G = -\frac{V}{\kappa}[e^{-\kappa(P_2 - P_1)} - 1]$$

If $\kappa\,\Delta P << 1$, $e^{-\kappa\Delta P} = 1 - \kappa\,\Delta P + \frac{1}{2}(\kappa\Delta P)^2 - \ldots$

$$\Delta G = \frac{V}{\kappa}[\kappa\Delta P - \frac{1}{2}\kappa^2(\Delta P)^2 + \ldots] = V\Delta P - \frac{1}{2}\kappa V(\Delta P)^2$$

4.10 The entropy of $O_2(g)$ at 1 bar is listed in Table A.2 in the Appendix as 205.138 J K^{-1} mol^{-1}, and the Gibbs energy of formation is listed as 0 kJ mol^{-1}. Assuming O_2 is a perfect gas, what will the molar entropy and molar Gibbs energy of formation be at 100 bar?

SOLUTION

$$S = S^\circ - R\ln\frac{P}{P^\circ} = 205.138 - 8.314\ln 100 = 166.849\ J\ K^{-1}\ mol^{-1}$$

$$\Delta_f G^\circ = 0 + RT \ln (P/P^\circ) = (8.314 \times 10^{-3})(298) \ln 100$$
$$= 11.41 \text{ kJ mol}^{-1}$$

4.11 A mole of helium is compressed isothermally and reversibly at 100 °C from a pressure of 2 to 10 bar. Calculate (a) q, (b) w, (c) ΔG, (d) ΔA, (e) ΔH, (f) ΔU, and (g) ΔS.

SOLUTION

(a) $\Delta U = q + W = 0$

$$q = -w = -RT \ln \frac{P_2}{P_1}$$

$$= -(8.314 \text{ J K}^{-1} \text{ mol}^{-1})(373.15 \text{ K}) \ln \frac{10 \text{ bar}}{2 \text{ bar}} = -4993 \text{ J mol}^{-1}$$

(b) $w = -q = 4993 \text{ J mol}^{-1}$

(c) $$\Delta G = \int_{P_1}^{P_2} V dP = RT \ln \frac{P_2}{P_1} = 4993 \text{ J mol}^{-1}$$

(d) $\Delta A = w_{max} = 4993 \text{ J mol}^{-1}$

(e) $\Delta H - \Delta U + \Delta(PV) = 0$

(f) $\Delta H = 0$

(g) $$\Delta S = \Delta H - \frac{\Delta G}{T} = 0 - \frac{4933 \text{ J mol}^{-1}}{373.15 \text{ K}} = -13.38 \text{ J K}^{-1} \text{ mol}^{-1}$$

4.12 (a) Calculate the work done against the atmosphere when 1 mol of toluene is vaporized at its boiling point, 111 °C. The heat of vaporization at this temperature is 361.9 J g^{-1}. For the vaporization of 1 mol, calculate (b) q, (c) ΔH, (d) ΔU, (e) ΔG, and (f) ΔS.

SOLUTION

(a) Assuming that toluene vapor is a perfect gas and that the volume of the liquid is negligible, the work done against the atmosphere is

$$w = P\Delta V = RT = (8.314\ \text{J K}^{-1}\ \text{mol}^{-1})(384\ \text{K})$$

$$= 3193\ \text{J mol}^{-1}$$

NOTE: The work done on the toluene is -3193 J mol^{-1}.

(b) $q_p = \Delta H_{vap} = (361.9\ \text{J g}^{-1})(92.13\ \text{g mol}^{-1}) = 33{,}342\ \text{J mol}^{-1}$.

(c) $\Delta H = 33{,}342\ \text{J mol}^{-1}$

(d) $\Delta U = q + w = 33{,}342 - 3193 = 30{,}149\ \text{J mol}^{-1}$

(e) $\Delta G = 0$ because the evaporation is reversible at the boiling point and 1 atm.

(f) $\Delta S = \dfrac{q_{rev}}{T} = \dfrac{33{,}342\ \text{J mol}^{-1}}{384\ \text{K}} = 86.8\ \text{J K}^{-1}\ \text{mol}^{-1}$

4.13 One mole of a perfect gas at 300 K has an initial pressure of 15 bar and is allowed to expand isothermally to a pressure of 1 bar. Calculate (a) the maximum work that can be obtained from the expansion, (b) ΔU, (c) ΔH, (d) ΔG, and (e) ΔS.

SOLUTION

(a) $w = RT \ln \dfrac{P_2}{P_1} = (8.314\ \text{J K}^{-1}\ \text{mol}^{-1})(300\ \text{K}) \ln \dfrac{1}{15}$

$$= -6.754\ \text{kJ mol}^{-1}$$

(b) $\Delta U = 0$ because $\partial U/\partial T = 0$ for a perfect gas

(c) $\Delta H = \Delta V + \Delta(PV) = 0$

(d) $\Delta G = RT \ln \dfrac{P_2}{P_1} = -6.754\ \text{kJ mol}^{-1}$

(e) $\Delta S = RT \ln \dfrac{P_1}{P_2} = (8.314\ \text{J K}^{-1}\ \text{mol}^{-1})(\ln 15)$

$$= 22.51\ \text{J K}^{-1}\ \text{mol}^{-1}$$

4.14 For $H_2(g)$ up to 100 bar the fugacity may be represented approximately by $f = Pe^{bP/RT}$ where $b = 0.01399\ L\ mol^{-1}$. What is the fugacity at 100 bar?

SOLUTION

$$f = (100\ bar) = \exp \frac{(0.01399\ L\ mol^{-1})(100\ bar)}{(0.08206\ L\ atm\ K^{-1}\ mol^{-1})(298\ K)}$$

$$= 105.9\ bar$$

4.15 Using the relation derived in Example 4.6, calculate the fugacity of $H_2(g)$ at 100 bar at 298 K.

SOLUTION

$$f = P \exp (b - \frac{a}{RT})\frac{P}{RT}$$

$$= (100\ bar) \exp [0.02661 - 0.2476/(.08314)(298)]100/(.08314)(298)$$

$$= 106.9\ bar$$

4.16 Show that if the compressibility factor is given by $Z = 1 + B'P$ the fugacity is given by $f = Pe^{Z-1}$. If Z is not very different from unity, $e^{Z-1} = 1 + (Z - 1) + \ldots \cong Z$ so that $f = PZ$. Using this approximation, what is the fugacity of $H_2(g)$ at 50 bar and 298 K using its van der Waals constants?

SOLUTION

$$\ln\left(\frac{f}{P^o}\right) = \ln\left(\frac{P}{P^o}\right) + \int_0^P \frac{Z - 1}{P}\, dP$$

$$= \ln\left(\frac{P}{P^o}\right) + \int_0^P B'dP = \ln\left(\frac{P}{P^o}\right) + B'P$$

$$f = Pe^{Z-1}$$

$$Z = 1 + (b - \frac{a}{RT})(\frac{P}{RT})$$

$$= 1 + [0.02661 - \frac{0.2476}{(0.08314)(298)}] \frac{50}{(0.08314)(298)} = 1.034$$

$$f = (50 \text{ bar})\ e^{0.034} = 51.7 \text{ bar}$$

4.17 As shown in Example 4.6, the fugacity of a van der Waals gas is given by a fairly simple expression if only the second virial coefficient is used. To this degree of approximation derive the expressions for G, S, A, U, H, and V.

SOLUTION

$$G = G^{\circ} + RT\ \ln\left(\frac{f}{P^{\circ}}\right) = G^{\circ} + RT\ \ln\left(\frac{P}{P^{\circ}}\right) + (b - \frac{a}{RT})P$$

$$S = -\left(\frac{\partial G}{\partial T}\right)_P = S^{\circ} - R\ \ln\left(\frac{P}{P^{\circ}}\right) + \frac{aP}{RT^2}$$

$$A = G - P\left(\frac{\partial G}{\partial P}\right)_T = G^{\circ} + RT\ \ln\left(\frac{P}{P^{\circ}}\right) - RT$$

$$U = G - T\left(\frac{\partial G}{\partial T}\right)_P - P\left(\frac{\partial G}{\partial P}\right)_T = G^{\circ} + TS^{\circ} - RT - \frac{aP}{RT}$$

$$= H^{\circ} - RT - \frac{aP}{RT} = U^{\circ} - \frac{aP}{RT}$$

Since the second term is small, we can use the perfect gas law to obtain $U = U^{\circ} - \frac{a}{V}$

Note that this agrees with the earlier result (Section 4.4) that $(\partial U/\partial V)_T = a/V^2$.

$$H = U + PV = U^{\circ} + \frac{aP}{RT} + RT + bP - \frac{aP}{RT}$$

$$= H^{\circ} + (b - \frac{2a}{RT})P$$

$$V = \left(\frac{\partial G}{\partial P}\right)_T = \frac{RT}{P} + b - \frac{a}{RT}$$

4.18 Using experimental data from Fig. 4.4 calculate $(\partial S/\partial l)_T$ and $(\partial U/\partial l)_T$ using only the data at 10 °C and 70 °C, assuming these quantities are independent of temperature. These thermodynamic quantities do depend on the elongation so calculate them for 200% and 400% elongation. (For f and T simply use averages for this temperature range.)

SOLUTION

At 200% elongation

$$\left(\frac{\partial S}{\partial l}\right)_T = -\left(\frac{\partial f}{\partial T}\right)_l = -\frac{(10.8 - 9.0)(10^5\ \mathrm{N\ m^{-2}})}{(70 - 10)\mathrm{K}}$$

$$= -3.0 \times 10^3\ \mathrm{N\ K^{-1}\ m^{-2}}$$

$$\left(\frac{\partial U}{\partial l}\right)_T = f - T\left(\frac{\partial f}{\partial T}\right)_l = 9.5 \times 10^5\ \mathrm{N\ m^{-2}} - (313\ \mathrm{K})(3.0 \times 10^3\ \mathrm{N\ K^{-1}\ m^{-2}})$$

$$= 0.1\ \mathrm{N\ m^{-2}}$$

At 400% elongation

$$\left(\frac{\partial S}{\partial l}\right)_T = -\frac{(26 - 21.3) \times 10^5\ \mathrm{N\ m^{-2}}}{60\ \mathrm{K}} = -7.8 \times 10^3\ \mathrm{N\ K^{-1}\ m^{-2}}$$

$$\left(\frac{\partial U}{\partial l}\right)_T = 23 \times 10^5\ \mathrm{N\ m^{-2}} - (313\ \mathrm{K})(7.8 \times 10^3\ \mathrm{N\ K^{-1}\ m^{-2}})$$

$$= -1.4\ \mathrm{N\ m^{-2}}$$

4.20 $$C_P - C_V = \frac{R^2 T V^3}{RTV^3 - 2a(V - b)^2}$$

4.21 $(\partial V/\partial V)_T = a/2T^{1/2}V(V - b)$

4.22 $-108.3\ \mathrm{J\ mol^{-1}}$

4.23 $89.5\ \mathrm{J\ mol^{-1}}$

4.24 $1.756\ \mathrm{kJ\ mol^{-1}}$

4.25 168.97 J K^{-1} mol^{-1}; 0, -11.42 kJ mol^{-1}

4.26 (a) 6820 J mol^{-1} (b) 6070 J mol^{-1}
(c) 0 J mol^{-1} (d) 75.7 J K^{-1} mol^{-1}

4.27 (a) -5230 J mol^{-1} (b) 5230 J mol^{-1}
(c) 0 J mol^{-1} (d) -5230 J mol^{-1}
(e) 19.142 J K^{-1} mol^{-1} (f) 0 J mol^{-1}
(g) 0 J mol^{-1} (h) 0 J mol^{-1}
(i) -5230 J mol^{-1} (j) 19.142 J K^{-1} mol^{-1}
(k) 0 J K^{-1} mol^{-1} (l) 19.142 J K^{-1} mol^{-1}

4.28 3100, -40,670, -40,670, -37,570, 0,
3100 J mol^{-1}
-108.8 J K^{-1} mol^{-1}

4.29 0, 19.16 J K^{-1} mol^{-1}

4.32 6.0 x 10^{25}, 5.8 x 10^{25} m^{-3}

CHAPTER 5: Chemical Equilibrium

5.1 Calculate ΔG and ΔS for the formation of a quantity of air containing 1 mol of gas by mixing nitrogen and oxygen at 298.15 K. Air may be taken to be 80% nitrogen and 20% oxygen by volume.

SOLUTION

$$\Delta G_{mix} = RT(y_1 \ln y_1 + y_2 \ln y_2)$$

$$= (8.314 \text{ J K}^{-1} \text{ mol}^{-1})(298.15 \text{ K})(0.8 \ln 0.8 + 0.2 \ln 0.2)$$

$$= -1239 \text{ J mol}^{-1}$$

$$\Delta S_{mix} = -R(y_1 \ln y_1 + y_2 \ln y_2) = 4.159 \text{ J K}^{-1} \text{ mol}^{-1}$$

5.2 A mole of gas A is mixed with a mole of gas B at 1 bar and 298 K. How much work is required to separate these gases to produce a container of each at 1 bar and 298 K?

SOLUTION

The mixture can be separated by diffusion through a perfect semi-permeable membrane. The highest partial pressure of each that can be reached is 1/2 bar. These gases then have to be compressed to 1 bar.

$$w = (2 \text{ mol}) \int_{1/2}^{1} V dP = (2 \text{ mol}) \int_{1/2}^{1} (RT/P) dP = (2 \text{ mol})\, RT \ln 2$$

$$= 2(8.314)(298) \ln 2$$

$$= 3.4 \text{ kJ}$$

The actual process will require more work.

5.3 For the reaction $N_2(g) + 3H_2(g) = 2NH_3(g)$, $K = 1.60 \times 10^{-4}$ at 400 °C. Calculate (a) ΔG° and (b) ΔG when the pressures of N_2 and H_2 are maintained at 10 and 30 bar, respectively, and NH_3 is removed at a partial pressure of 3 bar. (c) Is the reaction spontaneous under the latter conditions?

SOLUTION

(a) $\Delta G^\circ = -RT \ln K$

$= -(8.314 \text{ J K}^{-1} \text{ mol}^{-1})(673 \text{ K}) \ln 1.60 \times 10^{-4}$

$= 48.9 \text{ kJ mol}^{-1}$

(b) $\Delta G = \Delta G^\circ + RT \ln \dfrac{P_{NH_3}^2}{P_{N_2} P_{H_2}^3}$

$= 48.9 + (8.314 \times 10^{-3})(673) \ln \dfrac{3^3}{10(30)^3}$

$= 10.2 \text{ kJ mol}^{-1}$

(c) No

5.4 A 1:3 mixture of nitrogen and hydrogen was passed over a catalyst at 450 °C. It was found that 2.04% by volume of ammonia was formed when the total pressure was maintained at 10.13 bar [A. T. Larson and R. L. Dodge, J. Am. Chem. Soc., 45, 2918 (1923)]. Calculate the value of K for $\frac{3}{2} H_2(g) + \frac{1}{2} N_2(g) = NH_3(g)$ at this temperature.

SOLUTION

At equilibrium $P_{H_2} + P_{N_2} + P_{NH_3} = 10.13$ bar

$P_{N_2} = (10.13 \text{ bar})(0.0204) = 0.207$ bar

$P_{H_2} + P_{N_2} = 10.13 \text{ bar} - 0.207 \text{ bar} = 9.923$ bar

$P_{H_2} = 3P_{N_2}$ because this initial ratio is not changed by reaction

$$P_{N_2} = \frac{9.923\ \text{bar}}{4} = 2.481\ \text{bar}$$

$$P_{H_2} = \frac{3}{4}(9.923\ \text{bar}) = 7.442\ \text{bar}$$

$$K = \frac{\left(P_{NH_3}/P^0\right)}{\left(P_{H_2}/P^0\right)^{3/2}\left(P_{N_2}/P^0\right)^{1/2}}$$

$$= \frac{0.207}{7.442^{3/2}\ \ 2.482^{1/2}} = 6.47 \times 10^{-3}$$

5.5 Water vapor is passed over coal (assumed to be pure graphite in this problem) at 1000 K. Assuming that the only reaction occurring is the water gas reaction

$$C(\text{graphite}) + H_2O(g) = CO(g) + H_2(g) \qquad K = 2.52$$

calculate the equilibrium pressures of H_2O, CO, and H_2 at a total pressure of 1 bar. (Actually the water gas shift reaction

$$CO(g) + H_2O(g) = CO_2(g) + H_2(g)$$

occurs in addition, but it is considerably more complicated to take this subsequent reaction into account.)

SOLUTION

$$K = \frac{(P_{CO}/P^0)(P_{H_2}/P^0)}{(P_{H_2O}/P^0)} = \frac{x^2}{y} = \frac{x^2}{1-2x} = 2.52$$

$$2x + y = 1 \qquad x^2 = 2.52 - 5.04x$$

$$x^2 + 5.04x - 2.52 = 0$$

$$x = \frac{-5.04 \pm \sqrt{5.04^2 + 4(2.52)}}{2} = 0.458 = \frac{P_{CO}}{P^0} = \frac{P_{H_2}}{P^0}$$

$$1 - 2x = 0.084 = \frac{P_{H_2O}}{P^\circ}$$

$P_{H_2O} = 0.084$ bar $\qquad P_{CO} = 0.458$ bar $\qquad P_{H_2} = 0.458$ bar

5.6 An evacuated tube containing 5.96×10^{-3} mol L^{-1} of solid iodine is heated to 973 K. The experimentally determined pressure is 0.496 [M. L. Perlman and G. K. Rollefson, J. Chem. Phys., 9, 362 (1941)]. Assuming perfect-gas behavior, calculate K_P for $I_2(g) = 2I(g)$.

SOLUTION

$$P = \frac{n}{V} RT$$

$$0.496 \text{ bar} = (1 + \alpha)(5.96 \times 10^{-3} \text{ L}^{-1})(.08314 \text{ L bar K}^{-1} \text{ mol}^{-1})(973 \text{ K})$$

$$\alpha = \frac{0.496 \text{ bar}}{(5.96 \times 10^{-3} \text{ mol L}^{-1})(.08314 \text{ L bar}^{-1} \text{ K}^{-1} \text{ mol}^{-1})(973 \text{ K})} - 1$$

$$= 0.0287$$

$$K = \frac{4\alpha^2(P/P^\circ)}{1 - \alpha^2} = \frac{4(0.0287)^2(0.496)}{1 - 0.0287^2} = 1.64 \times 10^{-3}$$

5.7 At 55 °C and 1 bar the average molar mass of partially dissociated N_2O_4 is 61.2 g mol^{-1}. Calculate (a) α and (b) K for the reaction $N_2O_4(g) = 2NO_2(g)$. (c) Calculate α at 55 °C if the total pressure is reduced to 0.1 bar.

SOLUTION

$$(a) \quad \alpha = \frac{M_1 - M_2}{M_2 \; \Sigma \nu_i} = \frac{92.0 - 61.2}{61.2} = 0.503$$

$$(b) \quad K = \frac{4\alpha^2(P/P^\circ)}{1 - \alpha^2} = \frac{4\ (0.503)^2}{1 - 0.503^2} = 1.36$$

(c) $$\frac{\alpha^2}{1-\alpha^2} = \frac{K}{4(P/P^\circ)} = \frac{1.36}{4(0.1)}$$

$$\alpha = 0.879$$

5.8 A liter reaction vessel containing 0.233 mol of N_2 and 0.341 of PCl_5 is heated to 250 °C. The total pressure at equilibrium is 29.33 bar. Assuming that all the gases are ideal, calculate K for the only reaction that occurs.

$$PCl_5(g) = PCl_3(g) + Cl_2(g)$$

SOLUTION

$$n = \frac{PV}{RT} = \frac{(29.33\ \text{bar})(1\ \text{L})}{(0.08314\ \text{L bar K}^{-1}\ \text{mol}^{-1})(523.15\ \text{K})} = 0.674\ \text{mol}$$

Moles of reactants = 0.674 − 0.233 = 0.441 = (0.341 − x) + 2x

x = moles of PCl_3 = moles of Cl_2 = 0.100

$$K = \frac{(P_{PCl_3}/P^\circ)(P_{Cl_2}/P^\circ)}{P_{PCl_5}/P^\circ} = \frac{\left(\frac{0.100}{0.674}\right)^2 (29.33)^2}{\left(\frac{0.241}{0.679}\right)(29.33)} = 1.81$$

5.9 What are the percentage dissociations of $H_2(g)$, $O_2(g)$, and $I_2(g)$ at 2000 K and a total pressure of 1 bar?

SOLUTION

$$H_2(g) = 2H(g)$$

$$\Delta G^\circ = 2(106{,}653\ \text{J mol}^{-1})$$

$$= -RT \ln K$$

$$= -(8.31441\ \text{J K}^{-1}\ \text{mol}^{-1})(2000\ \text{K}) \ln K$$

$$K = 2.69 \times 10^{-6} = \frac{4\alpha^2}{1-\alpha^2}$$

$$\alpha = \left(\frac{K_P}{K_P + 4}\right)^{1/2} = \left(\frac{2.69 \times 10^{-6}}{4}\right)^{1/2} = 0.000819 \text{ or } 0.0819\%$$

$O_2(g) = 2O(g)$

$\Delta G° = 2(121{,}448 \text{ J mol}^{-1})$

$K = 4.53 \times 10^{-7}$ $\qquad \alpha = 0.0337\%$

$I_2(g) = 2I(g)$

$\Delta G° = 2(-29{,}637 \text{ J mol}^{-1})$

$K = 35.32$ $\qquad \alpha = 94.8\%$

5.10 Hydrogen is produced on a large scale from methane. Calculate the equilibrium constant K for the production of H_2 from CH_4 at 1000 K using the reaction $CH_4(g) + H_2O(g) = CO(g) + 3H_2(g)$.

SOLUTION

$$\Delta G° = -200.297 - 19.460 + 192.576 = -27.181 \text{ kJ mol}^{-1}$$

$$= -(8.31441 \times 10^{-3} \text{ kJ K}^{-1} \text{ mol}^{-1})(1000 \text{ K}) \ln K$$

$K = 26.3$

5.11 From the $\Delta_f G°$ of $Br_2(g)$ at 25 °C, calculate the vapor pressure of $Br_2(l)$. The pure liquid at 1 bar and 25 °C is taken as the standard state.

SOLUTION

$Br_2(l) = Br_2(g)$ $\qquad K = P_{Br_2}/P°$

$\Delta G° = -RT \ln (P_{Br_2}/P°)$

$$\frac{P_{Br_2}}{P°} = \exp\left[-\frac{3110 \text{ J mol}^{-1}}{(8.314 \text{ J K}^{-1} \text{ mol}^{-1})(298.15 \text{ K})}\right]$$

$P_{Br_2} = 0.258$ bar

5.12 In order to produce more hydrogen from "synthesis gas" ($CO + H_2$) the water gas shift reaction is used.

$$CO(g) + H_2O(g) = CO_2(g) + H_2(g)$$

Calculate K at 1000 K and the equilibrium extent of reaction starting with an equimolar mixture of CO and H_2O.

<u>SOLUTION</u>

$$\Delta G^\circ = -395{,}924 - (-200{,}297) - (-192{,}576) = -3051 \text{ J mol}^{-1}$$

$$= (8.314 \text{ J K}^{-1} \text{ mol}^{-1})(1000 \text{ K}) \ln K$$

$$K = 1.44 = \frac{P_{H_2} P_{CO_2}}{P_{CO} P_{H_2O}} = \frac{x^2}{(1 - x)^2}$$

$x = 0.545$, fractional conversion of reactants to products

(Note that this reaction is exothermic so that there will be a larger extent of reaction at lower temperatures. In practice this reaction is usually carried out at about 700 K.)

5.13 Acetic acid is produced on a large scale by the carbonylation of methanol at about 500 K and 25 bar using a rhodium catalyst. What is K_y under these conditions? ($\Delta_f G^\circ$ for acetic acid gas at 500 K is -335.28 kJ mol^{-1}.)

<u>SOLUTION</u>

$$CO(g) + CH_3OH(g) = CH_3CO_2H(g)$$

$$\Delta G^\circ = -335.28 + 155.68 + 133.52 = -46.00 \text{ kJ mol}^{-1}$$

$$K = 6.39 \times 10^4 = \frac{y_{AA} P}{y_{CO} y_{CH_3OH} P^2}$$

$$K_y = \frac{y_{AA}}{y_{CO}y_{CH_3OH}} = (6.39 \times 10^4)(25) = 1.60 \times 10^6$$

5.14 Calculate the degree of dissociation of $H_2O(g)$ into $H_2(g)$ and $O_2(g)$ at 2000 K and 1 bar. (Since the degree of dissociation is small, the calculation may be simplified by assuming that $P_{H_2O} = 1$ bar.)

SOLUTION

$$H_2O(g) = H_2(g) + \frac{1}{2} O_2(g)$$

$$\Delta G^\circ = 134{,}456 \text{ J}\cdot\text{mol}^{-1}$$

$$= -(8.3144 \text{ J K}^{-1} \text{ mol}^{-1})(2000 \text{ K}) \ln K$$

$$K = 3.079 \times 10^{-4}$$

$$= \frac{\left(P_{H_2}/P^\circ\right)\left(P_{O_2}/P^\circ\right)^{1/2}}{\left(P_{H_2O}/P^\circ\right)}$$

$$\left(\frac{P_{H_2}}{P^\circ}\right)\left(\frac{P_{H_2}}{2P^\circ}\right)^{1/2} = \frac{1}{\sqrt{2}}\left(\frac{P_{H_2}}{P^\circ}\right)^{3/2}$$

$$\left(\frac{P_{H_2}}{P^\circ}\right) = (\sqrt{2}\ \ K)^{2/3} = 0.0058$$

5.15 At 500 K CH_3OH, CH_4 and other hydrocarbons can be formed from CO and H_2. Until recently the main source of the CO mixture for the synthesis of CH_3OH was methane.

$$CH_4(g) + H_2O(g) = CO(g) + 3H_2(g)$$

When coal is used as the source, the "synthesis gas" has a different composition.

$$C(graphite) + H_2O(g) = CO(g) + H_2(g)$$

Suppose we have a catalyst that catalyzes only the formation of CH_3OH. (a) What pressure is required to convert 25% of the CO to CH_3OH at 500 K if the "synthesis gas" comes from CH_4? (b) If the synthesis gas comes from coal?

SOLUTION

(a)

	CO	+ $2H_2$	= CH_3OH	
Initial	1	3	0	
Equil.	$1-\xi$	$3-2\xi$	ξ	Total = $4-2\xi$

$$K = \frac{y_{CH_3OH}}{y_{CO}\, y_{H_2}^2\, P^2} = \frac{\xi(4-2\xi)^2}{(1-\xi)(3-2\xi)^2\, P^2}$$

$$\Delta G^\circ = -133.52 - (-155.68) = 22.16 \text{ kJ mol}^{-1}$$

$$K = 4.842 \times 10^{-3}$$

$$P = \left[\frac{\xi(4-2\xi)^2}{K_P(1-\xi)(3-2\xi)^2}\right]^{1/2}$$

$$= \left[\frac{(0.25)(3.5)^2}{4.842 \times 10^{-3}(0.75)(2.5)^2}\right]^{1/2} = 11.6 \text{ bar}$$

(b)

	CO	+ $2H_2$	= CH_3OH	
Initial	1	1	0	
Equil.	$1-\xi$	$1-2\xi$	ξ	Total = $2-2\xi$

$$K = \frac{\xi(2-2\xi)^2}{(1-\xi)(1-2\xi)^2\, P^2} \qquad P = \left[\frac{(0.25)(1.5)^2}{K(0.75)(0.5)^2}\right]^{1/2} = 24.9 \text{ bar}$$

5.16 Show that in going from a standard-state pressure of 1 atm to 1 bar, thermodynamic quantities for reactions are corrected as follows:

$$\Delta C_p^\circ(\text{bar}) = \Delta C_p^\circ(\text{atm})$$
$$\Delta H^\circ(\text{bar}) = \Delta H^\circ(\text{atm})$$
$$\Delta S^\circ(\text{bar}) = \Delta S(\text{atm}) + (0.109\ \text{J K}^{-1}\ \text{mol}^{-1})\ \Sigma\nu_i$$
$$\Delta G^\circ(\text{bar}) = \Delta G^\circ(\text{atm}) - (0.109 \times 10^{-3}\ \text{kJ K}^{-1}\ \text{mol}^{-1})T\Sigma\nu_i$$
$$K\ (\text{bar}) = K(\text{atm})(1.013\ 25)^{\Sigma\nu_i}$$

where $\Sigma\nu_i$ is the difference between the stoichiometric coefficients of gaseous products and gaseous reactants in the balanced chemical equation.

SOLUTION

For perfect gases the enthalpy and heat capacity are independent of pressure, and so the choice of standard state does not affect ΔH° or ΔC_p for a reaction.

As explained in connection with equation 3.75 the entropy decreases when a perfect gas is compressed from 1 bar to 1 atm.

$$\Delta S^\circ = -R \ln (101325\ \text{Pa}/10^5\ \text{Pa}) = -0.109\ \text{J K}^{-1}\ \text{mol}^{-1}$$

To apply this to a chemical reaction consider

A (1 atm)	+	B (1 atm)	=	C (1 atm)
$\uparrow$		$\uparrow$		$\downarrow$
A (1 bar)	+	B (1 bar)	=	C (1 bar)

$$\Delta S^\circ\ (\text{bar}) = -0.109 - 0.109 + \Delta S^\circ\ (\text{atm}) + 0.109$$
$$= \Delta S^\circ\ (\text{atm}) - 0.109\ \text{J K}^{-1}\ \text{mol}^{-1}$$

Since $\Sigma\nu_i = -1$, we can generalize this to the expression given above.

The expression for ΔG° (bar) can be derived from

$$\Delta G^\circ\ (\text{bar}) = \Delta H^\circ\ (\text{bar}) - T\Delta S^\circ\ (\text{bar})$$

but let us derive the expression for K first and use it.

$$K\ (\text{bar}) = \prod_i [P_i/(1\ \text{bar})]^{\nu_i} \text{ and } K\ (\text{atm}) = \prod_i [P_i/(1\ \text{atm}]^{\nu_i}$$

$$\frac{K\ (\text{bar})}{K\ (\text{atm})} = \prod_i \left[\frac{(1\ \text{atm})}{(1\ \text{bar})}\right]^{\nu_i} = \left[\frac{(1\ \text{atm})}{(1\ \text{bar})}\right]^{\Sigma\nu_i} = 1.01325^{\Sigma\nu_i}$$

$$\Delta G^\circ\ (\text{bar}) = -RT\ \ln\left\{\prod_i [P_i/(1\ \text{bar})]^{\nu_i}\right\}$$

$$= -RT\ \ln\left\{\prod_i [P_i/(1\ \text{atm})]^{\nu_i}[(1\ \text{atm})/(1\ \text{bar})]^{\Sigma\nu_i}\right\}$$

$$= -RT\ \ln K(\text{atm}) = RT\ \ln 1.01325^{\Sigma\nu_i}$$

$$= \Delta G^\circ\ (\text{atm}) - (0.109 \times 10^{-3}\ \text{kJ K}^{-1}\ \text{mol}^{-1})T\Sigma\nu_i$$

5.17 Calculate the total pressure that must be applied to a mixture of three parts of hydrogen and one part nitrogen to give a mixture containing 10% ammonia at equilibrium at 400 °C. At 400 °C, $K = 1.60 \times 10^{-4}$ for the reaction $N_2(g) + 3H_2(g) = 2NH_3(g)$.

SOLUTION

$$K = \frac{(P_{NH_3}/P^\circ)^2}{(P_{N_2}/P^\circ)(P_{H_2}/P^\circ)} = 1.60 \times 10^{-4} \qquad \text{OR}$$

$$K = \frac{P_{NH_3}^2}{P_{N_2}P_{H_2}^3} = 1.60 \times 10^{-4}\ \text{bar}^{-2}$$

$$= \frac{(0.1P)^2}{(\frac{1}{4}\ 0.9P)(\frac{3}{4}\ 0.9P)^3} = \frac{(0.01)4^4}{3^3(0.9)^4\ P^2}$$

$$P = 30.1\ \text{bar}$$

5.18 In the synthesis of methanol by $CO(g) + 2H_2(g) = CH_3OH(g)$ at 500 K, calculate the total pressure required for a 90% conversion to methanol if CO and H_2 are initially in a 1:2 ratio. Given: $K = 6.09 \times 10^{-3}$.

SOLUTION

	$CO(g)$	+ $2H_2(g)$	= $CH_3OH(g)$	
initial moles	1	2	0	
equil. moles	0.1	0.2	0.9	Total 1.2

$$K = \frac{\left(P_{CH_3OH}/P^o\right)}{\left(P_{CO}/P^o\right)\left(P_{H_2}/P^o\right)^2} = 6.09 \times 10^{-3}$$

$$= \frac{\frac{0.9}{1.2}\frac{P}{P^o}}{\frac{0.1}{1.2}\frac{P}{P^o}\left(\frac{0.2}{1.2}\frac{P}{P^o}\right)^2}$$

$$\frac{P}{P^o} = \sqrt{\frac{(0.9)(1.2)^2}{(0.1)(0.04)(6.09 \times 10^{-3})}} = 231$$

P = 231 bar = total pressure for 90% conversion to CH_3OH

5.19 At 1273 K and at a total pressure of 30.4 bar the equilibrium in the reaction $CO_2(g) + C(s) = 2CO(g)$ is such that 17 mole % of the gas is CO_2. (a) What percentage would be CO_2 if the total pressure were 20.3 bar? (b) What would be the effect on the equilibrium of adding N_2 to the reaction mixture in a closed vessel until the partial pressure of N_2 is 10 bar? (c) At what pressure of the reactants will 25% of the gas be CO_2?

SOLUTON

(a) P_{CO_2} = (30.4 bar)(0.17) = 5.2 bar

P_{CO} = (30.4 bar)(0.83) = 25.2 bar

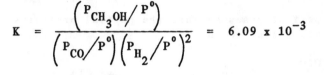

$$K = \frac{(25.2)^2}{5.2} = 122$$

Let x = mole fraction CO_2

$$K = \frac{[20.3(1 - x)]^2}{20.3x} = 122 \qquad x = 0.127$$

Percentage CO_2 at equilibrium = 12.7%

(b) No effect for perfect gases because the partial pressures of the reactants are not affected.

(c) $$K = \frac{[0.75(P/P^\circ)]^2}{0.25(P/P^\circ)} = 122$$

$$P = 54 \text{ bar}$$

5.20 At 2000 °C water is 2% dissociated into oxygen and hydrogen at a total pressure of 1 bar.

(a) Calculate K for $H_2O(g) = H_2(g) + \frac{1}{2} O_2(g)$

(b) Will the degree of dissociation increase or decrease if the pressure is reduced? (c) Will the degree of dissociation increase or decrease if argon gas is added, holding the total pressure equal to 1 bar? (d) Will the degree of dissociation change if the pressure is raised by addition of argon at constant volume to the closed system containing partially dissociated water vapor? (e) Will the degree of dissociation increase or decrease if oxygen gas is added while holding the total pressure constant at 1 bar?

SOLUTION

(a)	$H_2O(g)$	= $H_2(g)$	+ $\frac{1}{2} O_2(g)$	
Initially	1	0	0	
Equilibrium	$1-\xi$	ξ	$\xi/2$	Total = $1 + \xi/2$

$$P_{H_2O} = \frac{1 - \xi}{1 + \frac{\xi}{2}} P \qquad P_{H_2} = \frac{\xi}{1 + \frac{\xi}{2}} P \qquad P_{O_2} = \frac{\xi/2}{1 + \frac{\xi}{2}} P$$

$$K = \frac{\left[\frac{\xi/2}{1 + \xi/2}\frac{P}{P^\circ}\right]^{1/2}\left[\frac{\xi}{1 + \xi/2}\frac{P}{P^\circ}\right]}{\frac{1 - \xi}{1 + \xi/2}\frac{P}{P^\circ}} = \frac{\xi^{3/2}\,(P/P^\circ)^{1/2}}{\sqrt{2}\left(1 + \frac{\xi}{2}\right)^{1/2}(1 - \xi)}$$

$$= \frac{0.02^{3/2}\;1^{1/2}}{\sqrt{2}\;(1.01)^{1/2}\;(0.98)} = 2.03 \times 10^{-3}$$

(b) If the total pressure is reduced, the degree of dissociation will increase because the reaction will produce more molecules to fill the volume.

(c) If argon is added at constant pressure, the degree of dissociation will increase because the partial pressure due to the reactants will decrease.

(d) If argon is added at constant volume, the degree of dissociation will not be changed because the partial pressure due to the reactants will not change.

If oxygen is added at constant total pressure, the degree of dissociation of H_2O will decrease because the reaction will be pushed to the left.

5.21 At 250 °C PCl_5 is 80% dissociated at a pressure of 1.013 bar, and so K = 1.80. What is the percentage dissociation at equilibrium after sufficient nitrogen has been added at constant pressure to produce a nitrogen partial pressure of 0.9 bar? The total pressure is maintained at 1 bar.

SOLUTION

$$K = \frac{0.8^2\,(1.013)}{1 - 0.8^2} = 1.80$$

$$K = \frac{\alpha^2(0.1)}{1 - \alpha^2}$$

α = 0.973 or 97.3% dissociated.

5.22 The following data apply to the reaction $Br_2(g) = 2Br(g)$

T/K	1123	1172	1223	1273
$K/10^{-3}$	0.408	1.42	3.32	7.2

Determine by graphical means the enthalpy change when 1 mol of Br_2 dissociates completely at 1200 K.

SOLUTION

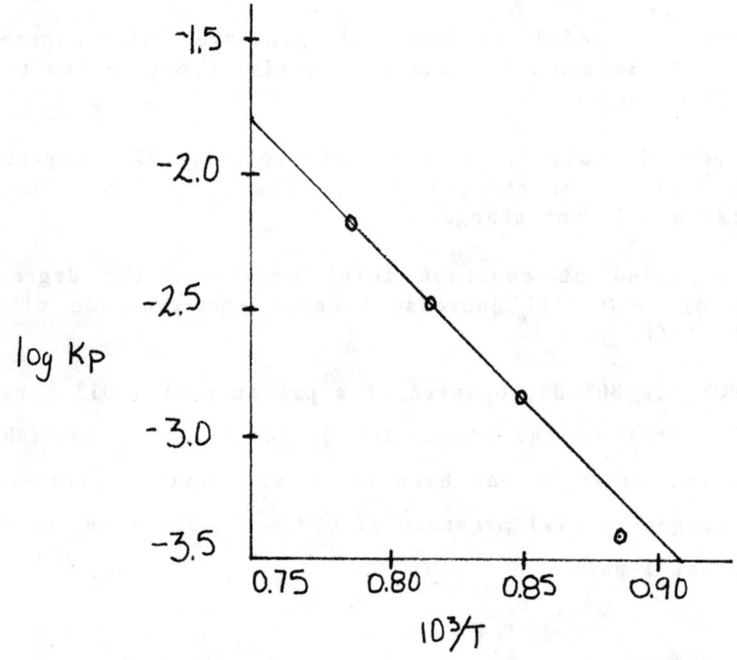

At T = 1200 K ($1/T = 0.833 \times 10^{-3}$) the slope is -10.4×10^3 K

$\Delta H^\circ = -2.303\ R\ (\text{slope})$

$= (2.303)(8.314\ \text{J K}^{-1}\ \text{mol}^{-1})(10.4 \times 10^3\ \text{K})$

$= 199.2\ \text{kJ mol}^{-1}$

5.23 The vapor pressure of water above mixtures of $CuCl_2 \cdot H_2O(cr)$ and $CuCl_2 \cdot 2H_2O(cr)$ is given as a function of temperature in the following table.

$t/^\circ C$	17.9	39.8	60.0	80.0
P/bar	0.0049	0.0250	0.122	0.327

(a) Calculate ΔH° for the reaction

$$CuCl_2 \cdot 2H_2O(cr) = CuCl_2 \cdot H_2O(cr) + H_2O(g)$$

(b) Calculate ΔG° for the reaction at 60 °C.

(c) Calculate ΔS° for the reaction at 60 °C.

SOLUTION

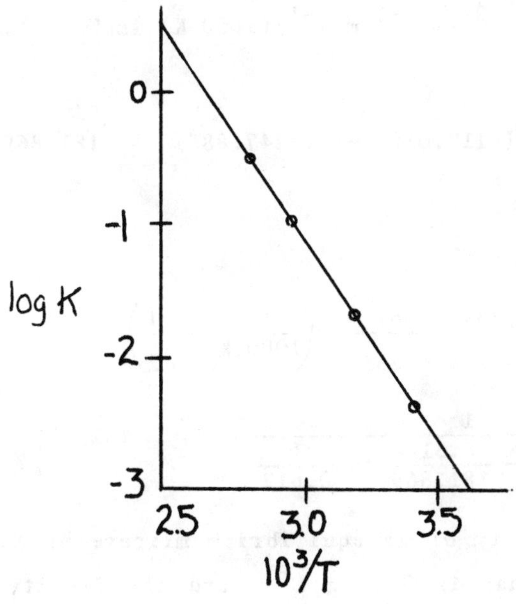

(a) $K = P_{H_2O}$ slope = -3060 K

$$\Delta H^\circ = -2.303\ R(\text{slope}) = 57.3\ \text{kJ mol}^{-1}$$

(b) $\Delta G^\circ = -RT \ln K = -(8.314)(333\ \text{K}) \ln 0.122 = 5.82\ \text{kJ mol}^{-1}$

(c) $\Delta S^\circ = (\Delta H^\circ - \Delta G^\circ)/T = 154.7\ \text{J K}^{-1}\ \text{mol}^{-1}$

5.24 The following reaction is nonspontaneous at room temperature and endothermic.

$$3C(graphite) + 2H_2O(g) = CH_4(g) + 2CO(g)$$

As the temperature is raised, the equilibrium constant will become equal to unity at some point. Estimate this temperature using data from Table A.2.

SOLUTION

At 1000 K

$$\Delta G^\circ = 19.460 + 2(-200.297) - 2(-192.576) = 4.018 \text{ kJ mol}^{-1}$$

$$= -(8.314 \times 10^{-3} \text{ kJ K}^{-1} \text{ mol}^{-1})(1000 \text{ K}) \ln K_P$$

$$K = 0.617$$

$$\Delta H^\circ = -89.881 + 2(-112.010) - 2(-247.885) = 181.869 \text{ kJ mol}^{-1}$$

$$\ln \frac{K_2}{K_1} = \frac{\Delta H^\circ}{R}\left(\frac{1}{T_1} - \frac{1}{T_2}\right)$$

$$\ln \frac{1}{0.617} = \frac{181{,}869 \text{ J mol}^{-1}}{8.314 \text{ J K}^{-1} \text{ mol}^{-1}}\left(\frac{1}{1000 \text{ K}} - \frac{1}{T_2}\right)$$

$$T_2 = \frac{1}{\dfrac{1}{1000 \text{ K}} - \dfrac{8.314}{181{,}869} \ln \dfrac{1}{0.617}} = 1023 \text{ K}$$

5.25 The measured density of an equilibrium mixture of N_2O_4 and NO_2 at 15 °C and 1.013 bar is 3.62 g L^{-1}, and the density at 75 °C and 1.013 bar is 1.84 g L^{-1}. What is the enthalpy change of the reaction $N_2O_4(g) = 2NO_2(g)$?

SOLUTION

At 15 °C

$$M = \frac{RT}{P}\frac{g}{V} = \frac{(0.08314)(288)(3.62)}{1.013} = 85.57 \text{ g mol}^{-1}$$

$$\alpha = \frac{M_1 - M_2}{M_2} = \frac{92.01 - 85.57}{85.57} = 0.0753$$

$$K = \frac{4\alpha^2 P}{1 - \alpha^2} = \frac{4(0.0753)^2(1.013)}{1 - 0.0753^2} = 0.0231$$

At 75 °C

$$M = \frac{(0.08314)(348)(1.84)}{1.013} = 52.55 \text{ g mol}^{-1}$$

$$= \frac{92.01 - 52.55}{52.55} = 0.751 \qquad K = 5.24$$

$$\Delta H = \frac{RT_1T_2}{(T_2 - T_1)} \ln\frac{K_2}{K_1} = \frac{(8.314)(288)(348)}{60} \ln\frac{5.24}{.0231}$$

$$= 75 \text{ kJ mol}^{-1}$$

5.26 (a) A mixture of one mole of hydrogen and one mole of nitrogen reacts to form ammonia at equilibrium at 500 K and 1 bar. The equilibrium extent of reaction is 0.0599. What is the effect of adding a small amount of nitrogen at constant pressure, as expressed by $\partial\xi/\partial n_2{}^o$? (b) A mixture of one mole of hydrogen and 1.2 mole of nitrogen reacts to form ammonia at equilibrium at 500 K and 1 bar. The equilibrium extent of reaction is 0.0597. What is the effect of adding a small amount of nitrogen at constant pressure, as expressed by $\partial\xi/\partial n_2{}^o$?

SOLUTION

(a) For the reaction $3H_2(g) + N_2(g) = 2NH_3(g)$, the numerator of equation 5.83 is equal to zero.

$$\frac{\upsilon}{n} - \frac{\upsilon_2}{n_2} = \frac{-2}{2 - 2\xi} + \frac{1}{1 - \xi} = 0$$

Thus the addition of a small amount of nitrogen at constant

pressure and temperature does not change the extent of reaction. The volume will of course increase slightly. Note that the mole fraction of nitrogen in the equilibrium mixture is $(1 - \xi)/(2 - 2\xi) = 0.5$.

(b) The numerator of equation 5.83 is negative.

$$\frac{\nu}{n} - \frac{\nu_2}{n_2} = \frac{-2}{2.2 - 2\xi} + \frac{1}{1.2 - \xi}$$

$$= \frac{-2}{2.2 - 2(.0597)} + \frac{1}{1.2 - .0597} = -.0843$$

Thus adding a small amount of nitrogen at constant pressure and temperature causes some ammonia to dissociate. This effect can also be seen in the extents of reaction given above. If the nitrogen addition is continued, the NH_3 will completely dissociate at equilibrium. This effect is not obtained when hydrogen is added to the equilibrium mixture at constant pressure and temperature.

5.27 The preceding problem shows that under certain conditions the addition of N_2 to the N_2, H_2, NH_3 equilibrium causes NH_3 to dissociate. (a) Why doesn't this happen with H_2? (b) Is NH_3 present in the limit of indefinitely large additions of hydrogen? (c) Nitrogen?

SOLUTION

This doesn't happen to H_2 because its stoichiometric coefficient is larger than that of N_2. Even when the ratio of hydrogen to nitrogen goes to infinity, there will still be some NH_3 if there is some N_2. As N_2 is added, however, all NH_3 dissociates in the limit. These same considerations apply to any gas reaction for which the reactants have different stoichiometric coefficients.

5.28 Mercuric oxide dissociates according to the reaction $2HgO(cr) = 2Hg(g) + O_2(g)$. At 420 °C the dissociation pressure is 5.16×10^4 Pa, and at 450 °C it is 10.8×10^4 Pa. Calculate (a) the equilibrium constants, and (b) the enthalpy of dissociation per mole of HgO.

SOLUTION

(a) $P_{Hg} = 2\,P_{O_2} \qquad P_{Hg} = \frac{2}{3}P \qquad P_{O_2} = \frac{1}{3}P$

$$K_{420} = P_{Hg}^2\,P_{O_2} = \left(\frac{2}{3}\right)^2\left(\frac{1}{3}\right)P^3$$

$$= \left(\frac{4}{27}\right)\left(\frac{5.16 \times 10^4\ \text{Pa}}{1.013 \times 10^5\ \text{Pa}}\right)^3 = 0.0196$$

$$K_{450} = \left(\frac{4}{27}\right)\left(\frac{10.8 \times 10^4\ \text{Pa}}{1.013 \times 10^5\ \text{Pa}}\right)^3 = 0.1794$$

(b)
$$\Delta H^\circ = \frac{RT_1T_2}{T_2 - T_1}\ln\frac{K_2}{K_1}$$

$$= \frac{(8.314\ \text{J K}^{-1}\ \text{mol}^{-1})(693\ \text{K})(723\ \text{K})}{30\ \text{K}}\ln\frac{0.1794}{0.0196}$$

$= 308\ \text{kJ mol}^{-1}$ for the reaction as written

$= 154\ \text{kJ mol}^{-1}$ of HgO(cr)

5.29 Calculate (a) K and (b) ΔG° for the following reaction at 20 °C.

$$CuSO_4\cdot 4NH_3(cr) = CuSO_4\cdot 2NH_3(cr) + 2NH_3(g)$$

The equilibrium pressure of NH_3 is 8.27 kPa.

SOLUTION

(a)
$$K = \left(P_{NH_3}/P^\circ\right)^2 = \left(\frac{8270\ \text{Pa}}{101{,}325\ \text{Pa}}\right)^2 = 6.66 \times 10^{-3}$$

(b) $\Delta G^{\circ} = -RT \ln K$

$= -(8.314 \text{ J K}^{-1} \text{ mol}^{-1})(293.15 \text{ K}) \ln 6.66 \times 10^{-3}$

$= 12.1 \text{ kJ mol}^{-1}$

5.30 The decomposition of silver oxide is represented by

$$2Ag_2O(cr) = 4Ag(cr) + O_2(g)$$

Using data from Table A.1 and assuming $\Delta C_P^{\circ} = 0$ calculate the temperature at which the equilibrium pressure of O_2 is 0.2 bar. This temperature is of interest because Ag_2O will decompose to yield Ag at temperatures above this value if it is in contact with air.

SOLUTION

$\Delta H^{\circ} = -2(-31.05) = 62.10 \text{ kJ mol}^{-1}$

$\Delta S^{\circ} = 4(42.55) + 205.138 - 2(121.3) = 132.7 \text{ J K}^{-1} \text{ mol}^{-1}$

$\Delta G = \Delta G^{\circ} + RT \ln P_{O_2}$

$0 = \Delta H^{\circ} - T\Delta S^{\circ} + RT \ln 0.2$

$\Delta H^{\circ} = T(\Delta S^{\circ} - R \ln 0.2)$

$$T = \frac{\Delta H^{\circ}}{\Delta S^{\circ} - R \ln 0.2} = \frac{62{,}100 \text{ J mol}^{-1}}{(132.7 - 8.3(4 \ln 0.2)) \text{J K}^{-1} \text{ mol}^{-1}}$$

$= 425 \text{ K}$

5.31 The dissociation of ammonium carbamate takes place according to the reaction

$$(NH_2)CO(ONH_4)(cr) = 2NH_3(g) + CO_2(g)$$

When an excess of ammonium carbamate is placed in a previously evacuated vessel, the partial pressure generated by NH_3 is twice the partial pressure of the CO_2, and the partial pressure of $(NH_2)CO(ONH_4)$ is negligible in comparison. Show that

$$K = \left(\frac{P_{NH_3}}{P^o}\right)^2\left(\frac{P_{CO_2}}{P^o}\right) = \frac{4}{27}\left(\frac{P}{P^o}\right)^3$$ where P is the total pressure.

SOLUTION

$P = P_{NH_3} + P_{CO_2} = 3\ P_{CO_2}$ since $P_{NH_3} = 2\ P_{CO_2}$

$P_{CO_2} = P/3$ $\qquad P_{NH_3} = \frac{2}{3}P$

$$K = \left(\frac{P_{NH_3}}{P^o}\right)^2\left(\frac{P_{CO_2}}{P^o}\right) = \left(\frac{2}{3}\frac{P}{P^o}\right)^2\left(\frac{1}{3}\frac{P}{P^o}\right) = \frac{4}{27}\left(\frac{P}{P^o}\right)^3$$

5.32 At 1000 K methane at 1 bar is in the presence of hydrogen. In the presence of a sufficiently high partial pressure of hydrogen, methane does not decompose to form graphite and hydrogen. What is this partial pressure?

SOLUTION

$CH_4(g) = C(\text{graphite}) + 2H_2(g)$

$\Delta G^o = -RT\ \ln K = -19.46\ \text{kJ mol}^{-1}$

$$K = 10.39 = \frac{\left(P_{H_2}/P^o\right)^2}{P_{CH_4}/P^o}$$

$P_{H_2} = P^o[(10.39)(1)]^{1/2} = 3.2\ \text{bar}$

5.33 A gaseous system contains CO, CO_2, H_2, H_2O, and C_6H_6 in chemical equilibrium. How many components are there? How many independent reactions? How many degrees of freedom are there?

SOLUTION

	CO	CO_2	H_2	H_2O	C_6H_6
C	1	1	0	0	6
O	1	2	0	1	0
H	0	0	2	2	6

Subtract the first row from the second and divide the third row by 2 to obtain

1	1	0	0	6
0	1	0	1	-6
0	0	1	1	3

Subtract the second row from the first

1	0	0	-1	12
0	1	0	1	-6
0	0	1	1	3

The rank of this matrix is 3, and so there are 3 independent components. The stoichiometric coefficients of 2 independent reactions are given by the last 2 columns.

$$CO_2 + H_2 = H_2O + CO$$

$$12CO + 3H_2 = C_6H_6 + 6CO_2$$

If the species are arranged in a different order in the matrix, a different pair of independent equations will be obtained.

$$F = c - p + 2$$

If the numbers of moles of 3 components and the temperature are specified, the system is completely determined; that is, the numbers of moles of the other two species and the pressure can be calculated--given values of the equilibrium constant. Alternatively, the mole fractions of 2 species and the temperature and pressure may be specified. We then have two mathematical expressions between 5 mole fractions, 4 of which ae independent since $\Sigma y_i = 1$. If 2 mole fractions are known, the other two can be calculated from the two simultaneous equations.

5.34 Superheated steam is passed over coal, represented in these calculations by graphite, at 1000 K to produce CO, CO_2, H_2, and CH_4. Since there are 6 species and 3 element balance equations, there are 3 independent chemical reactions which may be written as follows:

$$3(\text{graphite}) + H_2O(g) = CO(g) + H_2(g)$$

$$3C(\text{graphite}) + 2H_2O(g) = CH_4(g) + 2CO(g)$$

$$2C(\text{graphite}) + 2H_2O(g) = CH_4(g) + CO_2(g)$$

What are the equilibrium constants for these reactions? If the total pressure is raised to 100 bar, one of these reactions will predominate. Neglecting the other two reactions, calculate the equilibrium mole fractions of the gases present.

<u>SOLUTION</u>

Using Table A.2 the following equilibrium constants are calculated for 1000 K

$$C(\text{graphite}) + H_2O(g) = CO(g) + H_2(g) \qquad K = 2.5$$

$$3C(\text{graphite}) + 2H_2O(g) = CH_4(g) + 2CO(g) \qquad K = 0.61$$

$$2C(\text{graphite}) + 2H_2O(g) = CH_4(g) + CO_2(g) \qquad K = 0.35$$

At 100 bar the K_y values for these three reactions are 0.025, 0.0061, and 0.35, respectively. Thus the third reaction will predominate.

$$K_y = \frac{y_{CH_4}\, y_{CO_2}}{y_{H_2O}^2} = 0.35 \qquad y_{CH_4} = y_{CO_2} = y$$

$$1 = y_{H_2O} + y_{CH_4} + y_{CO_2} = y_{H_2O} + 2y \qquad y_{H_2O} = 1 - 2y$$

Substituting in K_y and solving the quadratic equation for y yields

$$y_{CH_4} = 0.27, \qquad y_{CO_2} = 0.27, \qquad y_{H_2O} = 0.46.$$

5.35 Acetic acid and ethanol react to form ethyl acetate and water according to

$$CH_3CO_2H(l) + C_2H_5OH(l) = CH_3CO_2C_2H_5(l) + H_2O(l)$$

Analytical data obtained by titrating the acetic acid remaining at equilibrium at 298 K indicates that K = 40 when mole fractions are used, and the solution is assumed to be ideal. The standard Gibbs energies of formation of ethanol, acetic acid, and water are given in Table A.1. Calculate the standard Gibbs energy of formation of ethyl acetate.

SOLUTION

$$\Delta G^\circ = -RT \ln K = -(8.314)(298) \ln 4.0 = -3.44 \text{ kJ mol}^{-1}$$

$$= \Delta_f G^\circ_{EtAc} - 228.57 + 389.9 + 174.78$$

$$\Delta_f G^\circ_{EtAc} = -339.6 \text{ kJ mol}^{-1}$$

5.36 The equilibrium constant for the association of benzoic acid to a dimer in dilute benzene solutions is as follows at 43.9 °C.

$$2C_6H_5COOH = (C_6H_5COOH)_2 \qquad K_c = 2.7 \times 10^2$$

Molar concentrations are used in expressing the equilibrium constant. Calculate ΔG°, and state its meaning.

SOLUTION

$$\Delta G^\circ = -RT \ln K_c = -(8.314 \text{ J K}^{-1} \text{ mol}^{-1})(317.1 \text{ K}) \ln (2.7 \times 10^2)$$

$$= -14.8 \text{ kJ mol}^{-1}$$

This is the decrease in Gibbs energy when 2 mol of monomer at unit activity on the molar scale is converted to 1 mol of dimer at unity activity on the molar scale.

5.37 Amylene, C_5H_{10}, and acetic acid react to give the ester according to the reaction

$$C_5H_{10}(1) + CH_3COOH(1) = CH_3CooC_5H_{11}(1)$$

What is the value of K_c if 0.006 45 mol of amylene and 0.001 mol of acetic acid, dissolved in a certain inert solvent to give a total volume of 0.845 L, react to give 0.000 784 mol of ester? Use the molar concentration scale.

SOLUTION

$$K = \frac{(7.84 \times 10^{-4})/0.845}{[(64.5 \times 10^{-4} - 7.84 \times 10^{-4})/0.845][10 \times 10^{-4} - 7.84 \times 10^{-4})/0.845]}$$

$= 541$

5.38 $\Delta S_{mix} = -R(x_1 \ln x_1 + x_2 \ln x_2 + x_3 \ln x_3)$

$\Delta G_{mix} = -T\Delta S_{mix}$

5.39 -4.73 kJ mol^{-1}, 15.88 J K^{-1} mol^{-1}

5.40 $\Delta G° > 17.2$ kJ mol^{-1}

5.41 0.696

5.42 1.87×10^{-4}, 1.64×10^{-4}, 1.90×10^{-4}, 0.364×10^{-4}
As the pressure is increased, the gases behave more imperfectly.

5.43 (a) 17.0 kJ mol^{-1} (b) 0.795 bar

5.44 (a) 9.63 g (b) 1.215, 0.032, 1.457 bar

5.45 $k = 3^{3/2}/4\ K = 0.0164$

5.46 0.351 bar

5.47 3.74 bar

5.48 (a) 2.48 (b) 0.54 mol

5.49	ξ = 0.800	K = 1.80	
5.50	(a) 0.0788	(b) 0.0565	
5.51	6.9		
5.52	1.81×10^{-2}		
5.53	2.99×10^{-3}	0.0273	0.0861
5.54	0.081	0.363	
5.55	0.53		
5.56	(a) 1.20×10^{10}	3.91×10^{-2}	
	(b) 8.25×10^{-2}	0.352	
5.57	18.2 bar		
5.58	(a) No	(b) 0.285	
5.59	(a) 5×10^{-3} bar	(b) 1.79%	(c) 0.0221 bar

5.60 (a)

$t/^{\circ}C$	25	45	65
α	0.1852	0.3775	0.6284
K	0.1451	0.6737	2.644

(b) 60.7 kJ mol^{-1} (c) 0.320 (d) 0.371

5.61	135.8 $^{\circ}C$	
5.62	153 kJ mol^{-1}	
5.63	1.94×10^{-13} bar	
5.64	6.88×10^{-24}	4.79×10^{-43}
5.65	452 kJ mol^{-1}	
5.66	(a) 19.70 bar	(b) 35.6 bar

5.67 (a) $y_{N_2} = 0.5000$, $y_{H_2} = 0.4363$, and $y_{NH_3} = 0.0637$

(b) $y_{N_2} = 0.5481$, $y_{H_2} = 0.3946$, and $y_{NH_3} = 0.0574$

5.69 Dissociation into atoms and solubility proportional to the partial pressure of atoms.

5.70 Fe_2O_3

5.71 5.6×10^{-5} bar

5.72 2.66×10^{-6} bar

5.73 7.55 bar

5.74 (a) 9.75 g (b) 19.75 g

5.75 3.31×10^{-16} bar of O_2 is required to oxidize Ni, but the partial pressure of O_2 from partially dissociation H_2O in the presence of 3 mole percent hydrogen is only 9.7×10^{-18} bar.

5.76 3 components, 1 independent reaction, variance of 4. If C is present, the answers are 3, 2, and 3.

5.77 16.06 kJ mol^{-1}

CHAPTER 6: Phase Equilibrium: One-Component Systems and Ideal Solutions

6.1 What is the maximum number of phases that can be in equilibrium at constant temperature and pressure in one-, two-, and three-component systems?

SOLUTION

$F = c - p + 2$

$0 = 1 - p + 2, \quad p = 3$

$0 = 2 - p + 2, \quad p = 4$

$0 = 3 - p + 2, \quad p = 5$

6.2 How many degrees of freedom are there for the following systems, and how might they be chosen?

(a) $CuSO_4 \cdot 5H_2O(cr)$ in equilibrium with $CuSO_4(cr)$ and $H_2O(g)$.

(b) N_2O_4 in equilibrium with NO_2 in the gas phase.

(c) CO_2, CO, H_2O, and H_2 in equilibrium in the gas phase.

(d) The system described in (c) is made up of specified amount of CO and H_2.

SOLUTION

(a) $c = s - r = 3 - 1 = 2$
$F = c - p + 2 = 2 - 3 + 2 = 1$
Only the temperature or pressure may be fixed.

(b) $c = s - r = 2 - 1 = 1$
$F = c - p + 2 = 1 - 1 + 2 = 2$
Temperature and pressure may be fixed.

(c) $c = 4 - 1 = 3$
$F = c - p + 2 = 3 - 1 + 2 = 4$
Temperature, pressure, and two mole fractions may be fixed.

(d) $c = s - r - m = 4 - 1 - 2 = 1$
$F = c - p + 2 = 1 - 1 + 2 = 2$
Only temperature and pressure may be fixed if the elemental composition is fixed.

6.3 How many degrees of freedom do the following systems have?

(a) $NH_4Cl(cr)$ is allowed to dissociate to $NH_3(g)$ and $HCl(g)$ until equilibrium is reached.

(b) A solution of alcohol in equilibrium with its vapor.

SOLUTION

(a) $NH_4Cl(cr) = NH_3(g) + HCl(g)$

$c = s - r - m = 3 - 1 - 1 = 1$

The additional constraint is that only equal numbers of moles of NH_3 and HCl can be formed.

$F = c - p + 2 = 1 - 2 + 2 = 1$
Only temperature or pressure may be fixed.

(b) $F = c - p + 2 = 2 - 2 + 2 = 2$
Temperature and pressure may be fixed.

6.4 Ice has the unusual property of a melting point that is lowered by increasing pressure. Thus, one can skate on ice provided that the pressure exerted by one's skates is great enough to liquefy the ice under them. Would a 75-kg skater whose skates contact the ice with an area of 0.1 cm^2 be able to skate at -3 °C?

SOLUTION

From Example 6.3, $\Delta T = -(7.5 \times 10^{-3}\ K\ bar^{-1})\Delta P$

$$\Delta T = \frac{-(7.5 \times 10^{-3}\ K\ bar^{-1})(75\ kg)(9.80\ m\ s^{-2})}{(0.1\ cm^2)(10^{-2}\ m\ cm^{-1})^2(10^5\ Pa\ bar^{-1})}$$

$$= -5.51\ ^{\circ}C$$ so the answer is yes.

6.5 The change in Gibbs energy for the conversion of aragonite to calcite at 25 °C is −1046 J mol^{-1}. The density of aragonite is 2.93 g cm^{-3} at 25 °C and the density of calcite is 2.71 g cm^{-3}. At what pressure at 25 °C would these two forms of $CaCO_3$ be in equilibrium?

SOLUTION

aragonite = calcite $\Delta G° = -1046 \text{ J mol}^{-1}$

$$\Delta V = \frac{100.09 \text{ g mol}^{-1}}{2.71 \text{ g cm}^{-3}} - \frac{100.09 \text{ g mol}^{-1}}{2.73 \text{ g cm}^{-3}}$$

$$= 2.77 \text{ cm}^3 \text{ mol}^{-1} = 2.77 \times 10^{-6} \text{ m}^3 \text{ mol}^{-1}$$

$$\left(\frac{\partial \Delta G}{\partial P}\right)_T = \Delta V \qquad \int_1^2 d\Delta G = \int_1^P \Delta V dP$$

$$\Delta G_2 - \Delta G_1 = \Delta V(P - 1)$$

$$0 + 1046 \text{ J mol}^{-1} = (2.77 \times 10^{-6} \text{ m}^3 \text{ mol}^{-1})(P - 1)$$

$$P = 3780 \text{ bar}$$

6.6 n-Propyl alcohol has the following vapor pressures.

$t/°C$	40	60	80	100
P_{sat}/kPa	6.69	19.6	50.1	112.3

Plot these data so as to obtain a nearly straight line, and calculate (a) the heat of vaporization, and (b) the boiling point at 1 bar.

SOLUTION

log P is plotted versus 1/T
(Please see top of page 85)

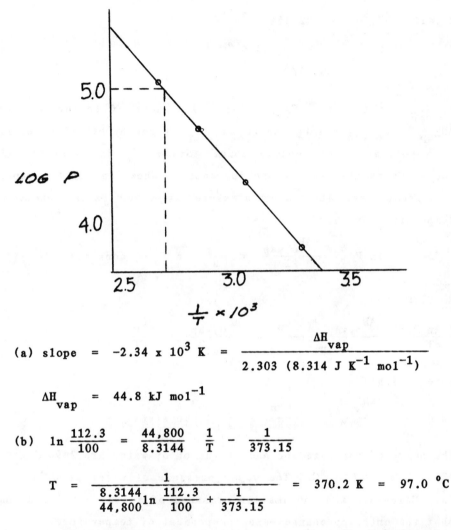

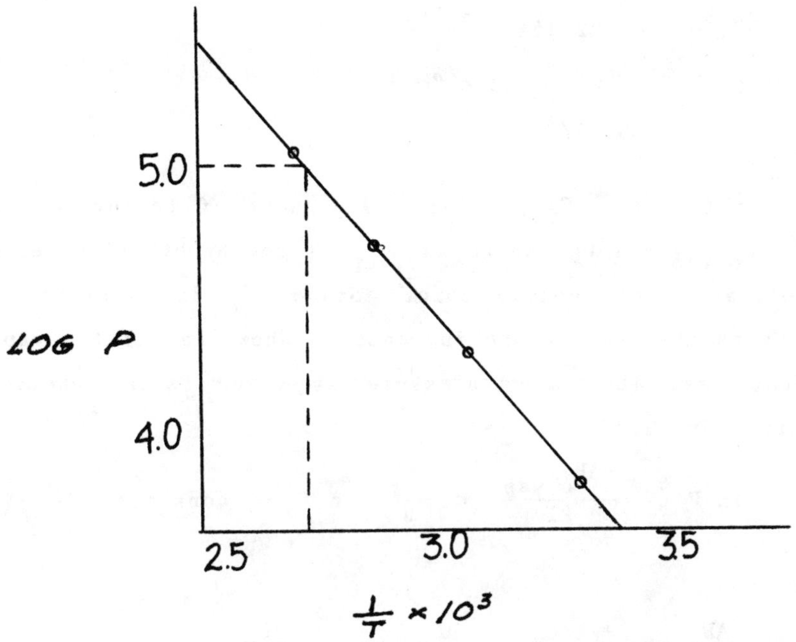

(a) slope $= -2.34 \times 10^3\ K = \dfrac{\Delta H_{vap}}{2.303\ (8.314\ J\ K^{-1}\ mol^{-1})}$

$$\Delta H_{vap} = 44.8\ kJ\ mol^{-1}$$

(b) $\ln \dfrac{112.3}{100} = \dfrac{44{,}800}{8.3144} \left(\dfrac{1}{T} - \dfrac{1}{373.15} \right)$

$$T = \frac{1}{\frac{8.3144}{44{,}800} \ln \frac{112.3}{100} + \frac{1}{373.15}} = 370.2\ K = 97.0\ ^{\circ}C$$

6.7 For uranium hexafluoride the vapor pressures (in Pa) for the solid and liquid are given by

$$\ln P_s = 29.411 - 5893.5/T$$

$$\ln P_l = 22.254 - 3479.9/T$$

Calculate the temperature and pressure of the triple point.

SOLUTION

$$29.411 - \frac{5893.5}{T} = 22.254 - \frac{3479.9}{T}$$

$$T = 337.2 \quad K = 64.0\ ^{\circ}C$$

$$P = e^{29.411 - 5893.5/337.2} = 152.2\ kPa$$

6.8 If $\Delta C_P = C_{P,vap} - C_{P,liq}$ is independent of temperature, then $\Delta H_{vap} = \Delta H_{o,vap} + T\Delta C_P$ where $\Delta H_{o,vap}$ is the hypothetical enthalpy of vaporization at abolute zero. Since ΔC_P is negative, ΔH_{vap} decreases as the temperature increases. Show that if the vapor is a perfect gas, the vapor pressure is given as a function of temperature by

$$\ln P = \frac{-\Delta H_{o,vap}}{RT} + \frac{\Delta C_P}{R} \ln T + \text{constant}$$

SOLUTION

$$\frac{d \ln P}{dT} = \frac{\Delta H_{o,vap} + T\Delta C_P}{RT^2} = \frac{\Delta H_{o,vap}}{RT^2} + \frac{\Delta C_P}{RT}$$

Integrating

$$\ln P = -\frac{\Delta H_{o,vap}}{RT} + \frac{\Delta C_P}{R} \ln T + \text{constant}$$

6.9 The heats of vaporization and of fusion of water are 2490 J g^{-1} and 333.5 J g^{-1} at 0 °C. The vapor pressure of water at 0 °C is 611 Pa. Calculate the sublimation pressure of ice at -15 °C, assuming that the enthalpy changes are independent of temperature.

SOLUTION

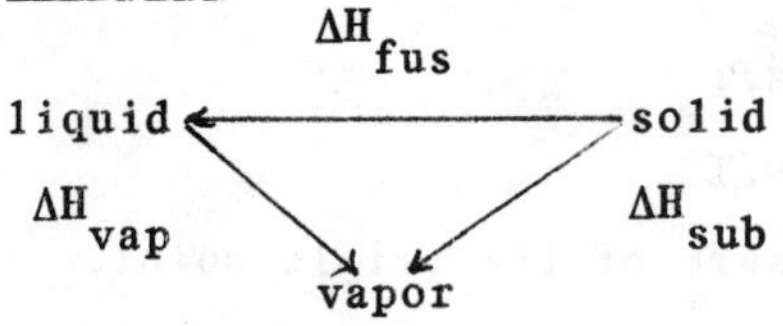

$$\Delta H_{sub} = \Delta H_{fus} + \Delta H_{vap}$$
$$= 333.5 + 2490$$
$$= 2824\ J\ g^{-1}$$

$$\ln \frac{P_2}{P_1} = \frac{\Delta H_{sub}(T_2 - T_1)}{RT_1T_2}$$

$$P_2 = P_1 \exp[\Delta H_{sub}(T_2 - T_1)/(RT_1T_2]$$

$$= (611 \text{ Pa}) \exp\left[\frac{(2824 \times 18 \text{ J mol}^{-1})(-15 \text{ K})}{(8.314 \text{ J K}^{-1} \text{ mol}^{-1})(273.15 \text{ K})(258.15 \text{ K})}\right]$$

$$= 166 \text{ Pa}$$

6.10 The vapor pressure of toluene is 8.00 kPa at 40.3 °C and 2.67 kPa at 18.4 °C. Calculate (a) the heat of vaporization, and (b) the vapor pressure at 25 °C.

SOLUTION

(a) $$\ln \frac{P_2}{P_1} = \frac{\Delta H_{vap}(T_2 - T_1)}{RT_1T_2}$$

$$\Delta H = \frac{(8.314 \text{ J K}^{-1} \text{ mol}^{-1})(291.6 \text{ K})(313.5 \text{ K})}{21.9 \text{ K}} \ln \frac{8.00}{2.67}$$

$$= 38.1 \text{ kJ mol}^{-1}$$

(b) $$\ln \frac{P_2}{2.67 \text{ kPa}} = \frac{(38{,}100 \text{ J mol}^{-1})(6.6 \text{ K})}{(8.314 \text{ J K}^{-1} \text{ mol}^{-1})(298.15 \text{ K})(291.6 \text{ K})}$$

$$P_2 = 3.78 \text{ kPa}$$

6.11 The sublimation pressures of solid Cl_2 are 352 Pa at -112 °C and 35 Pa at -126.5 °C. The vapor pressures of liquid Cl_2 are 1590 Pa at -110 °C and 7830 Pa at -80 °C. Calculate (a) ΔH_{sub}, (b) ΔH_{vap}, (c) ΔH_{fus}, and (d) the triple point.

SOLUTION

(a) $$\Delta H_{sub} = \frac{RT_1T_2}{T_2 - T_1} \ln \frac{P_2}{P_1}$$

$$= \frac{(8.314\ \mathrm{J\ K^{-1}\ mol^{-1}})(161.15\ \mathrm{K})(146.65\ \mathrm{K})}{14.5\ \mathrm{K}}\ \ln\frac{352}{35}$$

$$= 31.4\ \mathrm{kJ\ mol^{-1}}$$

(b) $$\Delta H_{vap} = \frac{(8.314\ \mathrm{J\ K^{-1}\ mol^{-1}})(173.15\ \mathrm{K})(193.15\ \mathrm{K})}{20\ \mathrm{K}}\ \ln\frac{7830}{1590}$$

$$= 22.1\ \mathrm{kJ\ mol^{-1}}$$

(c) $\Delta H_{fus} = \Delta H_{sub} - \Delta H_{vap} = 31.4 - 22.1 = 9.3\ \mathrm{kJ\ mol^{-1}}$

(d) For the solid

$$\ln P = \ln 352 + \frac{31{,}400}{8.314}\left(\frac{1}{161.15} - \frac{1}{T}\right) = 29.300 - \frac{3777}{T}$$

For the liquid

$$\ln P = \ln 1590 + \frac{22{,}100}{8.314}\left(\frac{1}{173.15} - \frac{1}{T}\right) = 22.723 - \frac{2658}{T}$$

At the triple point

$$29.300 - \frac{3777}{T} = 22.723 - \frac{2658}{T} \qquad T = \frac{1119}{6.577} = 170.1\ \mathrm{K}$$

$$T = 107.1\ \mathrm{K} - 273.15\ \mathrm{K} = -103.0\ ^{\circ}\mathrm{C}$$

6.12 The boiling point of n-hexane at 1 bar is 68.6 °C. Estimate (a) its molar heat of vaporization, and (b) its vapor pressure at 60 °C.

SOLUTION

(a) $\Delta H_{vap} = (88\ \mathrm{J\ K^{-1}\ mol^{-1}})(69 + 273)\mathrm{K} = 30.1\ \mathrm{kJ\ mol^{-1}}$

(b) $$\ln\frac{P_2}{P_1} = \frac{\Delta H_{vap}}{R}\left(\frac{1}{T_1} - \frac{1}{T_2}\right)$$

$$\ln\frac{1\ \mathrm{bar}}{P} = \frac{30{,}100\ \mathrm{J\ mol^{-1}}}{8.314\ \mathrm{J\ K^{-1}\ mol^{-1}}}\left(\frac{1}{333\ \mathrm{K}} - \frac{1}{341.8\ \mathrm{K}}\right)$$

$$P = 0.755\ \mathrm{bar}$$

6.13 According to Trouton's rule the entropy of vaporization of a liquid at its boiling point is 88 J K^{-1} mol^{-1}. What is the change in boiling point expected for a liquid with a boiling point of (a) 100 °C and (b) 200 °C at 101,325 Pa in going to a reference state of 1 bar?

SOLUTION

(a) $\Delta H_{vap} = (88 \text{ J K}^{-1} \text{ mol}^{-1})(373 \text{ K}) = 32{,}800 \text{ J mol}^{-1}$

$$\ln \frac{P_2}{P_1} = \frac{\Delta H_{vap}(T_2 - T_1)}{RT_1T_2}$$

$$\ln \frac{10^5}{101{,}325} = \frac{32{,}800\,(T_2 - T_1)}{(8.314)(373)^2} \qquad T_2 - T_1 = -0.46\ ^\circ\text{C}$$

(b) $\Delta H_{vap} = (88 \text{ J K}^{-1} \text{ mol}^{-1})(473 \text{ K}) = 41{,}600 \text{ J mol}^{-1}$

$$T_2 - T_1 = \frac{(8.314)(473)^2}{41{,}600} \ln \frac{10^5}{101{,}325} = -0.59\ ^\circ\text{C}$$

6.14 Calculate ΔG° for H_2O(g, 25 °C) = H_2O(1, 25 °C). The vapor pressure of water at 25 °C is 3168 Pa.

SOLUTION

H_2O(g, 101325 Pa, 25 °C) = H_2O(g, 3168 Pa, 25 °C) = H_2O(1, 25 °C)

Since ΔG = 0 for the last step

$$\Delta G = RT \ln(P_2/P_1) = (8.314)(298.15) \ln (3168/101325)$$

$$= -8.59 \text{ kJ mol}^{-1}$$

6.15 From tables giving ΔG_f°, ΔH_f°, and C_P for H_2O(1) and H_2O(g) at 298 K, calculate (a) the vapor pressure of H_2O(1) at 25 °C, and (b) the boiling point at 1 atm.

SOLUTION

(a) $H_2O(l) \; = \; H_2O(g)$

$$\Delta G^\circ \;=\; -228.572 - (-237.129) \;=\; 8.557 \text{ kJ mol}^{-1}$$

$$\Delta G^\circ \;=\; -RT \ln (P/P^\circ)$$

$$P/P^\circ \;=\; \exp \frac{-8557 \text{ J mol}^{-1}}{(8.314 \text{ J K}^{-1} \text{ mol}^{-1})(298.15 \text{ K})} \;=\; 3.168 \times 10^{-2}$$

$$P \;=\; (3.168 \times 10^{-2})(10^5 \text{ Pa}) \;=\; 3.17 \times 10^3 \text{ Pa}$$

(b) $\Delta H^\circ(298.15 \text{ K}) \;=\; -241.818 - (-285.830) \;=\; 44.012 \text{ kJ mol}^{-1}$

In the absence of data on the dependence of C_P on T we will calculate $\Delta H^\circ(T)$ from

$$\Delta H^\circ(T) \;=\; 44{,}012 + \int_{273.15}^{T} (33.577 - 75.291)\,dT$$

$$= \; 44.012 - 41.714\,(T - 273.15)$$

$$\frac{\partial(\Delta G^\circ/T)}{\partial_T} \;=\; -\frac{\Delta H^\circ(T)}{T^2}$$

$$\frac{\Delta G^\circ}{T} \;=\; -\int \left[\frac{-44{,}012}{T^2} - \frac{41.714}{T} + \frac{(41.714)(273.15)}{T^2}\right] dT$$

$$= \; \frac{44.012}{T} + 41.714 \ln T + \frac{(41.714)(298.15)}{T} + I$$

where I is an integration constant which can be evaluated from $\Delta G^\circ(298.15 \text{ K})$.

$$\frac{8590}{298.15} \;=\; \frac{44.012}{298.15} + 41.714 \ln 298.15 + \frac{(41.714)(298.15)}{(298.15)} + I$$

$$I \;=\; -398.191$$

At the boiling point $\Delta G^\circ = 0$ and so we need to solve the following equation by successive approximations.

$$0 = \frac{44,012}{T} + 41.714 \ln T + \frac{(41.714)(298.15)}{T} - 398.191$$

Trying T = 373 K	RHS = 0.163
Trying T = 374 K	RHS = -0.130

Therefore the standard boiling point calculated in this way is close to 373.5 K.

6.16 Calculate the partial molar volume of zinc chloride in 1-molal $ZnCl_2$ solution using the following data.

% by weight of $ZnCl_2$	2	6	10	14	18	20
Density/g cm^{-3}	1.0167	1.0532	1.0891	1.1275	1.1665	1.1866

SOLUTION

Taking the first solution as an example, 1 g of solution contains 0.02 g of $ZnCl_2$ ($M = 136.28$ g mol^{-1}) and 0.98 g of H_2O. The weight of solution containing 1000 g of water is

$$\frac{1000}{0.98} = 1020 \text{ g}$$

$$\text{molality } ZnCl_2 = \frac{(0.02)(1020 \text{ g})}{136.28 \text{ g mol}^{-1}} = 0.1497 \text{ m}$$

Volume of solution containing 1000 g of H_2O

$$\frac{1020 \text{ g}}{1.0167 \text{ g cm}^{-3}} = 1003.2 \text{ cm}^3$$

Wt %	**Molality**	**Volume Containing 1000 g of H_2O**
2	0.1497 m	1003.2
6	0.4683	1010.1
10	0.8152	1020.2
14	1.194	1031.3
18	1.610	1045.5
20	1.834	1053.4

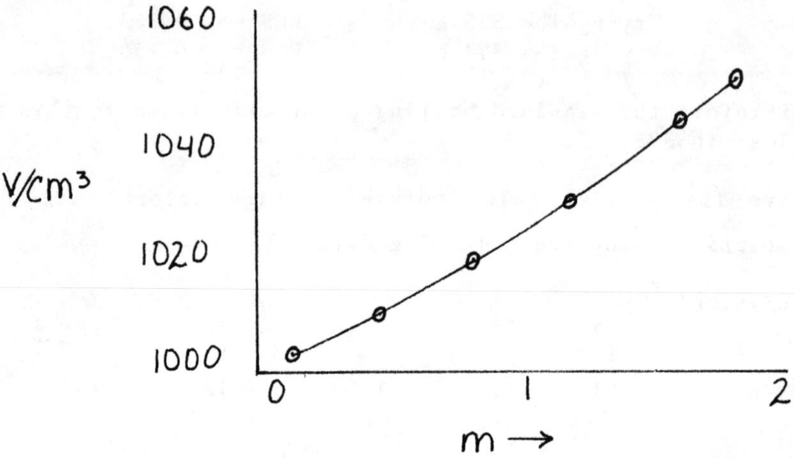

The slope of this plot at m = 1 molar is 29.3 cm^3 mol^{-1} and so this is the partial molar volume of $ZnCl_2$.

6.17 For a solution of ethanol and water at 20 °C which has a mole fraction of ethanol of 0.2, the partial molar volume of water is 17.9 cm^3 mol^{-1} and the partial molar volume of ethanol is 55.0 cm^3 mol^{-1}. What volumes of pure ethanol and water are required to make a liter of this solution? At 20 °C the density of ethanol is 0.789 g cm^{-3} and the density of water is 0.998 g cm^{-3}.

SOLUTION

Component 1 is water, and component 2 is ethanol.

$$V = n_1V_1 + n_2V_2 = 4n_2V_1 + n_2V_2 = n_2(4V_1 + V_2)$$

$$1000\ cm^3 = n_2[(4)(17.9\ cm^3\ mol^{-1}) + 55.0\ cm^3\ mol^{-1}]$$

$$n_2 = 7.90\ mol = \frac{g_2}{46.07\ g\ mol^{-1}}$$

$$g_2 = 363.9\ g$$

$$\text{Volume of pure ethanol} = \frac{363.9 \text{ g}}{0.789 \text{ g cm}^{-3}} = 461 \text{ cm}^3$$

$$n_1 = 4n_2 = 31.60 = \frac{g_1}{18.016} \qquad g_1 = 569.3 \text{ g}$$

$$\text{Volume of pure water} = \frac{569.3 \text{ g}}{0.998 \text{ g cm}^{-3}} = 570 \text{ cm}^3$$

6.18 Use the Gibbs-Duhem equation to show that if one component of a binary liquid solution follows Raoult's law, the other component will too.

SOLUTION

$$x_1 d\mu_1 + x_2 d\mu_2 = 0$$

If $\mu_1 = \mu_1^{o} + RT \ln x_1$

$$d\mu_1 = \frac{RT}{x_1} dx_1$$

Using the first equation

$$d\mu_2 = -\frac{x_1}{x_2} d\mu_1 = -\frac{RT}{x_2} dx_1$$

Since $x_1 + x_2 = 1$, $dx_2 = -dx_1$

$$d\mu_2 = \frac{RT}{x_2} dx_2 = RT \, d \ln x_2$$

$$\mu_2 = \text{const} + RT \ln x_2$$

If $x_2 = 1$, $\text{const} = \mu_2^{o}$

$$\mu_2 = \mu_2^{o} + RT \ln x_2$$

6.19 Show that the effect of external hydrostatic pressure P on the vapor pressure p of a liquid of partial molar volume V_1 is given by

$$RT \frac{d \ln p}{dP} = V_1$$

provided the vapor behaves as a perfect gas. **Hint:** When the pressure is changed at constant temperature, $dG_1 = dG_{vap}$, so that $V_1 dP = V_{gas} dP$. Integrate this equation to find an expression for the vapor pressure p of the liquid at some higher applied pressure P in terms of the vapor pressure p^o.

SOLUTION

The vapor pressure is represented by a lower case p.

$$\frac{dp}{dP} = \frac{\overline{V}_1}{V_{gas}} = \frac{\overline{V}_{1,p}}{RT} \qquad RT \frac{dp}{p} = \overline{V}_1 dP \qquad RT \frac{d \ln p}{dP} = \overline{V}_1$$

$$\int_{p^o}^{p} d \ln p = \frac{1}{RT} \int_{p^o}^{P} \overline{V}_1 dP$$

Assuming the partial molar volume of the liquid is independent of pressure,

$$\ln \frac{p}{p^o} = \frac{\overline{V}_1 (P - p^o)}{RT} \qquad p = p^o \exp [\overline{V} (P - p^o)/RT]$$

6.20 What is the difference in chemical potential of benzene (component 1) in toluene (component 2) at $x_1 = 0.5$ and $x_1 = 0.1$ at 25 °C, assuming ideal solutions?

SOLUTION

$$\mu_1 = \mu_1^* + RT \ln 0.5$$

$$\mu_1 = \mu_1^* + RT \ln 0.1$$

$$\Delta\mu_1 = (8.314 \text{ J K}^{-1} \text{ mol}^{-1})(298 \text{ K}) \ln \frac{0.5}{0.1}$$

$$= 3988 \text{ J mol}^{-1}$$

6.21 Ethanol and methanol form very nearly ideal solutions. The vapor pressure of ethanol is 5.93 kPa, and that of methanol is 11.83 kPa, at 20 °C. (a) Calculate the mole fraction of methanol and ethanol in a solution obtained by mixing 100 g of each. (b) Calculate the partial pressures and the total vapor pressure of the solution. (c) Calculate the mole fraction of methanol in the vapor.

SOLUTION

(a) $$x_{C_2H_5OH} = \frac{100/146}{100/146 + 100/32} = 0.410$$

$$x_{CH_3OH} = \frac{100/32}{100/146 + 100/32} = 0.590$$

(b) $$P_{C_2H_5OH} = x_{C_2H_5OH}\ P^{0}_{C_2H_5OH}$$

$$= (0.410)(5930\ \text{Pa}) = 2430\ \text{Pa}$$

$$P_{CH_3OH} = x_{CH_3OH}\ P^{0}_{CH_3OH}$$

$$= (0.590)(11{,}830\ \text{Pa}) = 6980\ \text{Pa}$$

$$P_{total} = 2430\ \text{Pa} + 6980\ \text{Pa} = 9410\ \text{Pa}$$

(c) $$y_{CH_3OH,\ vapor} = \frac{6980\ \text{Pa}}{9410\ \text{Pa}} = 0.742$$

6.22 Ethylene dibromide and propylene dibromide form very nearly ideal solutions. Plot the partial vapor pressure of ethylene dibromide (P^{sat} = 22.9 kPa), the partial vapor pressure of propylene dibromide (P^{sat} = 16.9 kPa), and the total vapor pressure of the solution versus the mole fraction of ethylene dibromide at 80 °C. (a) What will be the composition of the vapor in equilibrium with a solution containing 0.75 mole fraction of ethylene dibromide? (b) What will be the composition of the liquid phase in equilibrium with ethylene dibromide-propylene dibromide vapor containing 0.50 mole fraction of each?

SOLUTION

(a)

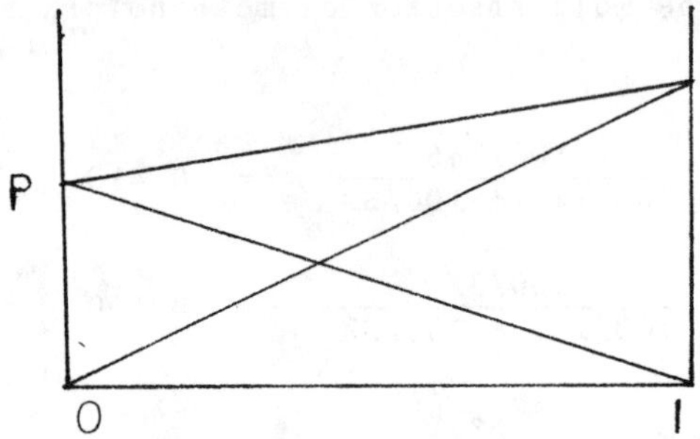

mole fraction ethylene dibromide

$$y_{EtBr_2} = \frac{(0.75)(22.9)}{(0.75)(22.9) + (0.25)(16.9)} = 0.802$$

$$y_{PrBr_2} = \frac{(0.25)(16.9)}{(0.75)(22.9) + (0.25)(16.9)} = 0.198$$

(b) $$y_{EtBr_2} = 0.5 = \frac{x_{EtBr_2} 22.9}{x_{EtBr_2} 22.9 + (1 - x_{EtBr_2})(16.9)}$$

$$x_{EtBr_2} = 0.425$$

6.23 The vapor pressure of a solution containing 13 g of a nonvolatile solute in 100 g of water at 28 °C is 3.6492 kPa. Calculate the molar mass of the solute, assuming that the solution is ideal. The vapor pressure of water at this temperature is 3.7417 kPa.

SOLUTION

$$x_2 = \frac{n_2}{n_1 + n_2} = \frac{P_1^{\circ} - P_1}{P_1^{\circ}} = \frac{3741.7 - 3649.2}{3741.7} = 0.0247$$

$$\frac{n_2}{n_1 + n_2} = 0.0247 = \frac{13/M}{(100/18) + (13/M)}$$

$$M = 92.4 \text{ g mol}^{-1}$$

6.24 One mole of benzene (component 1) is mixed with two moles of toluene (component 2). At 60 °C the vapor pressures of benzene and toluene are 51.3 and 18.5 kPa, respectively. (a) As the pressure is reduced at what pressure will boiling begin? (b) What will be the composition of the first bubble of vapor?

SOLUTION

(a)
$$P = P_2^{sat} + (P_1^{sat} - P_2^{sat})x_1$$

$$= 18.5 \text{ kPa} + (51.3 \text{ kPa} - 18.5 \text{ kPa})(0.333)$$

$$= 29.4 \text{ kPa}$$

(b)
$$y_1 = \frac{x_1 P_1^{sat}}{P_2^{sat} + (P_1^{sat} - P_2^{sat})x_1}$$

$$= \frac{(0.333)(51.3 \text{ kPa})}{18.5 \text{ kPa} + (32.8 \text{ kPa})(0.333)}$$

$$= 0.581 \text{ kPa}$$

6.25 The vapor pressures of benzene and toluene have the following values in the temperature range between their boiling points at 1 bar:

$t/°C$	79.4	88	94	100	110.0
$P^{sat}_{C_6H_6}$ / bar	1.000	1.285	1.526	1.801	
$P^{sat}_{C_7H_8}$ / bar		0.508	0.616	0.742	1.000

(a) Calculate the compositions of the vapor and liquid phases at each temperature and plot the boiling point diagram. (b) If a solution containing 0.5 mole fraction benzene and 0.5 mole fraction toluene is heated, at what temperature will the first bubble of vapor appear and what will be its composition?

SOLUTION

(a) At 88 °C

$$x_B(1.285) + (1 - x_B)\,0.508 = 1$$

$$x_B = 0.633$$

$$y_B = (0.633)(1.285)/1 = 0.814$$

At 94 °C

$$x_B(1.526) + (1 - x_B)\,0.616 = 1$$

$$x_B = 0.422$$

$$y_B = (0.4222)(1.526)/1 = 0.644$$

At 100 °C

$$x_B(1.801) + (1 - x_B)\,0.742 = 1$$

$$x_B = 0.244$$

$$y_B = (0.244)(1.801)/1 = 0.439$$

(b) Please see diagram at the top of page 99.

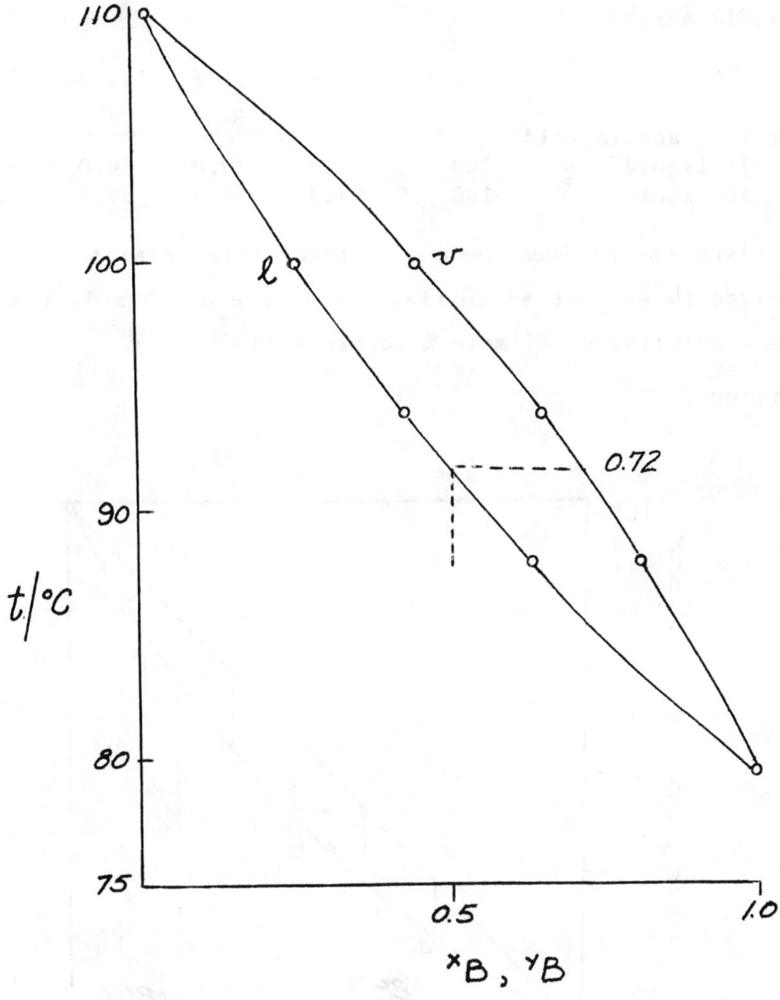

Bubble point = 92 °C y_B = 0.72

6.26 The following table gives mole % acetic acid in aqueous solutions and in the equilibrium vapor at the boiling point of the solution at 1.013 bar.

B.P., °C	118.1	113.8	107.5	104.4	102.1	100.0
Mole % of acetic acid						
In liquid	100	90.0	70.0	50.0	30.0	0
In vapor	100	83.3	57.5	37.4	18.5	0

Calculate the minimum number of theoretical plates for the column required to produce an initial distillate of 28 mole % acetic acid from a solution of 80 mole % acetic acid.

SOLUTION

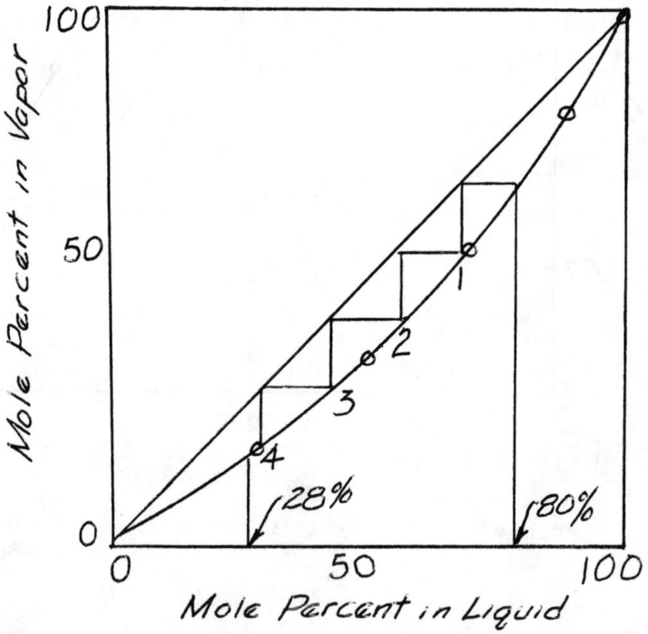

Since there are four steps, three theoretical plates are required in the column. The distilling pot counts as one plate.

6.27 If two liquids (1 and 2) are completely immiscible, the mixture will boil when the sum of the two partial pressures exceeds the applied pressure:

$$P = P_1^{sat} + P_2^{sat} .$$

In the vapor phase the ratio of the mole fractions of the two components is equal to the ratio of their vapor pressures.

$$\frac{P_1^{sat}}{P_2^{sat}} = \frac{x_1}{x_2} = \frac{g_1 M_2}{g_2 M_1}$$

where g_1 and g_2 are the masses of components 1 and 2 in the vapor phase, and M_1 and M_2 are their molar masses. The boiling point of the immiscible liquid system naphthalene-water is 98 °C under a pressure of 97.7 kPa. The vapor pressure of water at 98 °C is 94.3 kPa. Calculate the weight percent of naphthalene in the distillate.

SOLUTION

$$\frac{g_1}{g_2} = \frac{P_1^{\circ} M_1}{P_2^{\circ} M_2} = \frac{(97.7 - 94.3)(128)}{(94.3)(18)} = 0.261$$

$$\text{Weight percent naphthalene} = \frac{0.261}{1.261} = 0.207 \text{ or } 20.7\%$$

6.28 Calculate the solubility of p-dibromobenzene in benzene at 20 °C and 40 °C assuming ideal solutions are formed. The enthalpy of fusion of p-dibromobenzene is 13.22 kJ mol^{-1} at its melting point (86.9 °C).

SOLUTION

$$\text{At 20 °C} \quad \ln x_2 = \frac{\Delta H_{2f}(T - T_{2f})}{RTT_{2f}}$$

$$= \frac{(13{,}220 \text{ J mol}^{-1})(-66.9 \text{ K})}{(8.314 \text{ J K}^{-1} \text{ mol}^{-1})(360.1 \text{ K})(293.2 \text{ K})}$$

$$x_2 = 0.365$$

At 40 °C $\ln x_2 = \dfrac{(13{,}220 \text{ J mol}^{-1})(-46.9 \text{ K})}{(8.314 \text{ J K}^{-1} \text{ mol}^{-1})(360.1 \text{ K})(313.2 \text{ K})}$

$$x_2 = 0.516$$

6.29 Calculate the solubility of naphthalene at 25 °C in any solvent in which it forms an ideal solution. The melting point of naphthalene is 80 °C, and the heat of fusion is 19.29 kJ mol^{-1}. The actual measured solubility of naphthalene in benzene is $x_1 = 0.296$.

SOLUTION

$$-\ln x_1 = \frac{\Delta H_{fus,1}(T_{0,1} - T)}{RTT_{0,1}}$$

$$\ln x_1 = \frac{-(19{,}290 \text{ J mol}^{-1})(353 \text{ K} - 298 \text{ K})}{(8.314 \text{ J K}^{-1} \text{ mol}^{-1})(353 \text{ K})(298 \text{ K})}$$

$$x_1 = 0.297$$

6.30 If 68.4 g of sucrose (M = 342 g mol^{-1}) is dissolved in 1000 g of water: (a) What is the vapor pressure at 20 °C? (b) What is the freezing point? The vapor pressure of water at 20 °C is 2.3149 kPa.

SOLUTION

(a) $x_2 = \dfrac{\frac{68.4}{342}}{\frac{68.4}{342} + \frac{1000}{18}} = 3.59 \times 10^{-3}$

$$\frac{P_1^{\circ} - P_1}{P_1^{\circ}} = x_2 \qquad \frac{2.3149 - P_1}{2.3149} = 3.59 \times 10^{-3}$$

$$P_1 = 2.3149(1 - 3.59 \times 10^{-3}) = 2.3066 \text{ kPa}$$

(b) $\Delta T_f = K_f m = 1.86(0.2) = -0.372$

$$T_f = -0.372 \ ^{\circ}\text{C}$$

6.31 The protein human plasma albumin has a molar mass of 69,000 g mol^{-1}. Calculate the osmotic pressure of a solution of this protein containing 2 g per 100 cm^3 at 25 °C in (a) pascals, and (b) millimeters of water. The experiment is carried out using a salt solution for solvent and a membrane permeable to salt.

SOLUTION

$$\pi = \frac{cRT}{M} = \frac{(20 \times 10^{-3}\text{ kg L}^{-1})(10^3\text{ L m}^{-3})(8.314\text{ J K}^{-1}\text{mol}^{-1})(298.15\text{ K})}{69\text{ kg mol}^{-1}}$$

$$= 719 \text{ Pa}$$

$$h = \frac{\pi}{dg} = \frac{719\text{ Pa}}{(1\text{ g cm}^{-3})(10^{-3}\text{ kg g}^{-1})(10^2\text{ cm m}^{-1})(9.8\text{ m s}^{-2})}$$

$$= 7.34 \times 10^{-2}\text{ m} = 73.4 \text{ mm of water}$$

6.32 The following osmotic pressures were measured for solutions of a sample of polyisobutylene in benzene at 25 °C.

$c/10^{-2}$ g cm^{-3}	0.500	1.00	1.50	2.00
π/g cm^{-2}	0.505	1.03	1.58	2.15

Calculate the number average molar mass from the value of π/c extrapolated to zero concentration of the polymer.

SOLUTION

$$\frac{\pi}{c} = \frac{RT}{M} + Bc$$

The ratio π/c is plotted versus c and extrapolated to c = 0.

$c/g/100\ cm^3$	0.50	1.0	1.50	2.00
$\frac{\pi(100)}{c\ 1028} \Big/ \frac{atm\ cm^3}{g}$	0.0982	0.1002	0.1025	0.1046

$$\left(\frac{\pi}{c}\right)_{c=0} = (9.60 \times 10^{-2}\ atm\ cm^3\ g^{-1})(101325\ Pa\ atm^{-1}) \times (10^3\ g\ kg^{-1})(10^{-2}\ m\ cm^{-1})^3$$

$$= 9.73\ Pa(mol\ m^{-3})^{-1}$$

$$\left(\frac{\pi}{c}\right)_{c=0} = \frac{RT}{M}$$

$$M = \frac{RT}{\left(\frac{\pi}{c}\right)_{c=0}} = \frac{(8.314\ J\ K^{-1}\ mol^{-1})(298\ K)}{9.73\ Pa\ (mol\ m^{-3})^{-1}} = 255\ kg\ mol^{-1}$$

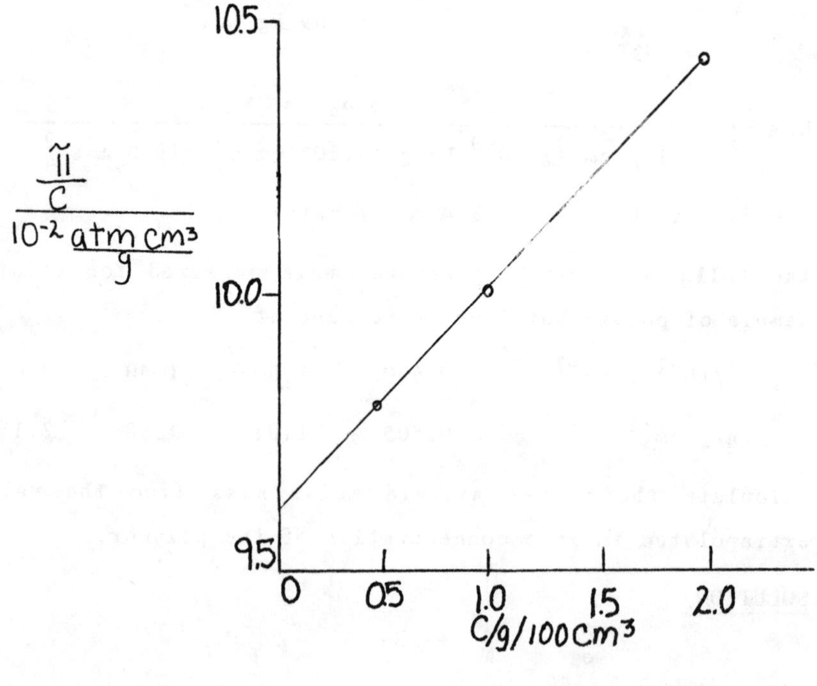

6.33 A sample of polymer contains 0.50 mole fraction with molar mass 100,000 g mol^{-1} and 0.50 mole fraction with molar mass 200,000 g mol^{-1}. Calculate (a) M_n, and (b) M_m.

SOLUTION

(a) $M_n = \frac{\Sigma n_i M_i}{\Sigma n_i} = \Sigma x_i M_i$ where x_i is mole fraction of the ith component

$= 0.5(10^5) + 0.5(2 \times 10^5) = 150{,}000$ g mol^{-1}

(b) $M_m = \frac{\Sigma n_i M_i^2}{\Sigma n_i M_i}$

Dividing numerator and denominator by Σn_i

$$M_m = \frac{\Sigma x_i M_i^2}{\Sigma x_i M_i} = \frac{0.5(10^{10}) + 0.5(4 \times 10^{10})}{0.5(10^5) + 0.5(2 \times 10^5)} = 167{,}000 \text{ g mol}^{-1}$$

6.34 Calculate the osmotic pressure of a 1 mol L^{-1} sucrose solution in water from the fact that at 30 °C the vapor pressure of the solution is 4.1606 kPa. The vapor pressure of water at 30 °C is 4.2429 kPa. The density of pure water at this temperature (0.99564 g cm^{-3}) may be used to estimate V_1 for a dilute solution. To do this problem, Raoult's law is introduced into equation 6.88.

SOLUTION

$V_1\pi = -RT \ln x_1$

Substituting Raoult's law $P_1 = x_1 P_1^{\circ}$

$$\pi = -\frac{RT}{V_1} \ln \frac{P_1}{P_1^{\circ}}$$

$$V_1 = \frac{18.02 \text{ g mol}^{-1}}{0.99564 \text{ g cm}^{-3}} = 1810 \text{ cm}^3 \text{ mol}^{-1} = 0.01810 \text{ L mol}^{-1}$$

$$\pi = \frac{(0.08314\ \text{L bar K}^{-1}\ \text{mol}^{-1})(303.15\ \text{K})}{0.0180\ \text{L mol}^{-1}} \ln \frac{4.2429\ \text{kPa}}{4.1606\ \text{kPa}}$$

$$= 27.3\ \text{bar}$$

6.35 4

6.36 3

6.37 (a) -38.81, (b) -16.6 °C

6.38 4.9 K

6.39 1.67 kPa
0.167 kPa

6.40 96 °C

6.41 (a) 2.82 kJ (b) 50.1 Pa K^{-1} (c) 360 Pa 387 Pa

6.42 (a) 1623 kPa (b) 1428 kPa

6.43 0.076 Pa

6.44 27.57 kJ mol^{-1}

6.45 (a) 5.8 °C 3.28 kPa (b) 9.90 kJ mol^{-1}

6.46 94 °C

6.47 0.79 cm^3 g^{-1}

6.48 (a) 0.204 cm^3 g^{-1} (b) 19.4 cm^3 mol^{-1}

6.49 26.01 cm^3

6.50 5.293 J K^{-1} mol^{-1} -1577 J mol^{-1}

6.51 (a) $x_{CHCl_3} = 0.653$ $x_{CCl_4} = 0.365$

(b) 20.92 kPa

6.52 (a) 3.1615 kPa (b) 3.1615 kPa
(c) 3.1557 kPa (d) 3.1615 kPa

6.53 (a) x_B = 0.72 x_T = 0.28 3.162 kPa

(b) x_B = 0.537

6.54 0.194 24.67 kPa

6.55 $\Delta G_{mix} = RT(x_1 \ln x_1 + x_2 \ln x_2) + wx_1x_2$

$\Delta S_{mix} = -R(x_1 \ln x_1 + x_2 \ln x_2)$

$\Delta H_{mix} = wx_1x_2$

$\Delta V_{mix} = 0$

6.56 (a) At −31.2 °C $x_{propane}$ = 0.560

At −16.3 °C $x_{propane}$ = 0.196

(b) At −31.2 °C $x_{propane,vap}$ = 0.885

At −16.3 °C $x_{propane,vap}$ = 0.578

6.57 x_B = 0.240 y_B = 0.434

6.58 562 g

6.59 0.108

6.60 x_{Cd} = 0.844 or 74.4 g/100 g of solution

6.61 251 °C

6.62 12.7 kJ mol^{-1}

6.63 122,000 g mol^{-1}

6.64 1810 Pa

6.65 (a) 36,800 g mol^{-1} (b) 48,500 g mol^{-1}

CHAPTER 7: Phase Equilibrium: Nonideal Solutions

7.1 For a solution of n-propanol and water, the following partial pressures in kPa are measured at 25 °C. Draw a complete pressure-composition diagram, including the total pressure. What is the composition of the vapor in equilibrium with a solution containing 0.5 mole fraction of n-propanol?

$x_{n\text{-propanol}}$	P_{H_2O}	$P_{n\text{-propanol}}$	$x_{n\text{-propanol}}$	P_{H_2O}	$P_{n\text{-propanol}}$
0	3.168	0.00	0.600	2.65	2.07
0.020	3.13	0.67	0.800	1.79	2.37
0.050	3.09	1.44	0.900	1.08	2.59
0.100	3.03	1.76	0.950	0.56	2.77
0.200	2.91	1.81	1.000	0.00	2.901
0.400	2.89	1.89			

SOLUTION

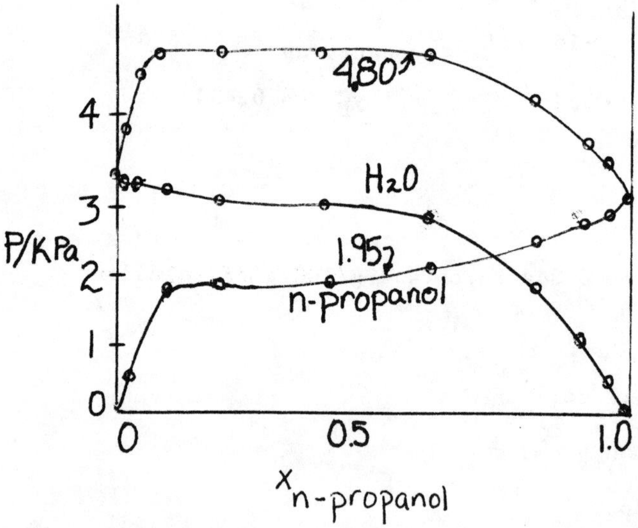

$$x_{vapor} = \frac{1.95\ \text{kPa}}{4.80\ \text{kPa}} = 0.406$$

7.2 Using the Henry law constants in Table 7.1, calculate the percentage (by volume) of oxygen and nitrogen in air dissolved in water at 25 °C. The air in equilibrium with the water at 1 bar pressure may be considered to be 20% oxygen and 80% nitrogen by volume.

<u>SOLUTION</u>

$$x_{N_2} = \frac{P_{N_2}}{K_{N_2}} = \frac{0.80 \times 10^5 \text{ Pa}}{8.68 \times 10^9 \text{ Pa}} = 9.22 \times 10^{-6}$$

$$x_{O_2} = \frac{P_{O_2}}{K_{O_2}} = \frac{0.20 \times 10^5 \text{ Pa}}{4.40 \times 10^9 \text{ Pa}} = 4.55 \times 10^{-6}$$

The percentage of oxygen in this dissolved air is given by

$$\frac{x_{O_2}}{x_{O_2} + x_{N_2}} \times 100 = \frac{4.55 \times 10^{-4}}{(4.55 + 9.22) \times 10^{-6}} = 33.0\%$$

The percentage of nitrogen is 100 − 33.0 = 67.0%

7.3 The following data on ethanol-chloroform solutions at 35 °C were obtained by G. Scatchard and C. L. Raymond [<u>J</u>. <u>Am</u>. <u>Chem</u>. <u>Soc</u>., <u>60</u>, 1278 (1938)]:

$x_{EtOH,liq}$	0	0.2	0.4	0.6	0.8	1.0
$x_{EtOH,vap}$	0.0000	0.1382	0.1864	0.2554	0.4246	1.0000
Total pressure, kPa	39.345	40.559	38.690	34.387	25.357	13.703

Calculate the activity coefficients of both in these solutions according to Convention I.

SOLUTION

$x_{EtOH} = 0.2 \qquad \gamma_{EtOH} = \dfrac{y_{EtOH}\,P}{x_{EtOH}\,P^{sat}_{EtOH}}$

$= \dfrac{(0.1382)(40.559)}{(0.2)(13.703)} = 2.045$

$x_{EtOH} = 0.4 \qquad \gamma_{EtOH} = \dfrac{(0.1864)(38.690)}{(0.4)(13.703)} = 1.316$

$x_{EtOH} = 0.6 \qquad \gamma_{EtOH} = \dfrac{(0.2554)(34.287)}{(0.6)(13.703)} = 1.065$

$x_{EtOH} = 0.8 \qquad \gamma_{EtOH} = \dfrac{(0.4246)(25.357)}{(0.8)(13.703)} = 0.982$

$x_{EtOH} = 1.0 \qquad \gamma_{EtOH} = \dfrac{(1.000)(13.703)}{(1.000)(13.703)} = 1.000$

$$\gamma_{CHCl_3} = \frac{y_{CHCl_3}\,P}{x_{CHCl_3}\,P^{sat}_{CHCl_3}}$$

$x_{CHCl_3} = 1 \qquad \gamma_{CHCl_3} = \dfrac{(1)(39.345)}{(1)(39.345)} = 1$

$x_{CHCl_3} = 0.8 \qquad \gamma_{CHCl_3} = \dfrac{(0.8618)(40.559)}{(0.8)(39.345)} = 1.111$

$x_{CHCl_3} = 0.6 \qquad \gamma_{CHCl_3} = \dfrac{(0.8136)(38.690)}{(0.6)(39.345)} = 1.333$

$x_{CHCl_3} = 0.4 \qquad \gamma_{CHCl_3} = \dfrac{(0.7446)(34.387)}{(0.4)(39.345)} = 1.627$

$x_{CHCl_3} = 0.2 \qquad \gamma_{CHCl_3} = \dfrac{(0.5754)(25.357)}{(0.2)(39.345)} = 1.854$

7.4 Using the data of the following table, which gives vapor pressures at 35.2 °C, calculate the activity coefficients of acetone and chloroform at 35.2 °C, following Convention I, for mole fractions of chloroform of 0.20, 0.40, 0.60, and 0.80.

x_{CHCl_3}	0	0.2	0.4	0.6	0.8	1.00
P_{CHCl_3}/kPa	0	4.5	10.9	19.7	30.0	39.1
$P_{acetone}$/kPa	45.9	36.0	24.4	13.6	5.6	0

SOLUTION

Component 1 is chloroform. Component 2 is acetone.

$$\gamma_i = \frac{P_i}{x_i P_i^{sat}}$$

$x_1 = 0.2$ $$\gamma_1 = \frac{4.5}{0.2(39.1)} = 0.58$$

$x_1 = 0.4$ $$\gamma_1 = \frac{10.9}{0.4(39.1)} = 0.70$$

$x_1 = 0.6$ $$\gamma_1 = \frac{19.7}{0.6(39.1)} = 0.84$$

$x_1 = 0.8$ $$\gamma_1 = \frac{30.0}{0.8(39.1)} = 0.96$$

$x_1 = 0.2$ $$\gamma_2 = \frac{36.0}{0.8(45.9)} = 0.98$$

$x_1 = 0.4$ $$\gamma_2 = \frac{24.4}{0.6(45.9)} = 0.89$$

$x_1 = 0.6$ $$\gamma_2 = \frac{13.6}{0.4(45.9)} = 0.74$$

$x_1 = 0.8$ $$\gamma_2 = \frac{5.6}{0.2(45.9)} = 0.61$$

7.5 Using the data in Problem 7.1, calculate the activity coefficients of water and n-propanol at 0.20, 0.40, 0.60, and 0.80 mole fraction n-propanol, using Convention II and considering n-propanol to be the solvent.

SOLUTION

$x_1 = 0.2 \qquad \gamma_1 = \dfrac{1.81}{(0.2)(2.901)} = 3.12$

$x_1 = 0.4 \qquad \gamma_1 = \dfrac{1.89}{(0.4)(2.901)} = 1.63$

$x_1 = 0.6 \qquad \gamma_1 = \dfrac{2.07}{(0.6)(2.901)} = 1.19$

$x_1 = 0.8 \qquad \gamma_1 = \dfrac{2.37}{(0.8)(2.901)} = 1.02$

To obtain the Henry law constant for H_2O (Component 2) plot P_2/x_2 versus x_2.

$$\text{At } x_2 = 0.05, \ \frac{P_2}{x_2} = \frac{0.56}{0.05} = 11.2$$

$$\text{At } x_2 = 0.10, \ \frac{P_2}{x_2} = \frac{1.08}{0.1} = 10.8$$

Thus as $x_2 \longrightarrow 0, P_2 = 11.6\ x_2$

$\left.\begin{array}{l} x_2 = 0.2 \\ x_1 = 0.8 \end{array}\right\} \ \gamma_2 = \dfrac{1.79}{0.2(11.6)} = 0.772$

$\left.\begin{array}{l} x_2 = 0.4 \\ x_1 = 0.6 \end{array}\right\} \ \gamma_2 = \dfrac{2.65}{0.4(11.6)} = 0.571$

$\left.\begin{array}{l} x_2 = 0.6 \\ x_1 \ \ 0.4 \end{array}\right\} \ \gamma_2 = \dfrac{2.89}{0.6(11.6)} = 0.415$

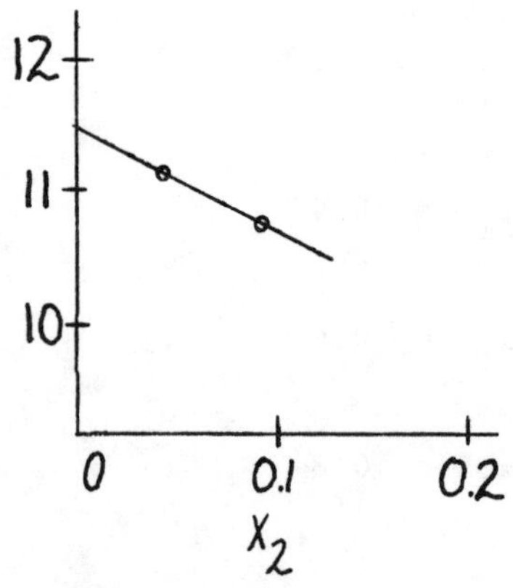

$$\left.\begin{array}{l} x_2 = 0.8 \\ x_1 \quad 0.2 \end{array}\right\} \gamma_2 = \frac{2.91}{0.8(11.6)} = 0.314$$

7.6 Show that the equations for the bubble-point line and dew-point line for nonideal solutions are given by

$$x_1 = \frac{P - \gamma_2 P_2^{sat}}{\gamma_1 P_1^{sat} - \gamma_2 P_2^{sat}}$$

$$y_1 = \frac{P\gamma_1 P_1^{sat} - \gamma_1 \gamma_2 P_1^{sat} P_2^{sat}}{P\gamma_1 P_1^{sat} - P\gamma_2 P_2^{sat}}$$

SOLUTION

$$P = P_1 + P_2 = x_1 \gamma_1 P_1^{sat} + (1 - x_1)\gamma_2 P_2^{sat}$$

$$= \gamma_2 P_2^{sat} + x_1(\gamma_1 P_1^{sat} - \gamma_2 P_2^{sat})$$

$$x_1 = \frac{P - \gamma_2 P_2^{sat}}{\gamma_1 P_1^{sat} - \gamma_2 P_2^{sat}}$$

$$y_1 = \frac{P_1}{P_1 + P_2} = \frac{x_1 \gamma_1 P_1^{sat}}{\gamma_2 P_2^{sat} + x_1(\gamma_1 P_1^{sat} - \gamma_2 P_2^{sat})}$$

Substituting the expression for x_1 yields the desired expression for y_1.

7.7 A regular binary solution is defined as one for which

$$\mu_1 = \mu_1^{o} + RT \ln x_1 + w x_2^2$$

$$\mu_2 = \mu_2^{o} + RT \ln x_2 + w x_1^2$$

(a) Calculate ΔG_{mix}, ΔS_{mix}, ΔH_{mix}, and ΔV_{mix} for the mixing of x_1 moles of Component 1 with x_2 moles of Component 2. Assume the coefficient w is independent of temperature and pressure. (b) Derive the expressions for the activity coefficients γ_1 and γ_2 in terms of w.

SOLUTION

(a) $$\Delta G_{mix} = x_1\mu_1 + x_2\mu_2 - RT(x_1 \ln x_1 + x_2 \ln x_2)$$

$$= x_1\mu_1^o + x_2\mu_2^o + wx_1x_2^2 + wx_1^2x_2$$

$$= x_1\mu_1^o + x_2\mu_2^o + wx_1x_2$$

$$\Delta S = -\left(\frac{\partial \Delta G_{mix}}{\partial T}\right)_P = x_1S_1^o + x_2S_2^o$$

$$G = x_1\mu_1 + x_2\mu_2 = x_1\mu_1^o + x_2\mu_2^o + RT(x_1 \ln x_1 + x_2 \ln x_2) + wx_1x_2$$

$$\Delta G_{mix} = G - (x_1\mu_1^o + x_2\mu_2^o)$$

$$= RT(x_1 \ln x_1 + x_2 \ln x_2) + wx_1x_2$$

$$\Delta S_{mix} = -\left(\frac{\partial \Delta G_{mix}}{\partial T}\right)_P = -R(x_1 \ln x_1 + x_2 \ln x_2)$$

$$\Delta H_{mix} = \Delta G_{mix} + T\Delta S_{mix} = wx_1x_2$$

$$\Delta V_{mix} = \left(\frac{\partial \Delta G_{mix}}{\partial P}\right)_T = 0$$

(b) $$\mu_1 = \mu_1^o + RT \ln \gamma_1 x_1$$

$$\mu_1 = \mu_1^o + RT \ln x_1 + wx_2^2$$

$$= \mu_1^o + RT \ln x_1 + RT \ln e^{wx_2^2/RT}$$

$$= \mu_1^o + RT \ln \left(e^{wx_2^2\ RT}\right)x_1$$

Therefore

$$\gamma_1 = e^{wx_2^2/RT} \qquad \gamma_2 = e^{wx_1^2/RT}$$

7.8 The expressions for the activity coefficients of the components of a regular binary solution were derived in the preceding problem. Derive the expression for γ_1 in terms of the experimentally measured total pressure P, the vapor pressures of the two components, and the composition of the solution for the case that the deviations from ideality are small.

SOLUTION

$$\gamma_1 = \exp\,(wx_2^2/RT) \qquad \gamma_2 = \exp\,(wx_1^2/RT)$$

$$P = P_1 + P_2 = \gamma_1 x_1 P_1^{sat} + \gamma_2 x_2 P_2^{sat}$$

$$= x_1 P_1^{sat} \exp\left(\frac{wx_2^2}{RT}\right) + x_2 P_2^{sat} \exp\left(\frac{wx_1^2}{RT}\right)$$

If the exponentials are not far from unity, the exponentials may be expanded to obtain

$$P = x_1 P_1^{sat}\left(1 + \frac{wx_2^2}{RT}\right) + x_2 P_2^{sat}\left(1 + \frac{wx_1^2}{RT}\right)$$

$$\frac{w}{RT} = \frac{P - \left(x_1 P_1^{sat} + x_2 P_2^{sat}\right)}{x_1 x_2\left[P_1^{sat} + x_1\left(P_2^{sat} - P_1^{sat}\right)\right]}$$

Thus from a measurement of the total pressure of a mixture of the two components, the activity coefficients of the components can be calculated over the entire concentration range using the first two equations given above. This is a particular example of the general situation that γ_1 and γ_2 can be calculated from the total pressure as a function of x_1.

7.9 Using the data in Problem 7.1, calculate the activity coefficients of water and n-propanol at 0.20, 0.40, 0.60, and 0.80 mole fraction of n-propanol, using Convention II and considering water to be the solvent.

SOLUTION

The data at the lowest mole fractions of n-propanol are used to calculate the Henry law constant for n-propanol in water.

x_2	$K_2^1 = P_2/x_2$
0.02	33.5
0.05	28.8
0.10	17.6

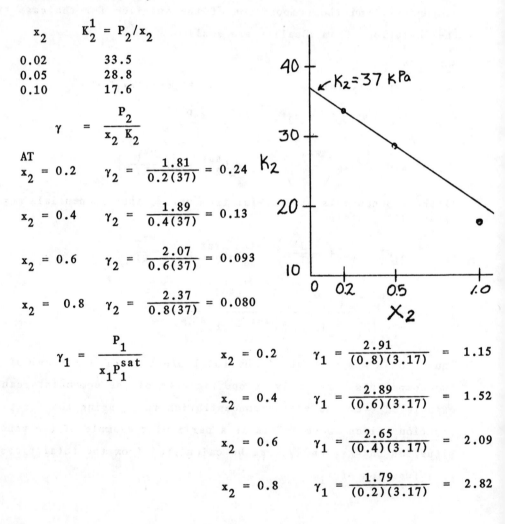

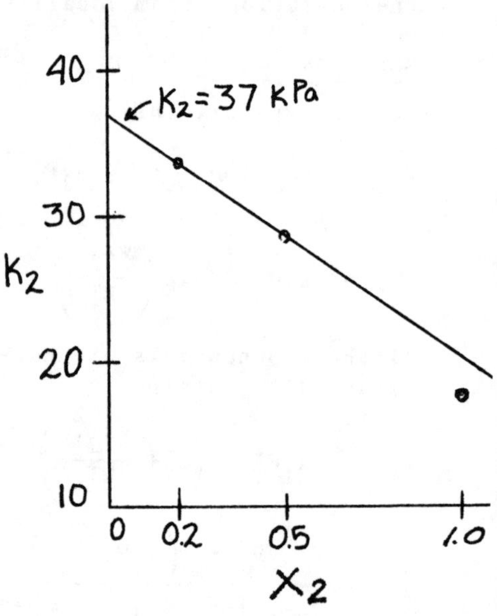

$$\gamma = \frac{P_2}{x_2 K_2}$$

AT

$x_2 = 0.2 \quad \gamma_2 = \dfrac{1.81}{0.2(37)} = 0.24$

$x_2 = 0.4 \quad \gamma_2 = \dfrac{1.89}{0.4(37)} = 0.13$

$x_2 = 0.6 \quad \gamma_2 = \dfrac{2.07}{0.6(37)} = 0.093$

$x_2 = 0.8 \quad \gamma_2 = \dfrac{2.37}{0.8(37)} = 0.080$

$$\gamma_1 = \frac{P_1}{x_1 P_1^{sat}}$$

$x_2 = 0.2 \quad \gamma_1 = \dfrac{2.91}{(0.8)(3.17)} = 1.15$

$x_2 = 0.4 \quad \gamma_1 = \dfrac{2.89}{(0.6)(3.17)} = 1.52$

$x_2 = 0.6 \quad \gamma_1 = \dfrac{2.65}{(0.4)(3.17)} = 2.09$

$x_2 = 0.8 \quad \gamma_1 = \dfrac{1.79}{(0.2)(3.17)} = 2.82$

7.10 The Gibbs-Duhem equation in the form

$$\left(\frac{\partial M}{\partial T}\right)_{P,x} dT + \left(\frac{\partial M}{\partial P}\right)_{T,x} dP - \sum (x_i dM_i) = 0$$

applies to any molar thermodynamic property M in a homogeneous phase. If this is applied to G^E, it may be shown* that if the

---

*J. M. Smith and H. C. Van Ness, Introduction to Chemical Engineering Thermodynamics, McGraw-Hill, New York, 1975, p. 346.

---

vapor is an ideal gas

$$x_1 \frac{d \ln (y_1 P)}{dx_1} + x_2 \frac{d \ln (y_2 P)}{dx_1} = 0 \quad \text{(constant T)}$$

Show that this can be rearranged to the coexistence equation

$$\frac{dP}{dy_1} = \frac{P(y_1 - x_1)}{y_1(1 - y_1)}$$

Thus if P versus y_1 is measured, there is no need for measurements of x_1.

SOLUTION

$$x_1 d \ln y_1 + x_1 d \ln P + x_2 d \ln y_2 + x_2 d \ln P = 0$$

$$d \ln P = -\frac{x_1}{y_1} dy_1 - \frac{x_2}{y_2} dy_2 = \left(-\frac{x_1}{y_1} + \frac{x_2}{y_2}\right) dy_1$$

$$= \frac{x_2 y_1 - x_1 y_2}{y_1 y_2} dy_1 = \frac{y_1 - x_1}{y_1(1 - y_1)} dy_1$$

7.11 From the data given in the following table construct a complete temperature-composition diagram for the system ethanol-ethyl acetate for 1.013 bar. A solution containing 0.8 mole fraction of ethanol, EtOH, is distilled completely at 1.013 bar. (a) What is the composition of the first vapor to come off? (b) That of the last drop of liquid to evaporate? (c) What would be the values of these quantities if the distillation were carried out in a cylinder provided with a piston so that none of the vapor could escape?

x_{EtOH}	y_{EtOH}	B.P., °C	x_{EtOH}	y_{EtOH}	B.P., °C
0	0	77.15	0.563	0.507	72.0
0.025	0.070	76.7	0.710	0.600	72.8
0.100	0.164	75.0	0.833	0.735	74.2
0.240	0.295	72.6	0.942	0.880	76.4
0.360	0.398	71.8	0.982	0.965	77.7
0.462	0.462	71.6	1.000	1.000	78.3

SOLUTION

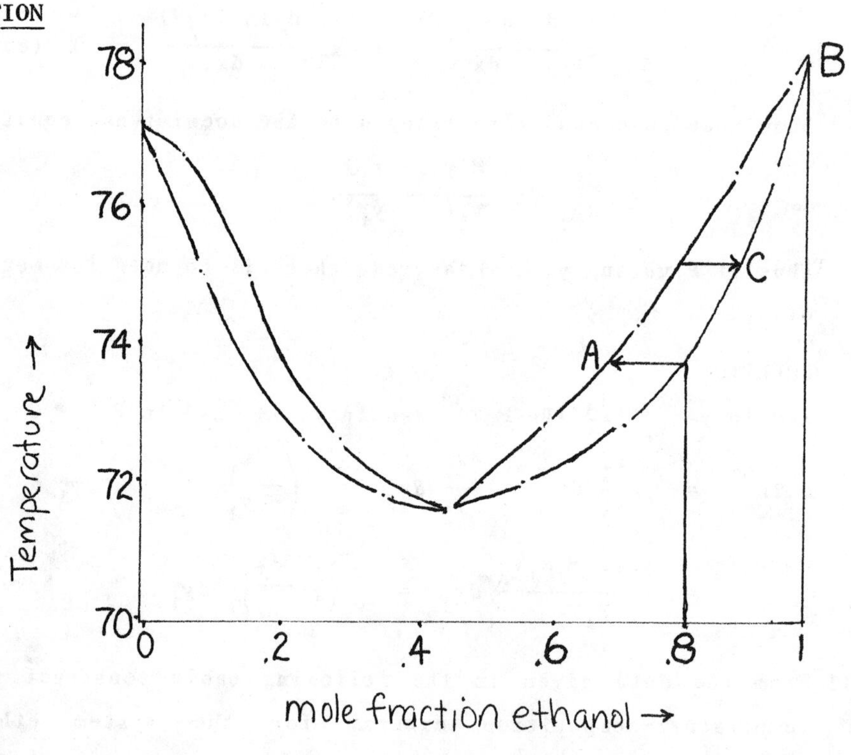

(a) The liquid will start to boil at 73.7 °C. The vapor in equilibrium with it is 0.69 mole fraction in ethanol (point A).

(b) As the boiling continues, more and more of the EtAc will be removed. As the mole fraction of EtOH increases, the boiling point will also increase. Finally, the very last vapors will be pure EtOH, as indicated by point B.

(c) The first vapor to come off will be the same as in part (a). However, when all of the mixture is in vapor state, the mole fraction of the vapor state will be .800, simply because that is the total mole fraction of EtOH in the entire mixture. Thus, to find the composition of the last drop to evaporate into the cylinder, simply note what liquid is in equilibrium with vapor which has .800 EtOH in it. This is given by point C.

7.12 The following table gives the mole percent of n-propanol ($M = 60.1$ g mol^{-1}) in aqueous solutions and in the vapor at the boiling point of the solution at 1.013 bar pressure.

B.P., °C	100.0	92.0	89.3	88.1	87.8	88.3	90.5	97.3
Mole % of n-propanol								
In liquid	0	2.0	6.0	20.0	43.2	60.0	80.0	100.0
In vapor	0	21.6	35.1	39.2	43.2	49.2	64.1	100.0

With the aid of a graph of these data, calculate the mole fraction of n-propanol in the first drop of distillate when the following solutions are distilled with a simple distilling flask that gives one theoretical plate: (a) 87 g of n-propanol and 211 g of water; (b) 50 g of n-propanol and 5.02 g of water.

<u>SOLUTION</u>

(a) $n_{Pr} = \dfrac{87}{60.1} = 1.44$

$n_{H_2O} = \dfrac{211}{18} = 11.7$

$x_{Pr} = \dfrac{1.44}{1.44 + 11.7} = 0.11$

From graph $y_{Pr} = 0.37$

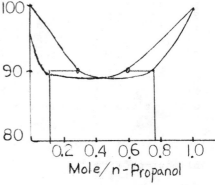

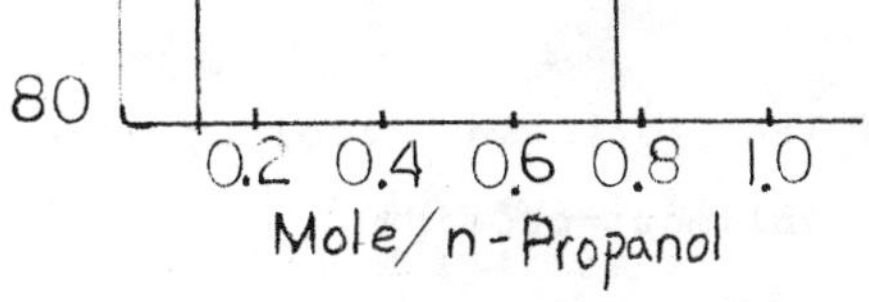

(b) $n_{Pr} = \frac{50}{60.1} = 0.831$

$n_{H_2O} = \frac{5.02}{18} = 0.279$

$x_{Pr} = \frac{0.831}{0.831 + 0.279} = 0.75$

$y_{Pr} = 0.59$ from graph

7.13 The <u>NBS Tables of Chemical Thermodynamic Properties</u> list $\Delta_f G^\circ$ for I_2 in C_6H_6:x as 7.1 kJ mol. The x indicates that the standard state for I_2 in C_6H_6 is on the mole fraction scale. What is the solubility of I_2 in C_6H_6 at 298 K on the mole fraction scale? A chemical handbook lists the solubility as 16.46 g I_2 in 100 cm^3 of C_6H_6. Are these solubilities consistent?

SOLUTION

$I_2\ (cr) + [C_6H_6] \quad = \quad I_2$ in C_6H_6

$\Delta G^\circ \quad = \quad -RT \ln x$

$7.1 \quad = \quad -(8.314 \times 10^{-3})(298) \ln x$

$x \quad = \quad 0.0569$ where x is the equilibrium mole fraction of I_2

The mole fraction calculated from the chemical handbook is

$$\frac{\frac{16.46}{253.8}}{\frac{16.46}{253.8} + \frac{(100)(.8787)}{78.12}} = 0.545$$

so the values are consistent.

7.14 The following cooling curves have been found for the system antimony-cadmium.

Cd, Wt. %	0	20	37.5	47.5	50	58	70	93	100
First break in curve, °C	--	550	461	--	419	--	400	--	--
Continuing constant temperature, °C	630	410	410	410	410	439	295	295	321

Construct a phase diagram, assuming that no breaks other than these actually occur in any cooling curve. Label the diagram completely and give the formula of any compound formed. How many degrees of freedom are there for each area and at each eutectic point?

SOLUTION

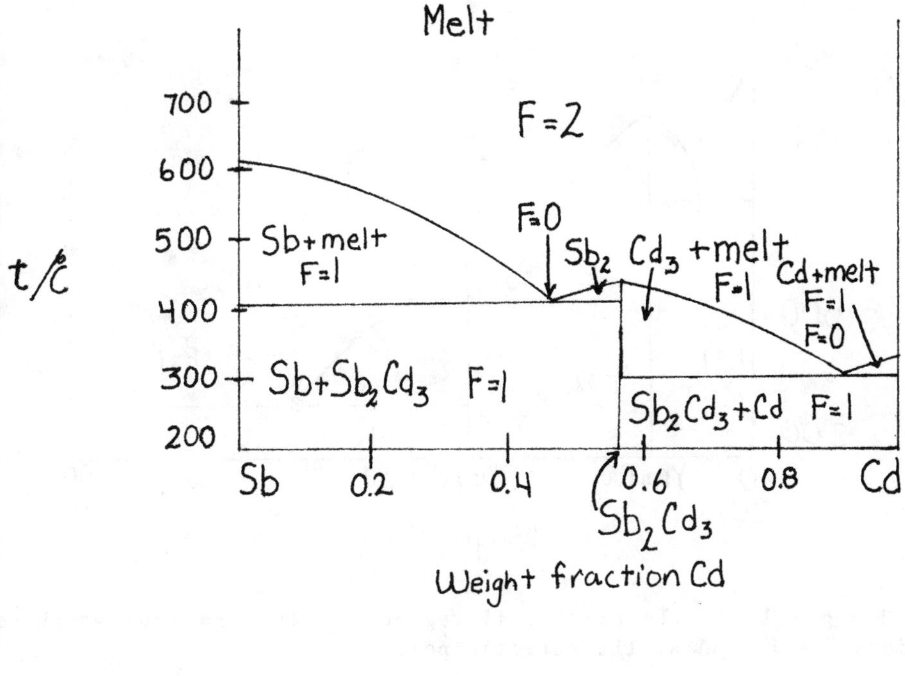

7.15 The phase diagram for magnesium-copper at constant pressure shows that two compounds are formed: $MgCu_2$ which melts at 800 °C, and Mg_2Cu which melts at 580 °C. Copper melts at 1085 °C, and Mg at 648 °C. The three eutectics are at 9.4% by weight Mg (680 °C), 34% by weight Mg (560 °C), and 65% by weight Mg (380 °C). Construct the phase diagram. How many degrees of freedom are there for each area and at each eutectic point?

SOLUTION

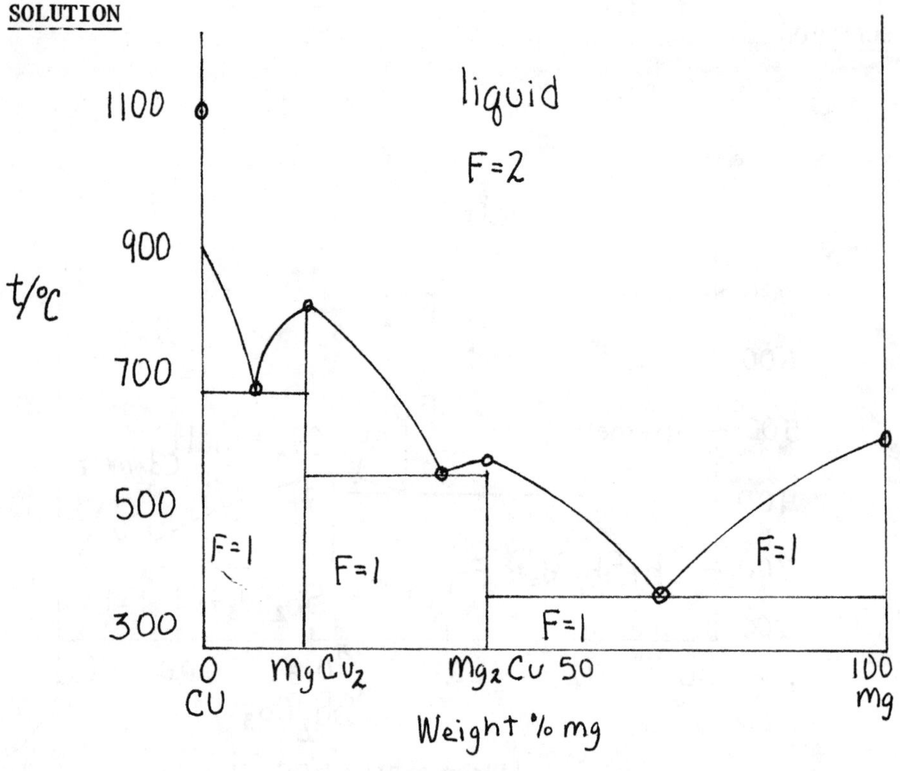

F = 2 - p + 1   In the liquid region F = 2; in the two-phase regions F = 1; and at the eutectic point F = 0.

7.16 For the ternary system benzene-isobutanol-water at 25 °C and 1 bar the following compositions have been obtained for the two phases in equilibrium.

Water-Rich Phase		**Benzene-Rich Phase**	
Isobutanol, wt.%	Water, wt.%	Isobutanol, wt.%	Benzene, wt.%
2.33	97.39	3.61	96.20
4.30	95.44	19.87	79.07
5.23	94.59	39.57	57.09
6.04	93.83	59.48	33.98
7.32	92.64	76.51	11.39

Plot these data on a triangular graph, indicating the tie lines. (a) Estimate the compositions of the phases that will be produced from a mixture of 20% isobutanol, 55% water, and 25% benzene. (b) What will be the composition of the principal phase when the first drop of the second phase separates when water is added to a solution of 80% isobutyl alcohol in benezene?

SOLUTION

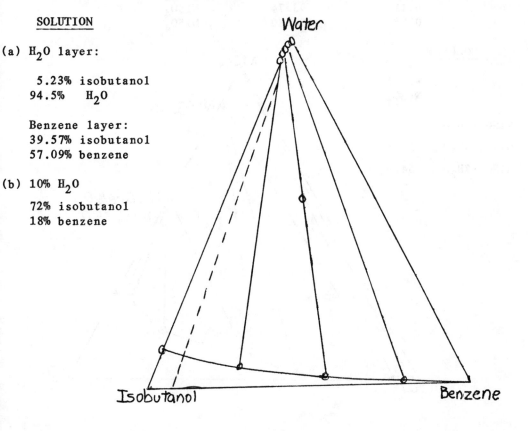

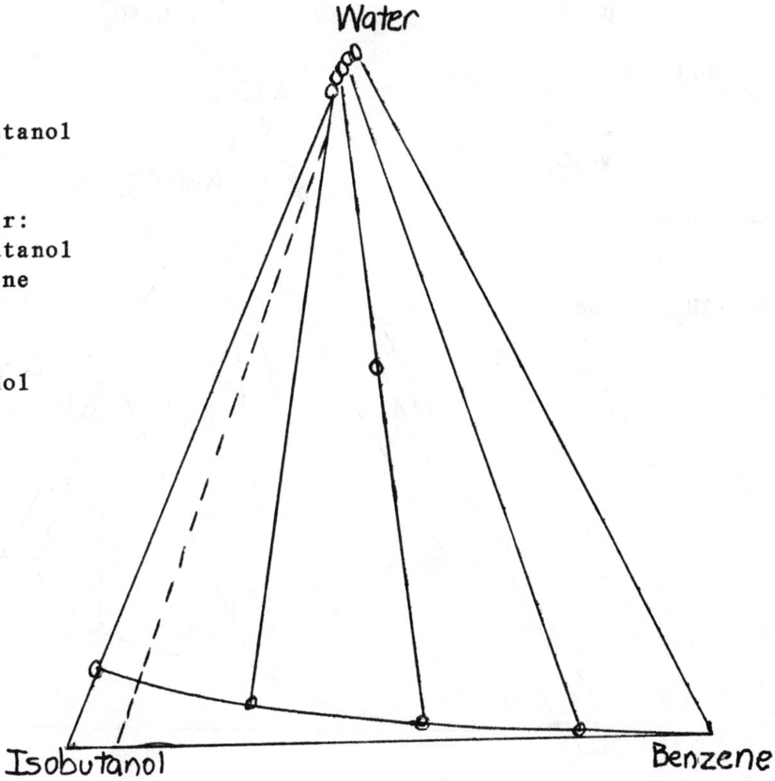

(a) H_2O layer:

5.23% isobutanol
94.5% H_2O

Benzene layer:
39.57% isobutanol
57.09% benzene

(b) 10% H_2O

72% isobutanol
18% benzene

7.17 The following data are available from the system nickel sulfate-sulfuric acid-water at 25 °C. Sketch the phase diagram on triangular coordinate paper, and draw appropriate tie lines.

Liquid Phase		Solid Phase
$NiSO_4$, wt.%	H_2SO_4, wt.%	
28.13	0	$NiSO_4 \cdot 7H_2O$
27.34	1.79	$NiSO_4 \cdot 7H_2O$
27.16	3.86	$NiSO_4 \cdot 7H_2O$
26.15	4.92	$NiSO_4 \cdot 6H_2O$
15.64	19.34	$NiSO_4 \cdot 6H_2O$
10.56	44.68	$NiSO_4 \cdot 6H_2O$
9.65	48.46	$NiSO_4 \cdot H_2O$
2.67	63.73	$NiSO_4 \cdot H_2O$
0.12	91.38	$NiSO_4 \cdot H_2O$
0.11	93.74	$NiSO_4$
0.08	96.80	$NiSO_4$

SOLUTION

	Wt. % $NiSO_4$
$NiSO_4 \cdot H_2O$	89.6
$NiSO_4 \cdot 6H_2O$	58.8
$NiSO_4 \cdot 7H_2O$	54.9

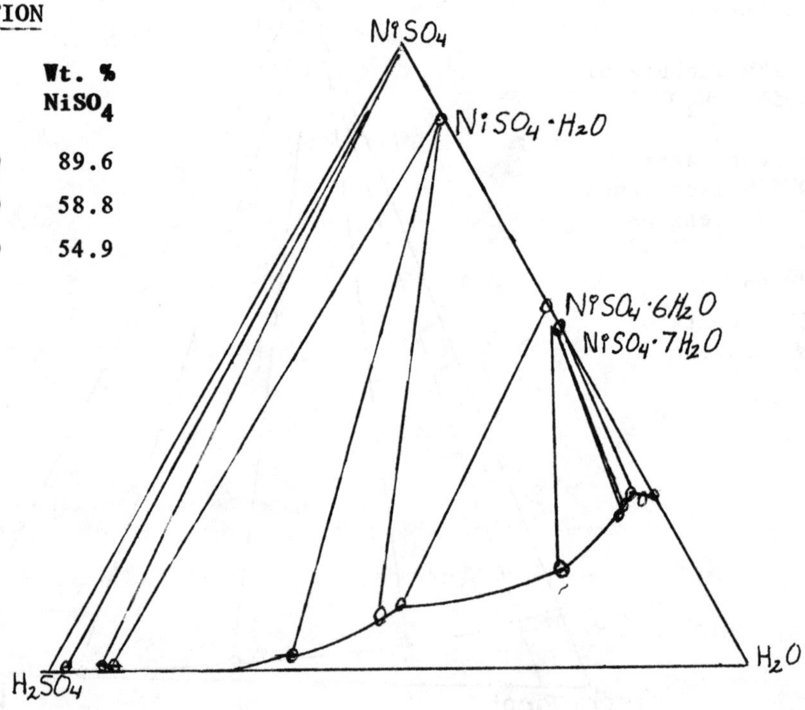

7.18 2.36×10^{-3} K

7.19 $\gamma_{acetone} = 1.67$ $\gamma_{CS_2} = 1.38$

7.20 (a) 19.9 kPa (b) 1.00, 1.13, 1.37, 1.65, 1.88, 1.96

7.22 0.125, 0.186, 0.182, 0.166, 0.120

7.23 $x_{C_6H_6} = 0.55$ Pure C_6H_6 may be obtained by distillation provided $x_{C_6H_6} > 0.55$.

7.24 11.17 kJ mol^{-1}

7.25

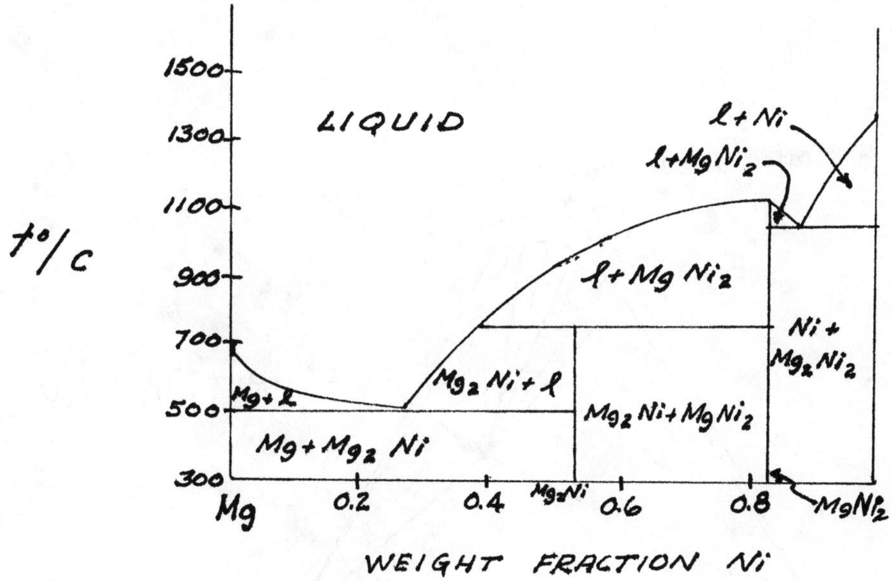

7.26

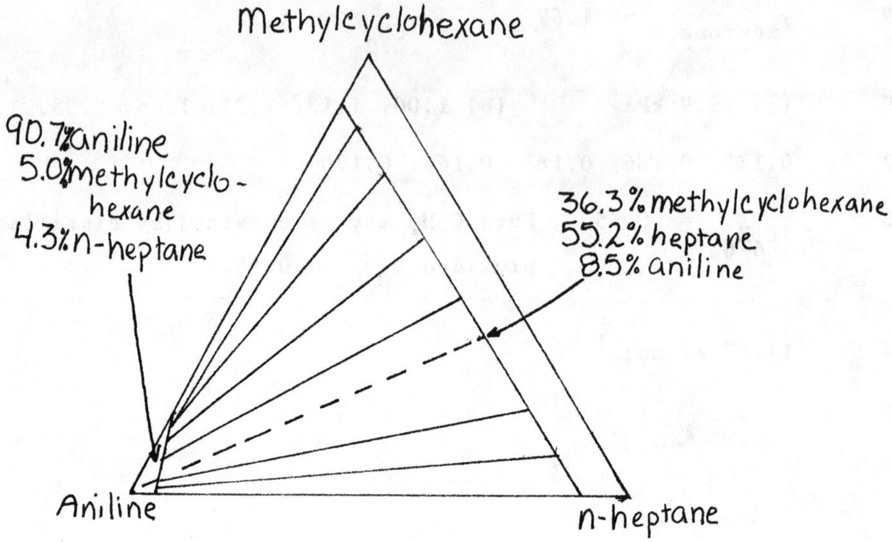
Methylcyclohexane
90.7% aniline
5.0% methylcyclo-
hexane
4.3% n-heptane
36.3% methylcyclohexane
55.2% heptane
8.5% aniline
Aniline
n-heptane

7.27

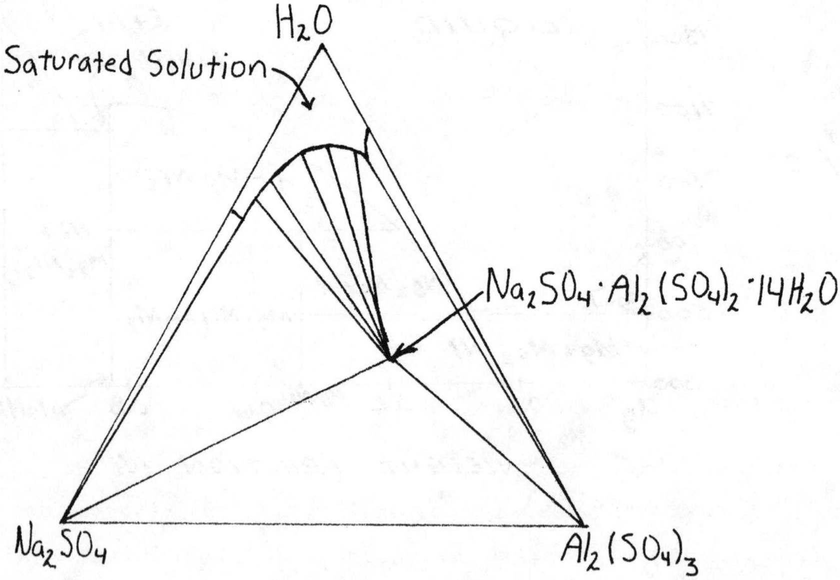
H2O
Saturated Solution
Na2SO4·Al2(SO4)2·14H2O
Na2SO4
Al2(SO4)3

CHAPTER 8: Electrochemical Equilibria

8.1 How much work is required to bring two protons from an infinite distance of separation to 0.1 nm? Calculate the answer in joules using the protonic charge 1.602×10^{-19} C. What is the work in kJ mol^{-1} for a mole of proton pairs?

SOLUTION

$$\text{Potential } \phi = \frac{Q_2}{4\pi\varepsilon_o r}$$

$$= \frac{(1.602 \times 10^{-19}\ C)(0.89875 \times 10^{10}\ N\ m^2\ C^{-2})}{10^{-10}\ m}$$

$$= 14.398\ J\ C^{-1}$$

$$= Q_1\phi = (1.602 \times 10^{-19}\ C)(14.398\ J\ C^{-1})$$

$$= 2.307 \times 10^{-18}\ J$$

$$= (2.307 \times 10^{-18}\ J)(6.022 \times 10^{23}\ mol^{-1})$$

$$= 1389.3\ kJ\ mol^{-1}$$

8.2 What is the electric field strength 0.5 nm from a proton?

SOLUTION

$$E = \frac{Q}{4\pi\varepsilon_o r^2}$$

$$= \frac{1.602 \times 10^{-19}\ C}{4\pi(8.854 \times 10^{-12}\ C^2\ N^{-1}\ m^{-2})(0.5 \times 10^{-9}\ m)^2}$$

$$= 5.759 \times 10^9\ V\ m^{-1}$$

8.3 A small dry battery of zinc and ammonium chloride weighing 85 g will operate continuously through a 4-Ω resistance for 450 min before its voltage falls below 0.75 V. The initial voltage is 1.60, and the effective voltage over the whole life of the battery is taken to be 1.00. Theoretically, how many kilometers above the earth could this battery be raised by the energy delivered under these conditions?

SOLUTION

$$I = \frac{E}{R} = \frac{1\ V}{4\ \Omega} = 0.25\ A$$

$$\text{Power} = I^2R = (0.25\ A)^2(4\ \Omega) = 0.25\ W$$

$$\text{Work} = (0.25\ W)(450 \times 60\ s) = 6.75 \times 10^3\ J$$

$$= (0.085\ kg)(9.80\ m\ s^{-2})h$$

$$h = \frac{6.75 \times 10^3\ J}{(0.085\ kg)(9.80\ m\ s^{-2})} = 8103\ m$$

$$= \frac{(8103 \times 10^5\ cm)}{(2.54\ cm\ in^{-1})(12\ in\ ft^{-1})(5280\ ft\ mile^{-1})} = 5.04\ miles$$

8.4 What is the expression for the activity of Na_2SO_4 in terms of the mean ionic activity coefficient and the molality.

SOLUTION

$$a_{Na_2SO_4} = a_{Na+}^2 \cdot a_{SO_4^=} \qquad \gamma_{Na^+} = \frac{a_{Na^+}}{2m}$$

$$= (2m\gamma_{Na^+})\ (\gamma_{SO_4^=}\ m) \qquad \gamma_{SO_4^=} = \frac{a_{SO_4^=}}{m}$$

$$= 4\ m^3\ \gamma_\pm^3 \qquad \text{Where } \gamma_\pm = (\gamma_{Na^+}^2 \cdot \gamma_{SO_4^=})^{1/3}$$

8.5 A solution of NaCl has an ionic strength of 0.24 mol kg^{-1}. (a) What is its molality? (b) What molality of Na_2SO_4 would have the same ionic strength? (c) What molality of $MgSO_4$?

SOLUTION

(a) $$I = \frac{1}{2}(m_1z_1^2 + m_2z_2^2)$$

$$0.24 = \frac{1}{2}(1^2 + 1^2)\ m \qquad m = 0.24 \text{ mol kg}^{-1}$$

(b) $$0.24 = \frac{1}{2}(2m \times 1^2 + m2^2) \qquad m = 0.08 \text{ mol kg}^{-1}$$

(c) $$0.24 = \frac{1}{2}(2^2 + 2^2)\ m \qquad m = 0.06 \text{ mol kg}^{-1}$$

8.6 For 0.002 mol kg^{-1} $CaCl_2$ at 25 °C use the Debye-Hückel limiting law to calculate the activity coefficients of Ca^{++} and Cl^-. What is the mean ionic activity coefficient for the electrolyte?

SOLUTION

$$I = \frac{1}{2}(0.002 \times 2^2 + 0.004 \times 1^2) = 0.006$$

$$\log \gamma_i = -0.509\ z_i^2\ I^{1/2}$$

$$\log \gamma_{Ca^{2+}} = -0.509(2^2)(0.006)^{1/2}$$

$$\gamma_{Ca^{2+}} = 0.695$$

$$\log \gamma_{Cl^-} = -0.509(1^2)(0.006)^{1/2}$$

$$\gamma_{Cl^-} = 0.913$$

$$\gamma_{\pm} = (\gamma_+\ \gamma_-^2)^{1/3} = (0.695 \times 0.913^2)^{1/3}$$

$$= 0.834$$

8.7 The cell Pt|H_2(1 bar)|HBr(m)|AgBr|Ag has been studied by H. S. Harned, A. S. Keston, and J. G. Donelson [J. Am. Chem. Soc., 58, 989 (1936)]. The following table gives the electromotive forces obtained at 25 °C.

m/mol kg^{-1}	0.01	0.01	0.05	0.10
E/V	0.3127	0.2786	0.2340	0.2005

Calculate (a) E^0 and (b) the activity coefficient for a 0.10 mol kg^{-1} solution of hydrogen bromide.

SOLUTION

(a) Plot $E + 0.1183 \log m - 0.0602 \sqrt{m}$ versus m.

The intercept at m = 0 is $E^\circ = 0.0710$ V.

(b) $E = 0.0710\ \text{V} - (0.05915\ \text{V}) \log \gamma_\pm^2 m^2$

$0.2005\ \text{V} = 0.0710\ \text{V} - (0.1183\ \text{V}) \log \gamma_\pm - (0.1183\ \text{V}) \log 0.1$

$$\log \gamma_\pm = -\frac{0.2005 - 0.1183 - 0.0710}{0.1183}$$

$$\gamma_\pm = 0.804$$

8.8 According to Table A.1 what is the value of the equilibrium constant for the reaction

$$\frac{1}{2} H_2(g) + AgCl(cr) = Ag(cr) + H^+(ao) + Cl^-(ao)$$

at 25 °C and how it is defined?

SOLUTION

$\Delta G^\circ = -131.228 - (-109.789) = -21.329\ \text{kJ mol}^{-1}$

$$K = e^{-\Delta G^\circ/RT} = 5.701 \times 10^3$$

$$= \frac{a(H^+)a(Cl^-)}{\left(P_{H_2}/P^\circ\right)^{1/2}} = \frac{a(HCl)}{\left(P_{H_2}/P^\circ\right)^{1/2}}$$

In writing the equilibrium constant expression the activities of the pure crystalline phases are taken equal to unity. Strictly speaking the pressure of hydrogen should be replaced by its fugacity.

8.9 The electromotive force of the cell

$$Pb\ (cr)|PbSO_4\ (cr)|Na_2SO_4\cdot 10H_2O(sat)|Hg_2SO_4\ (cr)|Hg(l)$$

is 0.9647 V at 25 °C. The temperature coefficient is 1.74×10^{-4} V K^{-1}. (a) What is the cell reaction? (b) What are the values of ΔG, ΔS, and ΔH?

SOLUTION

(a) $Pb(cr) + Hg_2SO_4(cr) = PbSO_4(cr) + 2Hg(l)$

(b) $\Delta G = -nFE = -(2)(96{,}485\ C\ mol^{-1})(0.9647\ V) = -186.16\ kJ\ mol^{-1}$

$$\Delta S = nF\left(\frac{\partial E}{\partial T}\right)_P = (2)(96{,}485\ C\ mol^{-1})(1.74 \times 10^{-4}\ V\ K^{-1})$$

$$= 33.58\ J\ K^{-1}\ mol^{-1}$$

$$\Delta H = -nFE + nFT\left(\frac{\partial E}{\partial T}\right)_P$$

$$= -2(96{,}485\ C\ mol^{-1})(0.9647\ V) + 2(96{,}485\ C\ mol^{-1})(298.15\ K)(1.74 \times 10^{-4}\ V\ K^{-1})$$

$$= -176.15\ kJ\ mol^{-1}$$

8.10 (a) Write the reaction that occurs when the cell $Zn|ZnCl_2(0.555\ mol\ kg^{-1})|AgCl|Ag$ delivers current and calculate (b) ΔG, (c) ΔS, and (d) ΔH at 25 °C for this reaction. At 25 °C, $E = 1.015$ V and $(\partial E/\partial T)_P = 4.02 \times 10^{-4}\ V\ K^{-1}$.

SOLUTION

(a)

R	$2AgCl + 2e^- = 2Ag + 2Cl^-$	
L	$Zn^{2+} + 2e^- = Zn$	
	$2AgCl + Zn = 2Ag + ZnCl_2(0.555\ mol\ kg^{-1})$	

(b) $\Delta G = -nFE = -2(96{,}485\ C\ mol^{-1})(1.015\ V) = -195.9\ kJ\ mol^{-1}$

(c) $$\Delta S = nF\left(\frac{\partial E}{\partial T}\right)_P = 2(96{,}485\ C\ mol^{-1})(-4.02 \times 10^{-4}\ V\ K^{-1})$$

$$= -77.6\ J\ K^{-1}\ mol^{-1}$$

(d) $\Delta H = \Delta G + T\Delta S = -195{,}900 + (298)(-77.6) = -219.0\ kJ\ mol^{-1}$

8.11 In problem 4.3 two equations were derived for calculating ΔG at another temperature if it is known at one. Compare the values of

ΔG°(323 K) and K_w calculated with these equations for

$$H_2O(l) = H^+(a0) + OH^-(ao)$$

SOLUTION

	$\Delta_f G^\circ$ (298 K)	ΔH° (298 K)	C_P° (298 K)
$H_2O(l)$	-237.129	-285.830	75.291
$H^+(ao)$	0	0	0
$OH^-(a0)$	-157.244	-229.994	-148.5

$\Delta G^\circ = -157.244 + 237.129 = 79.885$ kJ mol^{-1}

$\Delta H^\circ = -229.994 + 285.830 = 55.836$ kJ mol^{-1}

$\Delta C_P = -148.5 - 75.291 = -223.8$ J K^{-1} mol^{-1}

(a) $\Delta G_2^\circ = \Delta G_1^\circ \; T_2/T_1 + \Delta H(1 - T_2/T_1)$

$\Delta G^\circ(323) = (79.885)(323.15)/298.15 + 55.836\ (1 - 323.15/298.15)$

$= 81.902$ kJ mol^{-1}

$K_w = 6.94 \times 10^{-14}$

(b) $\Delta G_2 = \Delta G_1 \; T_2/T_1 + (\Delta H_1 - T_1 \Delta C_P)(1 - T_2/T_1) + T_2 \Delta C_P \ln (T_2/T_1)$

$= (79.885)(323.15)/298.15 + [55.836 - (298.15)(-0.2238)]$

$(1 - 323.15/298.15) + (323.15)(-0.2238) \ln (323.15/298.15)$

$= 70.483$ kJ mol^{-1}

$K_w = 4.046 \times 10^{-12}$

ΔC_P° is rather large for this reaction and so the simpler equation yields a misleading result.

8.12 For the electrochemical cell

$$Zn(s)|ZnO_2(aq)||CuO_2(aq)|Cu(s)$$

what are (a) the half cell reactions, (b) cell reaction, (c) the standard electromotive force, (d) the equilibrium constant for the cell reaction, and (e) equilibrium constant expression?

SOLUTION

(a) R $Cu^{2+}(aq) + 2e^- = Cu(cr)$ $\qquad E° = 0.337$ V

L $Zn^{2+}(aq) + 2e^- = Zn(cr)$ $\qquad E° = -0.763$ V

(b) $Zn(s) + Cu^{2+}(aq) = Zn^{2+}(aq) + Cu(s)$

(c) $E° = 0.337 - (-0.763) = +1.100$ V

(d) $K = e^{2(96,485)(1.1)/(8.314)(298)} = 1.62 \times 10^{37}$

(e) $$K = \frac{a_{ZnCl_2}}{a_{CuCl_2}} = \frac{4\, m^3_{ZnCl_2}\ \gamma^3_{\pm}(ZnCl_2)}{4\, m^3_{CuCl_2}\ \gamma^3_{\pm}(CuCl_2)}$$

If the $CuCl_2$ solution and $ZnCl_2$ solution have the same concentration, the activity coefficients will be approximately the same so that

$$\frac{m_{ZnCl_2}}{m_{CuCl_2}} \approx \left(1.62 \times 10^{37}\right)^{1/3}$$

8.13 Calculate the standard electromotive force of the cell Li|LiCl(ai)|Cl_2(g),Pt at 25 °C using (a) electrode potentials and (b) standard Gibbs energies of formation.

SOLUTION

R $Cl_2(g) + 2e^- = 2\ Cl^-(ao)$ $\qquad 1.360$ V

L $2\ Li^+ + 2e^- = 2\ Li$ $\qquad -3.045$ V

$Cl_2(g) + 2\ Li = 2\ LiCl(ai)$ $\qquad 4.405$ V

Using $\Delta_f G^\circ$ values

$$\Delta G^\circ = 2(-298.31 - 131.228) = -849.076 \text{ kJ mol}^{-1}$$

$$E = \frac{\Delta G^\circ}{nF} = \frac{849{,}076 \text{ J}}{2(96{,}485 \text{ C mol}^{-1})} = 4.400 \text{ V}$$

8.14 According to Table 8.1 what are the equilibrium constants for the following reactions at 25 °C?

(a) $H^+(ao) + Li(cr) = Li^+(ao) + \frac{1}{2} H_2(g)$

(b) $2H^+(ao) + Pb(cr) = Pb^{2+}(ao) + H_2(g)$

(c) $3H^+(ao) + Au(cr) = Au^{3+}(ao) + \frac{3}{2} H_2(g)$

SOLUTION

(a) R $H^+ + e^- = \frac{1}{2} H_2$ $\qquad E^\circ = 0$

L $Li^+ + e^- = Li$ $\qquad E^\circ = -3.045$ V

$H^+ + Li = Li^+ + \frac{1}{2} H_2$ $\qquad E^\circ = 3.045$ V

$$K = e^{nFE^\circ/RT}$$

$$= e^{(96{,}485 \text{ C mol}^{-1})(3.045 \text{ V})/(8.314 \text{ J K}^{-1} \text{ mol}^{-1})(298 \text{ K})}$$

$$= 3.1 \times 10^{51}$$

(b) R $2H^+ + 2e^- = H_2$ $\qquad E^\circ = 0$

L $Pb^{2+} + 2e^- = Pb$ $\qquad E^\circ = -0.126$ V

$2H^+ + Pb = Pb^{2+} + H_2$ $\qquad E^\circ = 0.126$

$$K = e^{2(96{,}485)(0.126)/(8.314)(298 \text{ K})} = 1.8 \times 10^4$$

(c) R $3H^+ + 3e^- = \frac{3}{2} H_2$ $\qquad E^\circ = 0$

L $Au^{3+} + 3e^- = Au$ $\qquad E^\circ = 1.50$ V

$3H^+ + Au = Au^{3+} + \frac{3}{2} H_2$ $\qquad E^\circ = -1.50$ V

$$K = e^{3(96,485)(-1.50)/(8.314)(298\ K)} = 7.8 \times 10^{-77}$$

The striking resistance of gold to corrosion by acid is evident.

8.15 Devise a cell for which the cell reaction is

$$H_2O(l) = H^+(ao) + OH^-(ao)$$

Calculate ΔG° at 25 °C from electrode potentials. What is the value of the equilibrium constant at this temperature?

SOLUTION

$Pt|H_2(g)|H^+(ao)||OH^-(ao)|H_2(g)|Pt$

$$\text{R}\quad H_2O + e^- = \frac{1}{2}H_2 + OH^- \qquad E^\circ = -0.8281\ V$$

$$\text{L}\quad H^+ + e^- = \frac{1}{2}H_2 \qquad E^\circ = 0$$

$$H_2O(l) = H^+(ao) + OH^-(ao) \qquad E^\circ = -0.8281\ V$$

$$\Delta G^\circ = -nFE^\circ = -(96,485\ C\ mol^{-1})(-0.8281\ V) = 79.899\ kJ\ mol^{-1}$$

$$K = e^{-\Delta G^\circ/RT} = e^{-79,899/(8.314)(298)} = 1.003 \times 10^{-14}$$

$$= \frac{(a_{H^+})(a_{OH^-})}{a_{H_2O}} = m^2\ \gamma_\pm^{\ 2}$$

The activity of water is taken to be unity since, in contrast with the ions, its activity is measured on the mole fraction scale (Section 5.17).

8.16 Devise an electromotive force cell for which the cell reaction is $AgBr(cr) = Ag^+ + Br^-$. Calculate the equilibrium constant (usually called the solubility product) for this reaction at 25 °C.

SOLUTION

$Ag|Ag^+||Br^-|AgBr|Ag$

$$\begin{array}{lllll} \text{R} & AgBr + e^- & = & Ag + Br^- & E^\circ = 0.095 \text{ V} \\ \text{L} & Ag^+ + e^- & = & Ag & E^\circ = 0.7991 \text{ V} \\ \hline & AgBr & = & Ag^+ + Br^- & E^\circ = -0.701 \text{ V} \end{array}$$

$$\log K = \frac{E^\circ}{0.05916} = \frac{-0.710 \text{ V}}{0.05916 \text{ V}} = -11.85$$

$$K = 10^{-11.85} = m^2 \gamma_\pm^2$$

8.17 We found in Section 8.8 that for $Pt|H_2|HCl|AgCl|Ag$ $E^\circ = 0.2224$ V at 25 °C. Using the value of $\Delta_f G^\circ[Cl^-(ao)]$ obtained in Example 8.6, what is the value of $\Delta_f G^\circ[AgCl(cr)]$?

SOLUTION

(a)

$$\begin{array}{lllll} \text{R} & AgCl + e^- & = & Ag + Cl^- & E^\circ = 0.2224 \text{ V} \\ \text{L} & H^+ + e^- & = & \frac{1}{2} H_2 & E^\circ = 0 \\ \hline & AgCl + \frac{1}{2} H_2 & = & Ag + H^+ + Cl^- & E^\circ = 0.2224 \text{ V} \end{array}$$

$$\Delta G^\circ = -(96{,}485 \text{ C mol}^{-1})(0.2224 \text{ V}) = -21{,}458 \text{ J mol}^{-1}$$

$$= \Delta_f G^\circ[Cl^-(ao)] - \Delta_f G^\circ[AgCl(cr)]$$

$$\Delta_f G^\circ[AgCl(cr)] = 21{,}458 - 131.263 = -109.805 \text{ kJ mol}^{-1}$$

8.18 From standard electrode potentials in Table 8.1 what are the standard Gibbs energies of formation at 25 °C for $Cl^-(ao)$ and $Na^+(ao)$?

SOLUTION

For $H_2(g)|HCl(ai)|Cl_2(g)$

$$\begin{array}{lllll} \text{R} & \frac{1}{2} Cl_2 + e^- & = & Cl^- & E^\circ = 1.3604 \text{ V} \\ \text{L} & H^+ + e^- & = & \frac{1}{2} H_2 & E^\circ = 0 \\ \hline & \frac{1}{2} Cl_2 + \frac{1}{2} H_2 & = & H^+(ao) + Cl^-(ao) & E^\circ = 1.3604 \text{ V} \end{array}$$

$\Delta G^\circ = -nFE^\circ = -(96{,}485\ C\ mol^{-1})(1.3604\ V) = -131.258\ kJ\ mol^{-1}$

This is $\Delta_f G^\circ[Cl^-(ao)]$ since ΔG° for reaction 1 is zero by convention. Table A.1 gives $-131.228\ kJ\ mol^{-1}$.

For $H_2(g)|H^+(ao)||OH^-(ao)|O_2(g)$

$$
\begin{array}{lllr}
R & \frac{1}{4} O_2 + \frac{1}{2} H_2O(l) + e^- = OH^-(ao) & E^\circ = & 0.401\ V \\
L & H^+(ao) + e^- = \frac{1}{2} H_2 & E^\circ = & 0 \\
\hline
 & \frac{1}{4} O_2 + \frac{1}{2} H_2 + \frac{1}{2} H_2O = H^+(ao) + OH^-(ao) & E^\circ = & 0.401
\end{array}
$$

$\Delta G^\circ = -(96{,}485\ C\ mol^{-1})(0.401\ V) = -38.69\ kJ\ mol^{-1}$

$= \Delta_f G^\circ[OH^-(ao)] - \frac{1}{2} \Delta_f G^\circ[H_2O(l)]$

$\Delta_f G^\circ\ [OH^-(ao)] = -38.69 + \frac{1}{2}(-237.129) = -157.26\ kJ\ mol^{-1}$

Table A.1 gives $-157.244\ kJ\ mol^{-1}$.

For $Na(cr)|Na^+(ao)||H^+(ao)|H_2(g)$

$$
\begin{array}{lllr}
R & H^+ + e^- = \frac{1}{2} H_2 & E^\circ = & 0 \\
L & Na^+ + e^- = Na & E^\circ = & -2.714\ V \\
\hline
 & Na(cr) + H^+(ao) = Na^+(ao) + \frac{1}{2} H_2(g) & E^\circ = & 2.714\ V
\end{array}
$$

$\Delta G^\circ = -(96{,}485\ C\ mol^{-1})(2.714\ V) = -261.86\ kJ\ mol^{-1} = \Delta_f G^\circ[Na^+(ao)]$

Table A.1 gives $-261.905\ kJ\ mol^{-1}$.

8.19 Using data from Table A.1 calculate the solubility of AgCl(cr) in water at 298.15 K. The salt is completely dissociated in the aqueous phase.

SOLUTION

$AgCl(cr) = Ag^+(ao) + Cl^-(ao)$

$\Delta G^\circ \quad = \quad 77.107 - 131.228 + 109.789 \quad = \quad 55.668 \text{ kJ mol}^{-1}$

$K \quad = \quad \exp(-55{,}668/8.3144 \times 298.15) \quad = \quad 1.767 \times 10^{-10}$

$\quad = \quad (a_{Ag^+})(a_{Cl^-}) \quad = \quad m^2 \gamma_\pm^2 \quad \approx \quad m^2$

$m \quad = \quad 1.33 \times 10^{-5} \text{ mol kg}^{-1}$

8.20 Calculate standard electrode potentials at 25 °C for the following electrodes using Table A.1.

(a) $Li^+(ao)|Li(cr)$

(b) $F^-(ao)|F_2(g)$

(c) $Pb^{2+}(ao)|PbO_2(cr)|Pb$

<u>SOLUTION</u>

$Li^+ + e^- = Li$

$\Delta G^\circ \quad = \quad -(-293.31) \quad = \quad 293.31 \text{ kJ mol}^{-1}$

$E^\circ \quad = \quad -\Delta G^\circ/nF = \quad -(293{,}310 \text{ J mol}^{-1})/(96{,}485 \text{ C mol}^{-1}) \quad = \quad -3.040 \text{ V}$

$\frac{1}{2} F_2(g) + e^- = F^-$

$\Delta G^\circ \quad = \quad -278.79 \text{ kJ mol}^{-1}$

$E^\circ \quad = \quad (278{,}790 \text{ J mol}^{-1})/(96{,}485 \text{ C mol}^{-1}) \quad = \quad 2.889 \text{ V}$

$\frac{1}{2} PbO_2 + 2H^+ + e^- = \frac{1}{2} Pb^{2+} + H_2O$

$\Delta G^\circ = -237.129 + \frac{1}{2}(-24.43) - \frac{1}{2}(-217.33) = -140.68 \text{ kJ mol}^{-1}$

$E^\circ = (140{,}680 \text{ J mol}^{-1})/(96{,}485 \text{ C mol}^{-1}) = 1.458 \text{ V}$

8.21 Using Table A.1 calculate the values of ΔG°, ΔH°, ΔS°, and ΔC_P° for the electrode reaction for the $Na^+|Na$ electrode.

SOLUTION

The shorthand notation for the electrode reaction is $Na^+ + e^- = Na$, but we must remember that we are really talking about the cell for which the cell reaction is

$$Na^+ + \frac{1}{2} H_2(g) = Na(cr) + H^+$$

$$\Delta G^o = -(-261.905) = 261.905 \text{ kJ mol}^{-1}$$

$$\Delta H^o = -(-240.12) = 240.12 \text{ kJ mol}^{-1}$$

$$\Delta S^o = 51.21 - 59.0 - \frac{1}{2}(130.684) = -73.132 \text{ J K}^{-1} \text{ mol}^{-1}$$

$$\Delta C_P{}^o = 28.24 - 46.4 - \frac{1}{2}(28.824) = -32.572 \text{ J K}^{-1} \text{ mol}^{-1}$$

According to Table 8.1

$$\Delta G^o = -nFE^o = -(96{,}485 \text{ C mol}^{-1})(-2.714 \text{ V}) = 261.860 \text{ kJ mol}^{-1}$$

8.22 Calculate the standard Gibbs energy of formation of $NO_3^-(ao)$ from its $\Delta_f H^o$ and S^o using data from Table A.2.

SOLUTION

$$\frac{1}{2} N_2(g) + \frac{3}{2} O_2(g) + \frac{1}{2} H_2(g) = NO_3^-(ao) + H^+(ao)$$

$$\Delta_f G^o = \Delta_f H^o - T[S^o(NO_3^-) - \frac{1}{2} S^o(N_2) - \frac{3}{2} S^o(O_2) - \frac{1}{2} S^o(H_2)]$$

$$= -205.0 - (298.15 \times 10^{-3})[146.4 - \frac{1}{2}(191.61) - \frac{3}{2}(205.138) - \frac{1}{2}(130.684)]$$

$$= -108.86 \text{ kJ mol}^{-1}$$

The tabulated value is -108.75 kJ mol^{-1}.

8.23 What are the differences between the standard electrode potentials for a standard-state pressure of 1 bar and 1 atm for the following electrodes:

	E°(1 atm)
$Cl^-|AgCl(cr)|Ag$	0.2224 V
$Cl^-|Cl_2(g)|Pt$	1.3604

SOLUTION

The change in potential with pressure should be calculated for the balanced reaction that includes the hydrogen electrode.

$$\text{R} \quad H^+ + e^- = Ag + Cl^-$$

$$\text{L} \quad AgCl + e^- = \frac{1}{2} H_2$$

$$\frac{1}{2} H_2 + AgCl = Ag + HCl(ao) \qquad \Sigma\nu_i = -\frac{1}{2}$$

$$E^\circ - E^* = -(\Delta_r G^\circ - \Delta_r G^*)/nF = \frac{1}{nF}[RT \ln (P^*/P^\circ)]\Sigma\nu_i$$

$$= (0.338\ \Sigma\nu_i/n)\ \text{mV at } 298.15\ \text{K} = -0.338/2 = -0.17\ \text{mV}$$

$$\frac{1}{2} Cl_2(g) + e^- = Cl^-$$

$$\frac{1}{2} H_2 = H^+ + e^-$$

$$\frac{1}{2} H_2 + \frac{1}{2} Cl_2 = HCl(ao) \qquad \Sigma\nu_i = -1$$

$$E^\circ - E^* = -0.34\ \text{mV}$$

8.24 (a) When one mole of methane is oxidized completely to $CO_2(g)$ and $H_2O(1)$ how much electrical energy can be produced using a fuel cell, assuming no electrical losses? What is the electromotive force of the fuel cell? (b) When one mole of methane is oxidized completely in a Carnot engine that operates between 500 K and 300 K, how much electrical energy can be produced, assuming that the mechanical energy can be converted completely into electrical energy?

SOLUTION

(a)

$$
\begin{array}{llllllll}
R & 2\ O_2 & + & 8\ e^- & + & 8\ H^+ & = & 4\ H_2O \\
L & CO_2 & + & 8\ H^+ & + & 8\ e^- & = & CH_4 + 2\ H_2O \\
\hline
 & 2\ O_2 & + & CH_4 & & & = & CO_2 + 2\ H_2O
\end{array}
$$

$$\Delta G^\circ = -394.359 + 2(-237.129) - (50.72) = -817.90\ \text{kJ mol}^{-1}$$

$$= -nFE^\circ$$

$$E^\circ = \frac{817.90 \times 10^3\ \text{J mol}^{-1}}{8(96{,}485\ \text{C mol}^{-1})} = 1.0596\ \text{V}$$

(b)

$$\Delta H^\circ = -393.509 + 2(-285.830) - (-74.81)$$

$$= -890.4\ \text{kJ mol}^{-1}$$

$$|w| = |q|(T_1 - T_2)/T_1 = 890.4\ \frac{200}{500} = 356\ \text{kJ mol}^{-1}$$

Thus the fuel cell would produce over twice as much electrical energy.

8.25 Calculate the electromotive force of $Li(l)|LiCl(l)|Cl_2(g)$ at 900 K for $P_{Cl_2} = 1$ bar. This high-temperature battery is attractive because of its high electromotive force and low atomic masses. Lithium chloride melts at 883 K and lithium at 453.69 K. [The $\Delta_f G^\circ$ for LiCl(l) at 900 K in JANAF Thermochemical Tables is -335.140 kJ mol^{-1}.]

SOLUTION

$$Li(l) + \frac{1}{2}\ Cl_2(g) = LiCl(l)$$

$$\Delta G^\circ = -335{,}140\ \text{J mol}^{-1} = -FE^\circ$$

$$E^\circ = \frac{-335{,}140 \text{ J mol}^{-1}}{-96{,}485 \text{ C mol}^{-1}} = 3.474 \text{ V}$$

8.26 Ammonia may be used as the anodic reactant in a fuel cell. The reactions occurring at the electrodes are

$$NH_3(g) + 3\ OH^-(ao) = \frac{1}{2} N_2(g) + 3\ H_2O(l) + 3\ e^-$$

$$O_2(g) + 2\ H_2O(l) + 4e^- = 4\ OH^-(ao)$$

What is the electromotive force of this fuel cell at 25 °C?

SOLUTION

$$\text{R} \quad 3\ O_2 + 6\ H_2O + 12\ e^- = 12\ OH^-$$

$$\text{L} \quad 2\ N_2 + 12\ H_2O + 12\ e^- = 4\ NH_3 + 12\ OH^-$$

$$3\ O_2 + 4\ NH_3 = 2\ N_2 + 6\ H_2O$$

Using Table A.1

$$\Delta G^\circ = 6(-237.129) - 4(-16.45) = -1356.97 \text{ kJ mol}^{-1} = -nFE^\circ$$

$$E^\circ = \frac{1{,}356{,}970 \text{ J mol}^{-1}}{(12)(96{,}485 \text{ C mol}^{-1})} = 1.1720 \text{ V}$$

8.27 When a hydrogen electrode and a normal calomel electrode are immersed in a solution at 25 °C a potential of 0.664 V is obtained. Calculate (a) the pH, and (b) the hydrogen-ion activity.

SOLUTION

$$\text{(a)} \quad E = E^\circ - \frac{RT}{nF} \ln a_{H^+} = E^\circ + 0.0591 \text{ pH}$$

$$\text{pH} = \frac{E - E^\circ}{0.0591} = \frac{0.664 - 0.2802}{0.0591} = 6.49$$

$$\text{(b)} \quad a_{H^+} = 10^{-6.49} = 3.24 \times 10^{-7}$$

8.28 Calculate the equilibrium constant at 25 °C for the reaction

$$2H^+ + D_2(g) = H_2(g) + 2D^+$$

from the electrode potential for $D^+|D_2|Pt$, which is -3.4 mV at 25 °C.

SOLUTION

$$Pt|D_2, D^+||H^+, H_2|Pt$$

$$E = 0 - (-0.0034) = 0.0034\ V$$

$$K = e^{nFE°/RT}$$

$$= \exp\left[\frac{2(96{,}500\ C\ mol^{-1})(0.0034\ V)}{(8.314\ J\ K^{-1}\ mol^{-1})(298\ K)}\right] = 1.30$$

8.29 A membrane permeable only by Na^+ is used to separate the following two solutions:

α	0.10 mol kg^{-1} NaCl	0.05 mol kg^{-1} KCl
β	0.05 mol kg^{-1} NaCl	0.10 mol kg^{-1} KCl

What is the membrane potential at 25 °C and which solution has the highest positive potential?

SOLUTION

$$\phi^\beta - \phi^\alpha = -\frac{RT}{z_i F}\ln\frac{a_i^\beta}{a_i^\alpha}$$

$$= -\frac{(8.314\ J\ K^{-1}\ mol^{-1})(298\ K)}{96{,}485\ C\ mol^{-1}}\ln\frac{0.05}{0.10} = 0.018\ V$$

The β phase is more positive because of the diffusion of Na^+ from α to β. Since the ionic strengths of the two solutions are the same, the activity coefficients of Na^+ in α and β are very nearly the same.

8.30 277.8 kJ mol^{-1}

8.31 461.1, 46.11, 5.77 kJ mol^{-1}

8.32 0.905

8.33 (a) $[(a_+)(a_-)]^{1/2}/m$ (b) $[(a_+)(a_-)^3]^{1/4}/3^{3/4}$ m

(c) $[(a_+)(a_-)]^{1/2}/m$

8.34 (a) 0.1 (b) 0.3 (c) 0.4 (d) 0.4

8.35 (a) $\frac{1}{2} H_2(g) + \frac{1}{2} Cl_2(g) = H^+(ao) + Cl^-(ao)$

-131.260, -167.127 kJ, -120.3 J K^{-1} mol^{-1}

(b) -131.260, -167.127 kJ, 56.5 J K^{-1} mol^{-1}

8.36 (a) The electrode with the higher percent thallium is negative.

(b) -1456 J mol^{-1} (c) 0.030462 V

8.37 -130.318 kJ mol^{-1} -125 J K^{-1} mol^{-1} -167.580 kJ mol^{-1}

8.38 (a) $Na|NaOH(m)|H_2$, Pt

At 25 °C $E = E° - 0.0591 \log (m^2 \gamma_\pm^2 P_{H_2}^{1/2})$

(b) $Pt|H_2|H_2SO_4(m)|Ag_2SO_4|Ag$

At 25 °C $E = E° - 0.0296 \log (4m^3 \gamma_\pm^3 P_{H_2}^{-1})$

8.39 $pK = 1.018 \sqrt{I} + \log \frac{m_1}{m_2} + \frac{(E - E°)F}{2.303\ RT} + \log m_3$

8.40 The equilibrium constants are 5.79×10^{-38}, 9.56×10^{45}, 2.01×10^8, 3.20×10^4, and 3.23.

8.41 0.828 V

8.42 -0.152 V

8.43 79.885, 55.835 kJ mol^{-1}, -80.668 J K^{-1} mol^{-1},

1.008 x 10^{-14}

8.45 -38.68 kJ mol^{-1}, -0.4009 V

8.46 6.39, 1.34 x 10^{-5} mol L^{-1}

8.47 -261.92 kJ mol^{-1}

8.48 -744.49 kJ mol^{-1}

8.49 1.239 V

8.50 (a) 1.229 V (b) 1.229 V

8.51 0.60

8.52 (a) 2.62 (b) 2.4 x 10^{-3}

8.53 15.3

CHAPTER 9: Equilibria of Biochemical Reactions

9.1 According to Table A.1, what are the values of ΔG°, ΔH°, and ΔS° at 298 K for

$$H_2O(1) \quad = \quad H^+(ao) \quad + \quad OH^-(ao)$$

Show that the same value of ΔS° is obtained from ΔG° and ΔH° as by using $\Delta S^\circ = \Sigma \nu_i S_i^\circ$.

SOLUTION

$$\Delta G^\circ = -157.244 + 237.129 = 79.885 \text{ kJ mol}^{-1}$$

$$\Delta H^\circ = -229.994 + 285.830 = 55.836 \text{ kJ mol}^{-1}$$

$$\Delta S^\circ = -10.75 - 69.91 = -80.66 \text{ J K}^{-1} \text{ mol}^{-1}$$

or

$$\Delta S^\circ = \frac{\Delta H^\circ - \Delta G^\circ}{T} = \frac{(55{,}836 - 79{,}885) \text{ J mol}^{-1}}{298.15 \text{ K}}$$

$$= -80.66 \text{ J K}^{-1} \text{ mol}^{-1}$$

9.2 Using Table A.1 calculate ΔH°, ΔG°, and ΔS° for

$$Ch_3CO_2H(aq) \quad = \quad H^+(ao) \quad + \quad CH_3CO_2^-(ao)$$

and compare the values in Table 9.1 for 298.15 K.

SOLUTION

$$\Delta H^\circ = -486.01 + 485.76 = -0.25 \text{ kJ mol}^{-1}$$

$$\Delta G^\circ = -369.31 + 396.46 = 27.15 \text{ kJ mol}^{-1}$$

$$\Delta S^\circ = 86.6 - 178.7 = -92.1 \text{ J K}^{-1} \text{ mol}^{-1}$$

9.3 For the acid dissociation of acetic acid ΔH° is approximately zero at room temperature in H_2O. For the acidic form of aniline, which is approximately as strong an acid as acetic acid, ΔH° is approximately 21 kJ mol^{-1}. Calculate ΔS° for each of the following reactions.

$$CH_3CO_2H \;=\; H^+ \;+\; CH_3CO_2^- \qquad pK \;=\; 4.75$$

$$C_6H_5NH_3^+ \;=\; H^+ \;+\; C_6H_5NH_2 \qquad pK \;=\; 4.63$$

How do you interpret these entropy changes? What compensates for the increase in entropy expected from the increase in number of molecules in the reaction?

SOLUTION

For acetic acid

$$\Delta G^\circ = -RT \ln K$$

$$= RT\ 2.303\ pK$$

$$= (8.314\ J\ K^{-1}\ mol^{-1})(298\ K)(2.303)(4.75)$$

$$= 27.1\ kJ\ mol^{-1}$$

$$\Delta S^\circ = (\Delta H^\circ - \Delta G^\circ)/T = -(27.1 \times 10^3\ J\ mol^{-1})/(298\ K)$$

$$= -91\ J\ K^{-1}\ mol^{-1}$$

This increase in order is due to the hydration of the ions that are formed.

For aniline

$$\Delta G^\circ = (8.314\ J\ K^{-1}\ mol^{-1})(298\ K)(2.303)(4.63)$$

$$= 26.4\ kJ\ mol^{-1}$$

$$\Delta S^\circ = [(21 - 26.4) \times 10^3\ J\ mol^{-1}]/(298\ K)$$

$$= -18\ J\ K^{-1}\ mol^{-1}$$

The entropy change is much smaller than for acetic acid because there is no change in the number of ions.

9.4 Estimate pK_3 and pK_2 for H_3PO_4 at 25 °C and 0.1 mol L^{-1} ionic strength. The values at zero ionic strength are $pK_3 = 2.148$ $pK_2 = 7.198$.

SOLUTION

$$pK_I = pK_{I=0} - \frac{(2n + 1)AI^{1/2}}{1 + I^{1/2}}$$

$A = 0.509$ at 25 °C

n is defined by $HA^{-n} = H^+ + A^{-(n+1)}$

For pK_3 of H_3PO_4, $n = 0$

$$pK_3 = 2.148 - \frac{(0.509)(0.1)^{1/2}}{1 + (0.1)^{1/2}} = 2.148 - 0.122 = 2.026$$

For pK_2 of H_3PO_4, $n = 1$

$$pK_2 = 7.198 - \frac{(3)(0.509)(0.1)^{1/2}}{1 + (0.1)^{1/2}}$$

$$= 7.198 - 0.369 = 6.831$$

9.5 At what pH is the average net charge on a lysine molecule zero? That is, what is its isoelectric point?

$pK_3 = 2.16(-CO_2H)$

$pK_2 = 9.18(\alpha - NH_3^+)$

$pK_1 = 10.79(\varepsilon - NH_3^+)$

(Simply set up the equation for calculating $[H^+]$, but do not attempt to calculate $[H^+]$ because the equation is cubic.)

SOLUTION

The three acid dissociation constants are represented by

$LH = L^- + H^+$ $K_1 = [L^-][H^+]/[LH]$

$LH_2^+ = LH + H^+$ $K_2 = [LH][H^+]/[LH_2^+]$

$LH_3^+ = LH_2^+ + H^+$ $K_3 = [LH_2^+][H^+]/[LH_3^{2+}]$

The net charge z is given by

$$z = \frac{-[L^-] + [LH_2^{\ +}] + 2[LH_3^{\ 2+}]}{[L^-] + [LH] + [LH_2] + [LH_3^{\ 2+}]}$$

$$= \frac{-K_1[LH]/H^+] + [LH][H^+]/K_2 + 2[LH][H^+]^2/K_2K_3}{\text{denominator}}$$

The net charge is zero when the numerator is zero so that

$$-\frac{K_1}{[H^+]} + \frac{[H^+]}{K_2} + \frac{2[H^+]^2}{K_2K_3} = 0$$

Multiplying each term by $K_2K_3[H^+]$ yields

$$-K_1K_2K_3 + K_3[H^+]^2 + 2[H^+]^3 = 0$$

9.6 At pH 7 and pMg 4 what value of pCa is required to put half the ATP in the form $CaATP^{-2}$? At 0.2 mol L^{-1} ionic strength and 25 °C the following constants are known.

$$HATP^{3-} = H^+ + ATP^{4-} \qquad pK = 6.95$$

$$MgATP^{2-} = Mg^{2+} + ATP^{4-} \qquad pK = 4.00$$

$$CaATP^{2-} = Ca^{2+} + ATP^{4-} \qquad pK = 3.60$$

SOLUTION

$$\frac{[CaATP^{2-}]}{[ATP^{4-}] + [HATP^{3-}] + [MgATP^{2-}] + [CaATP^{2-}]} = \frac{1}{2}$$

$$\frac{[CaATP^{2-}]/[ATP^{4-}]}{1 + \frac{[HATP^{3-}]}{[ATP^{4-}]} + \frac{[MgATP^{2-}]}{[ATP^{4-}]} + \frac{[CaATP^{2-}]}{[ATP^{4-}]}} = \frac{1}{2}$$

$$\frac{[Ca^{2+}]/10^{-3.60}}{1 + 10^{-0.05} + 1 + [Ca^{2+}]/10^{-3.60}} = \frac{1}{2}$$

$$[Ca^{2+}] = 7.25 \times 10^{-4} \text{ mol L}^{-1} \qquad pCa = 3.14$$

9.7 Show how the partition function method yields the number ν_A of A bound per P and the number ν_B of B are bound at the same site so that

$$PA = P + A \qquad K_A = [P][A]/[PA]$$

$$PB = P + B \qquad K_B = [P][B]/[PB]$$

Show that

$$\frac{\partial \nu_A}{\partial \ln [B]} = \frac{\partial \nu_B}{\partial \ln [A]}$$

for these linked bindings.

SOLUTION

$$Z = 1 + [A]/K_A + [B]/K_B$$

$$\nu_A = \frac{\partial \ln Z}{\partial \ln [A]} = \frac{[A]\partial Z}{Z\partial [A]} = \frac{[A]/K_A}{1 + [A]/K_A + [B]/K_B}$$

$$\nu_B = \frac{[K]/K_B}{1 + [A]/K_A + [B]/K_B}$$

$$\frac{\partial \nu_A}{\partial \ln [B]} = \frac{[B]\partial \nu_A}{\partial [B]} = \frac{[A][B]/K_AK_B}{\left(1 + [A]/K_A + [B]/K_B\right)^2}$$

$$\frac{\partial \nu_B}{\partial \ln [A]} = \frac{[A]\partial \nu_B}{\partial [A]} = \frac{[A][B]/K_AK_B}{\left(1 + [A]/K_A + [B]/K_B\right)^2}$$

9.8 Will 0.01 mol L^{-1} creatine phosphate react with 0.01 mol L^{-1} adenosine diphosphate to produce 0.04 mol L^{-1} creatine and 0.02 mol L^{-1} adenosine triphosphate at 25 °C, pH7, pMg4? What concentration of ATP can be formed if the other reactants are maintained at the indicated concentration?

SOLUTION

$$\text{Creatine P} + H_2O = \text{Creatine} + P \qquad \Delta G^{o\prime} = -43.5 \text{ kJ mol}^{-1}$$

$$\text{ADP} + P = \text{ATP} + H_2O \qquad \Delta G^{o\prime} = 39.8 \text{ kJ mol}^{-1}$$

$$\text{Creatine P} + \text{ADP} = \text{Creatine} + \text{ATP} \qquad \Delta G^{o\prime} = -3.7 \text{ kJ mol}^{-1}$$

$$\Delta G = \Delta G^{o} + RT \ln \frac{[\text{Creatine}][\text{ATP}]}{[\text{Creatine P}][\text{ADP}]}$$

$$= -3700 + (8.314)(298) \ln \frac{(0.04)\ (0.02)}{(0.01)\ (0.01)}$$

$$= 1340 \text{ J mol}^{-1}$$

Therefore the answer to the first question is no.

$$K = e^{-\Delta G^{o}/RT} = e^{1340/(8.314)(298)} = 4.5$$

$$= \frac{[\text{Creatine}][\text{ATP}]}{[\text{Creatine P}][\text{ADP}]}$$

If the reactants are maintained at the indicated concentrations,

$$[\text{ATP}] = \frac{4.5\ [\text{Creatine P}][\text{ATP}]}{[\text{Creatine}]} = \frac{4.5\ (0.01)(0.01)}{0.04}$$

$$= 1.1 \times 10^{-2} \text{ mol L}^{-1}$$

9.9 In a series of biochemical reactions the product in one reaction is a reactant in the next. This has the effect that spontaneous reactions drive nonspontaneous reactions. For example, reaction 2 follows reaction 1.

1. L-malate = fumarate + H_2O $\qquad \Delta G^{o\prime} = 2.9 \text{ kJ mol}^{-1}$
2. fumarate + ammonia = aspartate $\qquad \Delta G^{o\prime} = -15.6 \text{ kJ mol}^{-1}$

The $\Delta G^{o\prime}$ values are for pH 7 and 37 °C, and the state of ionization of the reactants is ignored. In reaction 1, the activity of H_2O is to be taken as 1. If the ammonia concentration is 10^{-2} mol L^{-1}, calculate [aspartate]/[L-malate] at equilibrium.

SOLUTION

L-malate + ammonia = aspartate + H_2O $\qquad \Delta G^{o\prime} = -15.6 \text{ kJ mol}^{-1}$

$$\Delta G = \Delta G^{o} + RT \ln \frac{[\text{aspartate}]}{[\text{L-malate}][\text{ammonia}]}$$

$$0 = -15{,}600 + (8.314)(298) \ln \frac{[\text{aspartate}]}{[\text{L-malate}](10^{-2})}$$

$$\frac{[\text{aspartate}]}{[\text{L-malate}]} = 1.3$$

9.10 Biochemistry textbooks give $\Delta G^{o\prime} = -20.1$ kJ mol^{-1} for the hydrolysis of ethyl acetate at pH 7 and 25 °C. Experiments in acid solution show that

$$\frac{[CH_3CH_2OH][CH_3CO_2H]}{[CH_3CO_2CH_2CH_3]} = 14$$

where concentrations are in moles per liter. What is the value of $\Delta G^{o\prime}$ obtained from this equilibrium quotient? The pK of acetic acid = 4.60 at 25 °C.

SOLUTION

$$K' = \frac{[CH_3CH_2OH]\left([CH_3CO_2H] + [CH_3CO_2^-]\right)}{[CH_3CO_2CH_2CH_3]}$$

$$= \frac{[CH_3CH_2OH][CH_3CO_2H]}{[CH_3CO_2CH_2CH_3]}\left(1 + \frac{[CH_3CO_2^-]}{[CH_3CO_2H]}\right)$$

$$= 14 \text{ mol L}^{-1}\left(1 + \frac{K_{CH_3CO_2H}}{[H^+]}\right)$$

At pH 7

$$K' = 14\left(1 + \frac{10^{-4.6}}{10^{-7}}\right) = 3530$$

$$\Delta G^{\circ\prime} = -RT \ln K = -(8.314 \text{ J K}^{-1} \text{ mol}^{-1})(298.15 \text{ K}) \ln 3540$$

$$= -20.3 \text{ kJ mol}^{-1}$$

9.11 The cleavage of fructose 1,6-diphosphate (FDP) to dihydroxyacetone phosphate (DHP) and glyceraldehyde 3-phosphate (GAP) is one of a series of reactions most organisms use to obtain energy. At 37 °C and pH 7, $\Delta G^{\circ\prime}$ for the reaction FDP = DHP + GAP is 23.97 kJ mol^{-1}. What is $\Delta G'$ in an erythrocyte in which [FDP] = 3×10^{-6} mol L^{-1}, [DHP] = 138×10^{-6} mol L^{-1}, and [GAP] = 18.5×10^{-6} mol L^{-1}?

SOLUTION

$$\text{FDP} + H_2O = \text{DHP} + \text{GAP}$$

$$\Delta G^{\circ\prime} = \Delta G^{\circ\prime} + RT \ln \frac{[\text{DHP}][\text{GAP}]}{[\text{FDP}]}$$

$$= 23{,}970 + (8.314)(310) \ln \frac{(138 \times 10^{-6})(18.5 \times 10^{-6})}{(3 \times 10^{-6})}$$

$$= 5770 \text{ J mol}^{-1}$$

9.12 Fumarase catalyzes the reaction fumarate + H_2O = L-malate . At 25 °C and pH 7 $K' = 4.4 = [\text{L-malate}]/[\text{fumarate}]$. What is the value of K' at pH 4? Given:

For fumaric acid $K_1 = 10^{-4.18}$

For L-malic acid $K_1 = 10^{-4.73}$

SOLUTION

$$\frac{[\text{L-malate}]}{[\text{fumarate}]} = \frac{[\text{M}] + [\text{HM}]}{[\text{F}] + [\text{HF}]} = \frac{[\text{M}][1 + [\text{H}^+]/K_{1M}]}{[\text{F}][1 + [\text{H}^+]/K_{1F}]}$$

$$= 4.4 \frac{[1 + 10^{-4}/10^{-4.73}]}{[1 + 10^{-4}/10^{-4.18}]} = 4.4 \frac{6.37}{2.51} = 11.2$$

9.13 Given $\Delta G^\circ = 49.4\ \text{kJ mol}^{-1}$ for

$$ATP^{4-} + H_2O = AMP^{2-} + P_2O_4{}^{7} + 2H^+$$

calculate $\Delta G^{\circ\prime}$ at pH 7 and 25 °C and 0.2 mol L^{-1} ionic strength. See Table 9.4

<u>SOLUTION</u>

$$\begin{array}{llllll}
ATP^{4-} & + \; H_2O \rightleftharpoons & AMP^{2-} & + & P_2O_7{}^{4-} & + \; 2H^+ \\
\Big\Updownarrow 10^{-6.95} & & \Big\Updownarrow 10^{-6.45} & & \Big\Updownarrow 10^{-8.95} & \\
HATP^{3-} & & HAMP^{1-} & & HP_2O_7{}^{3-} & \\
 & & & & \Big\Updownarrow 10^{-6.12} & \\
 & & & & H_2P_2O_7{}^{2-} &
\end{array}$$

$$K' = \frac{\left([AMP^{2-}] + [HAMP^-]\right)\left([P_2O_7{}^{4-}] + [HP_2O_7{}^{3-}] + [H_2P_2O_7{}^{2-}]\right)}{\left([ATP^{4-}] + [HATP^{3-}]\right)}$$

$$= \frac{[AMP^{2-}][P_2O_7{}^{4-}]\left(1 + \dfrac{[HAMP^-]}{[AMP^{2-}]}\right)\left(1 + \dfrac{[HP_2O_7{}^{3-}]}{[P_2O_7{}^{4-}]} + \dfrac{[H_2P_2O_7{}^{2-}]}{[P_2O_7{}^{4-}]}\right)}{[ATP^{4-}]\left(1 + \dfrac{[HATP^{3-}]}{[ATP^{4-}]}\right)}$$

$$= \frac{[AMP^{2-}][P_2O_7{}^{4-}][H^+]^2}{[ATP^{4-}]} \cdot \frac{\left(1 + \dfrac{[H^+]}{K_{1AMP}}\right)\left(1 + \dfrac{[H^+]}{K_{1PP}} + \dfrac{[H^+]^2}{K_{1PP}K_{2PP}}\right)}{[H^+]^2\left(1 + \dfrac{[H^+]}{K_{1ATP}}\right)}$$

$$= e^{\frac{-49,400}{(8.314)(298)}} \cdot \frac{\left(1+\frac{10^{-7}}{10^{-6.45}}\right)\left(1+\frac{10^{7}}{10^{-8.95}}+\frac{(10^{-7})^2}{10^{-8.95}\,10^{-6.12}}\right)}{(10^{-7})^2 \left(1+\frac{10^{-7}}{10^{-6.95}}\right)}$$

$$= 1.53 \times 10^7$$

$$\Delta G^{o\prime} = -RT \ln K' = -(8.314 \text{ J K}^{-1} \text{ mol}^{-1})(298 \text{ K}) \ln (1.53 \times 10^7)$$

$$= -41.0 \text{ kJ mol}^{-1}$$

9.14 The hydrolysis of adenosine triphosphate ATP to adenosine diphosphate ADP and inorganic phosphate at pH 8 and 25 °C

$$ATP^{-4} + H_2O = ADP^{-3} + HPO_4^{-2} + H^+$$

has a standard enthalpy change of -13 kJ mol^{-1}. The standard enthalpy changes of acid dissociation of $HATP^{-3}$, $HADP^{-2}$, and $H_2PO_4^{-1}$ are -8, 0, and $+8$ kJ mol^{-1}, respectively. Calculate the standard enthalpy change for the reaction

$$HATP^{-3} + H_2O = HADP^{-2} + H_2PO_4^{-}$$

SOLUTION

$ATP^{4-} + H_2O = ADP^{3-} + HPO_4^{2-} + H^+$	$\Delta H^o = -13$ kJ mol^{-1}
$HATP^{3-} = H^+ + ATP^{4-}$	-8
$H^+ + ADP^{3-} = HADP^{2-}$	0
$H^+ + HPO_4^{2-} = H_2PO_4^{-1}$	-8
$HATP^{3-} + H_2O = HADP^{2-} + H_2PO_4^{-1}$	$\Delta H^o = -29$ kJ mol^{-1}

9.15 Ethyl acetate is hydrolyzed in an aqueous buffer at pH 7 in a calorimeter. The enthalpy of hydrolysis as measured in the calorimeter does not correspond with what is calculated from the following standard enthalpies of formation from a chemical thermodynamic table.

	$\Delta_f H^\circ$ (298 K)/kJ mol^{-1}
acetic acid (1)	-484.5
ethanol (1)	-277.0
ethyl acetate (1)	-479.0
H_2O (1)	-285.8

Please explain why. What additional information would you need to calculate the heat absorbed in this experiment?

SOLUTION

ethyl acetate (1) + H_2O(1) = ethanol (1) + acetic acid (1)

$\Delta H^\circ = 3.3$ kJ mol^{-1}

This heat absorption would be obtained in acidic solution (where the acetic acid is undissociated) if the reactants and products form ideal solutions. At pH 7 H^+ is produced and reacts with the buffer. Therefore, it is necessary to know the heat of acid dissociation of the buffer. To calculate an accurate value for the heat absorbed, $\Delta_f H^\circ$ is needed for

acetic acid dissolved in water
ethanol dissolved in water
ethyl acetate dissolved in water

9.16 If n molecules of a ligand A combine with a molecule of protein to form PA_n without intermediate steps, derive the relation between the fractional saturation Y and the concentration of A.

SOLUTION

$$P + nA = PA_n$$

$$K = \frac{[P][A]^n}{[PA_n]} \qquad [P]_o = [P] + [PA_n]$$

$$\frac{\left([P]_o - [PA_n]\right)[A]^n}{[PA_n]} = K \qquad [P]_o[A]^n = [PA_n]\left(K + [A]^n\right)$$

$$Y = \frac{[PA_n]}{[P]_o} = \frac{1}{1 + K/[A]^n} = \frac{[A]^n/K}{1 + [A]^n/K}$$

This equilibrium represents a cooperative effect in that as soon as one ligand molecule is bound, the other (n - 1) ligand molecules are also bound.

9.17 The percent saturation of a sample of human hemoglobin was measured at a series of oxygen partial pressures at 20 $^{\circ}$C, pH 7.1, 0.3 mol L^{-1} phosphate buffer, and 3 x 10^{-4} mol L^{-1} heme.

P_{O_2} /Pa	Percent Saturation
393	4.8
787	20
1183	45
2510	78
2990	90

Calculate the values of h and K_h in the Hill equation. (See problem 9.19.)

SOLUTION

log P	$\log\left(\frac{Y}{1-Y}\right)$
2.594	-1.31
2.896	-0.602
3.074	-0.087
3.400	+0.550
3.476	+0.954

(Please see graph, top of page 158)

$$\log\left(\frac{Y}{1-Y}\right) = -\log K_h + n \log P$$

n = slope = 2.4 $\qquad K_h = 4 \times 10^7$

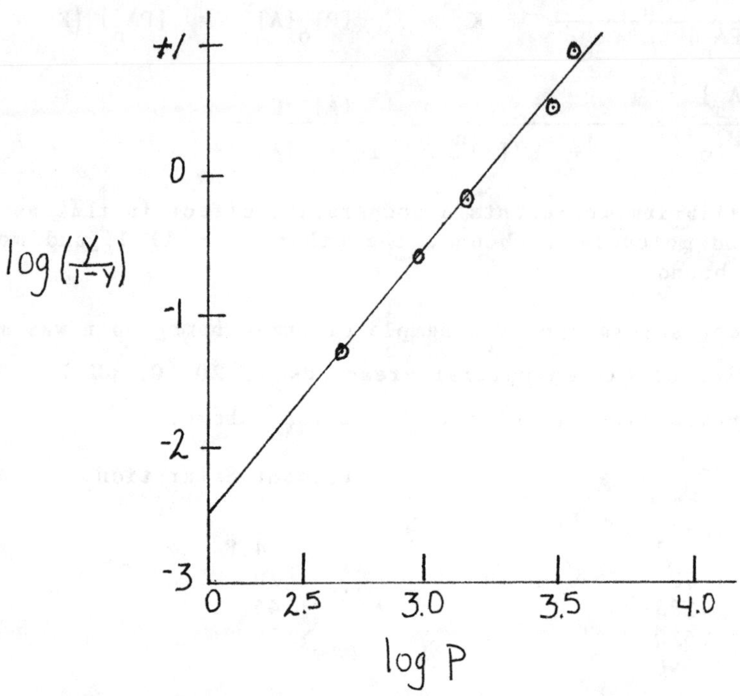

9.18 When myoglobin is in contact with air how many parts per million of CO are required to tie up 10% of the myoglobin? The partial pressure of oxygen required to half saturate myoglobin at 25 °C is 3.7 kPa. The partial pressure of CO required to half saturate myoglobin in the absence of oxygen is 0.009 kPa.

SOLUTION

$$\frac{P_{CO}/K_{CO}}{1 + P_{O_2}/K_{O_2} + P_{CO}/K_{CO}} = 0.1$$

$K_{O_2} = 3.7$ kPa $\qquad K_{CO} = 0.009$ kPa

$P_{O_2} = (0.2)(101.325 \text{ kPa}) = 20.3$ kPa

$$\frac{P_{CO}/0.009 \text{ kPa}}{1 + 20.3 \text{ kPa}/3.7 \text{ kPa} + P_{CO}/0.009 \text{ kPa}} = 0.1$$

$$P_{CO} = 0.0065 \text{ kPa} \qquad \frac{P_{CO}}{101.3 \text{ kPa}} = 64 \text{ ppm}$$

9.19 Since it is difficult to determine the values of the four dissociation constants in equation 9.67, the empirical Hill equation

$$Y = \frac{1}{1 + K_h/P_{O_2}^h}$$

is frequently used to characterize binding. Show that the Hill coefficient h may be obtained by plotting log [Y/(1 - Y)] versus P_{O_2}.

SOLUTION

$$\frac{Y}{1-Y} = \frac{P_{O_2}^h}{K_h} \qquad \log \frac{Y}{1-Y} = -\log K_h + h \log P_{O_2}$$

For a variety of binding systems relatively linear Hill plots are obtained for values of Y in the range 0.1 to 0.9, but deviations usually occur at the extremes unless h = 1. At the extremes the plot usually approaches a slope of unity.

9.20 $K_2 = 5.884 \times 10^{-4}$ $\quad pK_2 = 13.230$ $\quad pH = 6.615$

9.21 The pH of a solution of monosodium aspartate is $(pK_1 + pK_2)/2 = 6.9$. The pH of a solution of disodium aspartate is greater than 10.

9.23 4.710

9.25 4.35

9.26 $$Z = 1 + \frac{[H^+]}{K_{1PP}} + \frac{[H^+]^2}{K_{1PP}K_{2PP}} + \frac{[Mg^{2+}]}{K_{MgPP}} + \frac{[Mg^{2+}][H^+]}{K_{MgPP}K_{1PP}}$$

$$\nu_H = \left(\frac{[H^+]}{K_{1PP}} + \frac{2[H^+]^2}{K_{1PP}K_{2PP}} + \frac{[Mg^{2+}][H^+]}{K_{1PP}K_{MgPP}}\right) \Big/ Z$$

$$\nu_{Mg} = \left(\frac{[Mg^{2+}]}{K_{MgPP}} + \frac{[H^+][Mg^{2+}]}{K_{1PP}K_{MgPP}}\right) \Big/ Z$$

9.27 0.063

9.28 $\Delta G = 9.2$ kJ mol^{-1} No

9.29 133 g

9.30 3.88×10^{-2} mol L^{-1}

9.31 -42.3 kJ mol^{-1}

9.32 -30.33 kJ mol^{-1}

9.33 2.93×10^{7}

9.34

	pH 8, pMg 4	pH 8, pMg 8
$\Delta H^{o\prime}$	-26.57 kJ mol^{-1}	-19.7 kJ mol^{-1}
$\Delta G^{o\prime}$	-42.7 kJ mol^{-1}	-44.4 kJ mol^{-1}

9.35 (a) $K' = K_1(1 + K_{CH}/[H^+])/(1 + K_{AH}/[H^+])$

(b) $K_2K_{CH}/K_1K_{AH} = 1$

9.36 0.158 mol L^{-1} .3 25.6%

9.38 -10.75 J K^{-1} mol^{-1}

CHAPTER 10: Surface Thermodynamics

10.1 The surface tension of toluene at 20 °C is 0.0284 N m^{-1}, and its density at this temperature is 0.866 g cm^{-3}. What is the radius of the largest capillary that will permit the liquid to rise 2 cm?

SOLUTION

$$\gamma = \frac{1}{2} h\rho g r$$

$$r = \frac{2\gamma}{\rho g h} = \frac{2(0.284 \text{ N m}^{-1})}{(0.866 \times 10^3 \text{ kg m}^{-3})(9.80 \text{ m s}^{-2})(0.02 \text{ m})}$$

$$= 3.35 \times 10^{-4} \text{ m} = 3.35 \times 10^{-2} \text{ cm}$$

10.2 Mercury does not wet a glass surface. Calculate the capillary depression if the diameter of the capillary is (a) 0.1 mm. The density of mercury is 13.5 g cm^{-3}. The surface tension of mercury at 25 °C is 0.520 N m^{-1}.

SOLUTION

$r = hg\rho r/2$ $\qquad\qquad$ $h = 2\gamma/g\rho r$

(a) $h = (2)(0.520 \text{ N m}^{-1})/(9.80 \text{ m s}^{-2})(13.5 \times 10^3 \text{ kg m}^{-3})(0.05 \times 10^{-3} \text{ m})$

$= 15.7$ cm

(b) $h = (2)(0.520)/(9.80)(13.5 \times 10^3 \text{ kg m}^{-3})(10^{-3} \text{ m}) = 7.86$ mm

10.3 If the surface tension of a soap solution is 0.050 N m^{-1}, what is the pressure difference inside and outside for (a) a soap bubble 2 mm in diameter and (b) a bubble 2 cm in diameter.

SOLUTION

The effective surface tension is 2γ because the soap film has 2 surfaces.

(a) $\Delta P = \dfrac{2\gamma}{R} = \dfrac{4(0.050 \text{ N m}^{-1})}{0.002 \text{ m}} = 100 \text{ Pa}$

(b) $\Delta P = \dfrac{4(0.050 \text{ N m}^{-1})}{0.02 \text{ m}} = 10 \text{ Pa}$

10.4 Calculate the vapor pressure of a water droplet at 25 °C that has a radius of 2 nm. The vapor pressure of a flat surface of water is 3167 Pa at 25 °C.

SOLUTION

$$\ln\left(\frac{P}{P^{o}}\right) = \frac{2V_m\gamma}{r\,RT}$$

$$\ln\left(\frac{P}{3167 \text{ Pa}}\right) = \frac{2(18.016 \times 10^{-3} \text{ kg mol}^{-1})(0.07197 \text{ N m}^{-1})}{(2 \times 10^{-9} \text{ m})(8.314 \text{ J K}^{-1} \text{ mol}^{-1})(298.15 \text{ K})} \times \frac{1}{(10^{3} \text{ kg m}^{-3})}$$

$P = 5350 \text{ Pa}$

10.5 A solution of palmitic acid ($M = 256 \text{ g mol}^{-1}$) in benzene contains 4.24 g of acid per liter. When this solution is dropped on a water surface, the benzene evaporates and the palmitic acid forms a monomolecular film of the solid type. If we wish to cover an area of 500 cm^2 with a monolayer, what volume of solution should be used? The area occupied by one palmitic acid molecule may be taken to be $21 \times 10^{-20} \text{ m}^2$.

SOLUTION

The number of molecules in a cubic centimeter of solution is

$$\frac{(4.24 \text{ g L}^{-1})(6.022 \times 10^{23} \text{ mol}^{-1})}{(256 \text{ g mol}^{-1})(1000 \text{ cm}^3 \text{ L}^{-1})} = 9.97 \times 10^{18} \text{ cm}^{-3}$$

The number of molecules required to cover 500 cm^2 is

$$\frac{(500\ cm^2)(10^{-2}\ m\ cm^{-1})^2}{21 \times 10^{-20}\ m^2} = 2.38 \times 10^{17}$$

$$\frac{2.38 \times 10^{17}}{9.97 \times 10^{18}\ cm^{-3}} = 0.0239\ cm^3$$

10.6 A protein with a molar mass of 60,000 g mol^{-1} forms a perfect gaseous film on water. What area of film per milligram of protein will produce a pressure of 0.005 N m^{-1} at 25 °C?

SOLUTION

$$A_s = \frac{RT}{\pi} = \frac{(8.314\ J\ K^{-1}\ mol^{-1})(298\ K)}{0.005\ N\ m^{-1}} = 4.96 \times 10^5\ m^2\ mol^{-1}$$

$$n = \frac{10\ \ kg}{60\ kg\ mol^{-1}} = 1.67 \times 10^{-8}\ mol$$

Area per milligram

$$nA_s = (1.67 \times 10^{-8}\ mol)(4.96 \times 10^5\ m^2\ mol^{-1}) = 8.28 \times 10^{-3}\ m^2$$

$$= 82.8\ cm^2$$

10.7 The acid $CH_3(CH_2)_{13}CO_2H$ forms a nearly perfect gaseous monolayer on water at 25 °C. Calculate the mass of acid per 100 cm^2 required to produce a film pressure of 10^{-3} N m^{-1}.

SOLUTION

$$A_s = \frac{RT}{\pi} = \frac{(8.314\ J\ K^{-1}\ mol^{-1})(298\ K)}{10^{-3}\ N\ m^{-1}} = 2.48 \times 10^6\ m^2\ mol^{-1}$$

$$\frac{m}{M} = \frac{(100\ cm^2)(0.01\ m\ cm^{-1})^2}{2.48 \times 10^6\ m^2\ mol^{-1}}$$

$$m = \frac{(100\ cm^2)(0.01\ m\ cm^{-1})\ (244 \times 10^{-3}\ kg\ mol^{-1})}{2.48 \times 10^6\ m\ mol^{-1}}$$

$$= 9.85 \times 10^{-10}\ kg = 9.85 \times 10^{-4}\ mg$$

10.8 What volume of oxygen, measured at 25 °C and 1 bar, is required to form an oxide film on 1 m^2 of a metal with atoms in a square array 0.1 nm apart?

SOLUTION

Assuming one oxygen atom combines with each metal atom, the number of moles of O_2 adsorbed per m^2 is

$$\frac{(1\ m^2)}{(2)(0.1 \times 10^{-9}\ m)^2}$$

The volume of oxygen is given by

$$\frac{(1\ m^2)(8.314\ J\ K^{-1}\ mol^{-1})(298\ K)}{(2)(0.1 \times 10^{-9}\ m)^2(6.022 \times 10^{23}\ mol^{-1})(101\ 325\ Pa)}$$

$$= 2 \times 10^{-6}\ m^3 = 2 \times 10^{-3}\ L$$

10.9 The following table gives the number of milliliters (v) of nitrogen (reduced to 0 °C and 1 bar) adsorbed per gram of active carbon at 0 °C at a series of pressures.

P/Pa	524	1731	3058	4534	7497
$v/cm^3\ g^{-1}$	0.987	3.04	5.08	7.04	10.31

Plot the data according to the Langmuir isotherm, and determine the constants.

SOLUTION

$$\frac{1}{v} = \frac{1}{v_m} + \frac{1}{v_m KP} \qquad \text{Intercept} = 0.025\ g\ cm^{-3}$$

$$v_m = 1/(0.025\ g\ cm^{-3}) = 40\ cm^3\ g^{-1}$$

$$\text{Slope} = \frac{(1.00 - 0.025)\ \text{g cm}^{-3}}{1.9 \times 10^{-3}\ \text{Pa}^{-1}} = 513\ \text{g cm}^{-3}\ \text{Pa}$$

$$= \frac{1}{v_m K} = \frac{1}{(40\ \text{cm}^3\ \text{g}^{-1})K}$$

$$K = [(513\ \text{g cm}^{-3}\ \text{Pa})(40\ \text{cm}^3\ \text{g}^{-1})]^{-1} = 4.8 \times 10^{-5}\ \text{Pa}^{-1}$$

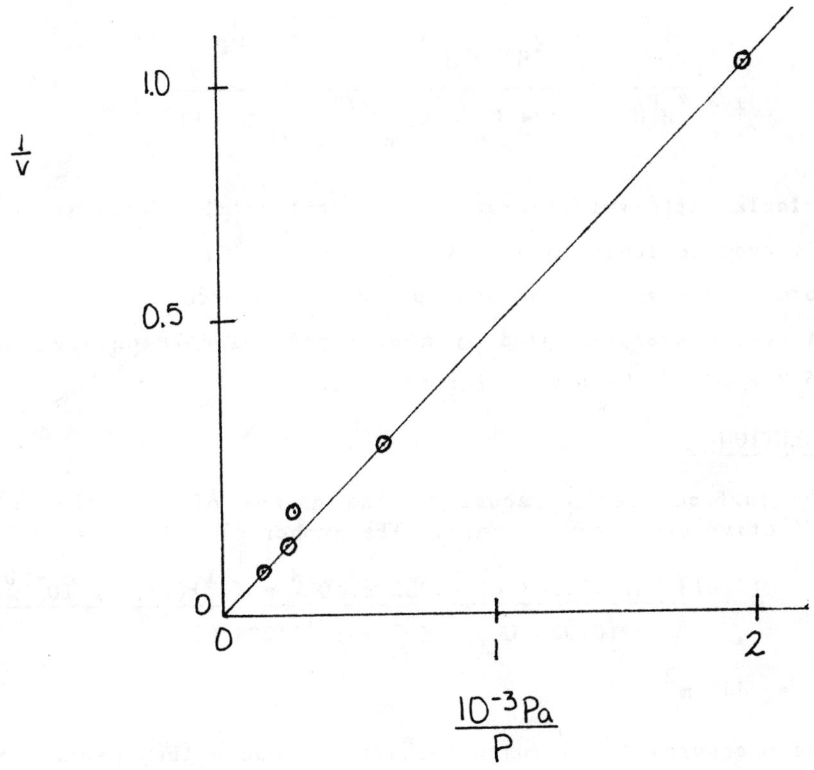

10.10 Hydrogen gas is adsorbed on a metal surface as atoms. Show that the fractional saturation θ of the surface is given by

$$\theta = \frac{KP_{H_2}^{1/2}}{1 + KP_{H_2}^{1/2}}$$

SOLUTION

$$H_2(g) = 2H(g) \qquad K_P = \frac{P_H^2}{P_{H_2}} \qquad P_H = (K_P P_{H_2})^{1/2}$$

$$\theta = \frac{K_H P_H}{1 + K_H P_H} = \frac{K_H (K_P P_{H_2})^{1/2}}{1 + K_H (K_P P_{H_2})^{1/2}} = \frac{KP_{H_2}^{1/2}}{1 + KP_{H_2}^{1/2}}$$

10.11 Calculate the surface area of a catalyst that adsorbs 103 cm^3 of nitrogen (calculated at 1.013 bar and 0 °C) per gram in order to form a monolayer. The adsorption is measured at −195 °C, and the effective area occupied by a nitrogen molecule on the surface is 16.2×10^{-20} m^2 at this temperature.

SOLUTION

The surface area is equal to the number of molecules times the effective area per molecule. The number of molecules is PVN_A/RT.

$$A_s = \frac{(1.013 \text{ bar})(0.103 \text{ L})(6.022 \times 10^{23} \text{ mol}^{-1})(16.2 \times 10^{-20} \text{ m}^2)}{(0.083 \text{ L bar K}^{-1} \text{ mol}^{-1})(273 \text{ K})}$$

$$= 449 \text{ m}^2$$

10.12 The pressures of nitrogen required to cause the adsorption of 1.0 cm^3 g^{-1} (25 °C, 1.013 bar) of gas on P-33 graphitized carbon black are 24 Pa at 77.5 K and 290 Pa at 90.1 K. Calculate the enthalpy of adsorption at this fraction of surface coverage using the Clausius-Clapeyron equation.

SOLUTION

$$\Delta H = \frac{RT_1T_2 \ln (P_2/P_1)}{T_2 - T_1}$$

$$= \frac{(8.314 \text{ J K}^{-1} \text{ mol}^{-1})(77.5 \text{ K})(90.1 \text{ K}) \ln (290/24)}{(90.1 \text{ K} - 77.5 \text{ K})}$$

$$= 11.6 \text{ kJ mol}^{-1}$$

10.13 0.0233 N m^{-1}

10.14 (a) 1.00288 (b) 1.0288 bar

10.15 (a) 15.2×10^{-10} m (b) 97

10.16 $4.04 \times 10^{-6} \text{ mol m}^{-2}$

10.17 244 m

10.18 20.3 L at 25 °C and 1 bar

10.19 (a) $v_m = 38.2 \text{ mm}^3$ $K = 1.22 \times 10^{-3} \text{ Pa}^{-1}$

(b) 30.4 mm^3

10.20 39.6 L

10.21 $$v = \frac{v_m(x_AK_A + x_BK_B)P}{1 + (x_AK_A + x_BK_B)P}$$

$$x_{A,ads} = \frac{\theta_A}{\theta_A + \theta_B} = \frac{x_AK_A}{x_AK_A + x_BK_B}$$

10.22 36.2 kJ mol^{-1}

10.23 29.5 kJ mol^{-1}

PART TWO

QUANTUM CHEMISTRY

CHAPTER 11: Quantum Theory

11.1 A hollow box with an opening of 1 cm^2 area is heated electrically. (a) What is the total energy emitted per second at 800 K? (b) How much energy is emitted per second if the temperature is 1600 K? (c) How long would it take the radiant energy emitted from the box at this temperature, 1600 K, to melt 1000 g of ice?

SOLUTION

(a) $I = \sigma T^4 = (5.67 \times 10^{-8}\ J\ s^{-1}\ m^{-2}\ K^{-4})(800\ K)^4$

$= 2.32 \times 10^4\ J\ M^{-2}\ s^{-1}$

(b) $I = \sigma T^4 = (5.67 \times 10^{-8}\ J\ s^{-1}\ m^{-2}\ K^{-4})(1600\ K)^4$

$= 37.2 \times 10^4\ J\ m^{-2}\ s^{-1}$

(c) $\Delta H = (333\ J\ g^{-1})(1000\ g\ kg^{-1}) = 3.33 \times 10^5\ J$

$$t = \frac{3.33 \times 10^5\ J}{37.2\ J\ s^{-1}} = 8950\ s$$

11.2 At what temperature does M_λ for cavity radiation have a maximum at (a) 800 nm and (b) 400 nm? (c) What is λ_{max} for cavity radiation at 25 °C?

SOLUTION

(a) $T = 2.898 \times 10^{-3}\ K\ m/\lambda_{max}$

$$= \frac{2.898 \times 10^{-3}\ K\ m}{800 \times 10^{-9}\ m} = 3623\ K$$

(b) $T = \dfrac{2.898 \times 10^{-3}\ \text{K m}}{400 \times 10^{-9}\ \text{m}} = 7245\ \text{K}$

(c) $\lambda_{max} = \dfrac{2.898 \times 10^{-3}\ \text{K m}}{298\ \text{K}} = 9.72 \times 10^{-6}\ \text{m}$

This wavelength is in the middle of the near infrared range.

11.3 Derive the value of the constant in the Wien displacement law (equation 11.7).

SOLUTION

$$M_\lambda = \frac{2\pi hc^2}{\lambda^4} \frac{1}{e^{hc/\lambda kT} - 1}$$

Setting $dM_\lambda/d\lambda = 0$ yields $\dfrac{xe^x}{e^x - 1} = 5$ where $x = \dfrac{hc}{kT\lambda}$

Substituting successive values of x shows that x = 4.965. Thus

$$\lambda_{max} T = \frac{hc}{4.965\ k} = \frac{(6.626 \times 10^{-34}\ \text{J s})(2.998 \times 10^{8}\ \text{m s}^{-1})}{(4.965)(1.3807 \times 10^{-23}\ \text{J K}^{-1})}$$

$$= 2.898 \times 10^{-3}\ \text{K m}$$

11.4 Calculate the wavelengths (in micrometers) of the first three lines of the Paschen series for atomic hydrogen.

SOLUTION

$$\lambda = \frac{1}{R} \frac{n_1^2\ n_2^2}{(n_2^2 - n_1^2)} = \frac{1}{R} \frac{9\ n_2^2}{(n_2^2 - 9)}$$

For $n_2 = 4$ $\quad \lambda = \dfrac{(9)(16)}{(109677.58\ \text{cm}^{-1})(7)} = 1.8756\ \mu\text{m}$

For $n_2 = 5$ $\quad \lambda = \dfrac{(9)(25)}{(109677.58\ \text{cm}^{-1})(16)} = 1.2822\ \mu\text{m}$

For $n_2 = 6 \quad \lambda = \dfrac{(9)(36)}{(109677.58\ \text{cm}^{-1})(27)}$

11.5 In the Balmer series for atomic hydrogen what is the wavelength of the series limit?

SOLUTION

$$\lim_{n_2 \to \infty} \frac{1}{\lambda} = \lim_{n_2 \to \infty} R\left(\frac{1}{2^2} - \frac{1}{n_2^2}\right) = \frac{R}{4}$$

$$\lambda = \frac{4}{R} = \frac{4}{1.0967758 \times 10^7\ \text{m}^{-1}} = 364.7054\ \text{nm}$$

11.6 Calculate the wavelength of light emitted when an electron falls from the n = 100 orbit to the n = 99 orbit of the hydrogen atom. Such species are known as high Rydberg atoms. They are detected in astronomy and are more and more studied in the laboratory.

SOLUTION

$$E_{100} - E_{99} = -(1.0967 \times 10^7\ \text{m}^{-1})(10^{-2}\ \text{m cm}^{-1})\left(\frac{1}{100^2} - \frac{1}{99^2}\right)$$

$$= 0.2227\ \text{cm}^{-1}$$

$$\lambda = 4.49\ \text{cm}$$

11.7 In the photoelectric effect an electron is emitted from a metal as the result of the absorption of a photon of light. Part of the energy of the photon is required to release the electron from the metal; this energy ϕ is called the work function. The kinetic energy of the ejected electron is given by

$$\frac{1}{2}mv^2 = h\nu - \phi$$

where m and v are the mass and velocity of the electron. For the

100 face (see Section 25.7) of silver the work function is 4.64 eV. What is v if the wavelength of the light is 300 nm?

SOLUTION

$$h\nu = \frac{(6.626 \times 10^{-34}\ \mathrm{J\ s})(2.998 \times 10^{8}\ \mathrm{m\ s^{-1}})}{200 \times 10^{-9}\ \mathrm{m}} = 9.93 \times 10^{-19}\ \mathrm{J}$$

$$\phi = (4.64\ \mathrm{V})(1.602 \times 10^{-19}\ \mathrm{C}) = 7.43 \times 10^{-19}\ \mathrm{J}$$

$$\frac{1}{2}mv^2 = h\nu - \phi = 9.93 \times 10^{-19}\ \mathrm{J} - 7.43 \times 10^{-19}\ \mathrm{J}$$

$$= 2.50 \times 10^{-19}\ \mathrm{J}$$

$$v = \left(\frac{2 \times 2.50 \times 10^{-19}\ \mathrm{J}}{9.10 \times 10^{-31}\ \mathrm{kg}}\right)^{1/2} = 7.42 \times 10^{5}\ \mathrm{m\ s^{-1}}$$

11.8 What potential difference is required to accelerate a singly charged gas ion in a vacuum so that it has (a) a kinetic energy equal to that of an average gas molecule at 25 °C; (b) an energy equivalent to 83.7 kJ mol^{-1}?

SOLUTION

(a) It has been shown earlier that the kinetic energy of an average gas molecule is (3/2)kT.

$$Ee = \frac{3}{2}kT$$

$$E = \frac{3(1.381 \times 10^{-23}\ \mathrm{J\ K^{-1}})(298\ \mathrm{K})}{2(1.602 \times 10^{-19}\ \mathrm{C})} = 0.0385\ \mathrm{V}$$

(b) $$Ee = \frac{83{,}700\ \mathrm{J\ mol^{-1}}}{6.022 \times 10^{23}\ \mathrm{mol^{-1}}} = 1.3896 \times 10^{-19}\ \mathrm{J}$$

$$E = \frac{1.3896 \times 10^{-19}\ \mathrm{J}}{1.602 \times 10^{-19}\ \mathrm{C}} = 0.867\ \mathrm{V}$$

11.9 Electrons are accelerated by a 1000 V potential drop. (a) Calculate the de Broglie wavelength. (b) Calculate the wavelength of the x-rays that could be produced when these electrons strike a solid.

SOLUTION

(a) $Ee = \frac{1}{2} mv^2$

$$v = \left(\frac{2Ee}{m}\right)^{1/2} = \left[\frac{2(1000\ V)(1.602 \times 10^{-19}\ C)}{9.110 \times 10^{-31}\ kg}\right]^{1/2}$$

$$= 1.875 \times 10^{7}\ m\ s^{-1}$$

$$\lambda = \frac{h}{mv} = \frac{6.626 \times 10^{-34}\ J\ s}{(9.110 \times 10^{-31}\ kg)(1.875 \times 10^{7}\ m\ s^{-1})}$$

$$= 0.0387\ nm$$

(b) $Ee = hc/\lambda \qquad \lambda = \frac{hc}{Ee}$

$$\lambda = \frac{(6.626 \times 10^{-34}\ J\ s)(2.998 \times 10^{8}\ m\ s^{-1})}{(1000\ V)(1.602 \times 10^{-19}\ C)} = 1.24\ nm$$

11.10 An ultraviolet photon (λ = 58.4 nm) from a helium gas discharge tube is absorbed by a hydrogen molecule which is at rest. Since momentum is conserved, what is the velocity of the hydrogen molecule after absorbing the photon? What is the kinetic energy of the hydrogen molecule in J mol^{-1}?

SOLUTION

$$p = \frac{h}{\lambda} = \frac{6.626 \times 10^{-34}\ J\ s}{58.4 \times 10^{-9}\ m} = 1.135 \times 10^{-26}\ kg\ m\ s^{-1} = mv$$

$$= \frac{2(1.0079 \times 10^{-3}\ kg\ mol^{-1})}{6.022 \times 10^{23}\ mol^{-1}}\ v$$

$$v = \frac{(1.135 \times 10^{-26}\ \text{kg m s}^{-1})(6.022 \times 10^{23}\ \text{mol}^{-1})}{2(1.0079 \times 10^{-3}\ \text{kg mol}^{-1})} = 3.39\ \text{m s}^{-1}$$

$$E = \frac{1}{2} mv^2 N_A = 0.012\ \text{J mol}^{-1}$$

11.11 What is the de Broglie wavelength of an oxygen molecule at room temperature? What is the average distance between oxygen molecules at 1 bar at room temperature?

SOLUTION

$$\frac{1}{2} mv^2 = \frac{3}{2} kT \qquad v = \left(\frac{3kT}{m}\right)^{1/2} = \frac{h}{mv}$$

$$\lambda = h/(3mkT)^{1/2}$$

$$= \frac{6.63 \times 10^{-34}\ \text{J s}}{[(3)(5.31 \times 10^{-26}\ \text{kg})(1.38 \times 10^{-23}\ \text{J K}^{-1})(298\ \text{K})]^{1/2}}$$

$$= 0.0259\ \text{nm}$$

$$\text{Since } m = (32 \times 10^{-3}\ \text{kg mol}^{-1})/(6.022 \times 10^{23}\ \text{mol}^{-1})$$

$$= 5.31 \times 10^{-26}\ \text{kg}$$

To calculate the average distance between molecules we calculate the length of the side of a cube containing one molecule.

$$V = RT/P = (8.314\ \text{J K}^{-1}\ \text{mol}^{-1})(298\ \text{K})/(10^5\ \text{Pa})$$

$$= 0.0247\ \text{m}^3\ \text{mol}^{-1}$$

$$l^3 = (0.0247\ \text{m}^3\ \text{mol}^{-1})/(6.022 \times 10^{23}\ \text{mol}^{-1})$$

$$l = 3.46\ \text{nm}$$

Since the de Broglie wavelength is much shorter than the average distance between molecules, translational motion can be treated classically.

11.12 What is the wavelength of a thermal neutron at 300 K?

SOLUTION

$$E = \frac{3}{2}kT = \frac{1}{2}m_n v^2 = \frac{p^2}{2m_n}$$

$$p = \sqrt{3kT\, m_n}$$

$$= \sqrt{3(1.381 \times 10^{-23}\ \mathrm{J\ K^{-1}})(300\ \mathrm{K})(1.675 \times 10^{-27}\ \mathrm{kg})}$$

$$= 4.563 \times 10^{-24}\ \mathrm{kg\ m\ s^{-1}}$$

$$x = \frac{h}{p} = \frac{6.626 \times 10^{-34}\ \mathrm{J\ s}}{4.563 \times 10^{-24}\ \mathrm{kg\ m\ s^{-1}}} = 1.452 \times 10^{-10}\ \mathrm{m}$$

$$= .1452\ \mathrm{nm}$$

11.13 Calculate the approximate quantum number corresponding with the translational energy of a 1-g bullet fired with a velocity of 300 $\mathrm{m\ s^{-1}}$ at a target 100 m away. What is the de Broglie wavelength?

SOLUTION

$$E = \frac{1}{2}mv^2 = \frac{1}{2}(10^{-3}\ \mathrm{kg})(300\ \mathrm{m\ s^{-1}})^2 = 45.00\ \mathrm{J}$$

$$E = \frac{n^2h^2}{8ma^2}$$

$$n = \left(\frac{a}{h}\right)(8mE)^{1/2} = \left(\frac{100\ \mathrm{m}}{6.626 \times 10^{-34}\ \mathrm{J\ s}}\right)[(8)(10^{-3}\ \mathrm{kg})(45\ \mathrm{J})]^{1/2}$$

$$= 9.05 \times 10^{34}$$

$$\lambda = \frac{h}{mv} = \frac{6.626 \times 10^{-34}\ \mathrm{J\ s}}{(10^{-3}\ \mathrm{kg})(300\ \mathrm{m\ s^{-1}})} = 2.21 \times 10^{-33}\ \mathrm{m}$$

11.14 A molecule exists in a certain electronic state for only 10^{-10} s. What is the uncertainty in energy of the molecule? What is the uncertainty in the energy of a mole of these molecules?

SOLUTION

$$\Delta E \geq \frac{\hbar}{2\Delta t} = \frac{6.626 \times 10^{-34} \text{ J s}}{4\pi(10^{-10} \text{ s})} = 5.27 \times 10^{-25} \text{ J}$$

$$= (5.27 \times 10^{-25} \text{ J})(6.022 \times 10^{23} \text{ mol}^{-1}) = 0.318 \text{ J mol}^{-1}$$

11.15 For a particle in a one-dimensional box, the wave function is

$$\psi = \left(\frac{2}{a}\right)^{1/2} \sin(\pi x/a)$$

(a) What is the probability that the particle is in the right half of the box? (b) What is the probability that the particle is in the middle third of the box?

SOLUTION

(a) 1/2, by symmetry

$$\text{(b) probability} = \int_{a/3}^{2a/3} \psi^2 dx = \frac{2}{a}\int_{a/3}^{2a/3} \sin^2(\pi x/a)dx$$

$$= \frac{2}{a}\left[\frac{1}{2}x - \frac{a}{4\pi}\sin(2\pi x/a)\right]_{a/3}^{2a/3}$$

$$= \frac{2}{a}\left[\frac{a}{3} - \frac{a}{6} - \frac{a}{4\pi}\sin\frac{2\pi 2a}{3a} + \frac{a}{4\pi}\sin\frac{2\pi a}{3a}\right]$$

$$= \frac{2}{a}\left[\frac{a}{6} - \frac{a}{4\pi}\sin\frac{4\pi}{3} + \frac{a}{4\pi}\sin\frac{2\pi}{3}\right]$$

$$= \frac{1}{3} - \frac{1}{2\pi}\sin\frac{4\pi}{3} + \frac{1}{2\pi}\sin\frac{2\pi}{3}$$

$$= \frac{1}{3} - \frac{1}{2\pi}\left(-\frac{\sqrt{3}}{2}\right) + \frac{1}{2\pi}\left(\frac{\sqrt{3}}{2}\right) = \frac{1}{3} + \frac{\sqrt{3}}{2\pi} = 0.609$$

11.16 Show that the function $\varphi = 8e^{5x}$ is an eigenfunction of the operator d/dx. What is the eigenvalue?

SOLUTION

$$\frac{d\varphi}{dx} = \frac{d}{dx}\,8e^{5x} = 40\,e^{5x} = 5\varphi$$ Thus the eigenvalue is 5.

11.17 What are the results of operating on the following functions with the operator d/dx?

(a) e^{-ax^2} (b) cos x (c) e^{ikx}

What eigenvalues are obtained?

SOLUTION

(a) $$\frac{de^{-ax^2}}{dx} = -2axe^{-ax^2}$$ (b) $$\frac{d\cos x}{dx} = -\sin x$$

(c) $$\frac{de^{ikx}}{dx} = ike^{ikx}$$ An eigenvalue, ik, is obtained only with the last function.

11.18 Show that the operators for the x coordinate and for the momentum p_x in the x-direction do not commute.

SOLUTION

$\hat{x} = x\cdot$ and $\hat{p}_x = \dfrac{\hbar}{i}\dfrac{\partial}{\partial x}$

$\hat{x}\hat{p}_x f(x) = x \cdot \dfrac{\hbar}{i} f'(x)$ where $f'(x)$ is $\partial f(x)/\partial x$.

$$\hat{p}_x\hat{x}f(x) = \frac{\hbar}{i}\frac{\partial}{\partial x}(xf(x)) = \frac{\hbar}{i}[f(x) + xf'(x)]$$

Since these operators do not commute, we cannot specify precise values of both x and p_x. The relationship between the uncertainties in these two variables is given by the Heisenberg uncertainty principle.

11.19 (a) Calculate the energy levels for n = 1 and n = 2 for an electron in a potential well of width 0.5 nm with infinite barriers on either side. The energies should be expressed in kJ mol^{-1}. (b) If an electron makes a transition from n = 2 to n = 1, what will be the wavelength of the radiation emitted?

SOLUTION

$$(a)\ E = \frac{n^2h^2}{8ma^2} = \frac{(6.626 \times 10^{-34}\ J\ s)^2}{8(9.110 \times 10^{-31}\ kg)(5 \times 10^{-10}\ m)^2}$$

$$= 2.4096 \times 10^{-19}\ J$$

$$= (2.4096 \times 10^{-19}\ J)(6.022 \times 10^{23}\ mol^{-1})$$

$$= 145.1\ kJ\ mol^{-1} \qquad \text{for } n = 1$$

For n = 2

$$E = 4(145.1\ kJ\ mol^{-1}) = 580.4\ kJ\ mol^{-1}$$

$$(b)\ \Delta E = (2.4096 \times 10^{-19}\ J)(4 - 1)$$

$$\Delta E = \frac{hc}{\lambda}$$

$$2.4096 \times 10^{-19}\ J)(4 - 1) = \frac{(6.626 \times 10^{-34}\ J\ s)(2.998 \times 10^{8}\ m\ s^{-1})}{\lambda}$$

$$\lambda = 274.7\ nm$$

11.20 For a hydrogen atom in a one-dimensional box calculate the value of the quantum number of the energy level for which the energy is equal to (3/2) kT at 25 °C (a) for a box 1 nm long, and (b) for a box 1 cm long.

SOLUTION

$$(a)\ E = \frac{3}{2}kT = \frac{3}{2}(1.38 \times 10^{-23}\ J\ K^{-1})(298\ K) = 6.17 \times 10^{-21}\ J$$

$$n = \frac{a}{h}(8mE)^{1/2}$$

$$= \frac{10^{-9}\ \text{m}}{6.63 \times 10^{-34}\ \text{J s}} \left[\frac{8(1.008 \times 10^{-3}\ \text{kg mol}^{-1})(6.17 \times 10^{-21}\ \text{J})}{6.02 \times 10^{23}\ \text{mol}^{-1}}\right]^{1/2}$$

$= 13.7$ or approximately 14

(b) $n = 1.37 \times 10^{8}$

11.21 Show that the wave functions for a particle in a one-dimensional box are orthogonal using

$$\sin\alpha \sin\beta = \frac{1}{2}\cos(\alpha - \beta) - \frac{1}{2}\cos(\alpha + \beta)$$

SOLUTION

$$\psi_n(x) = \left(\frac{2}{a}\right)^{1/2} \sin\frac{n\pi x}{a} \qquad n = 1, 2, \ldots$$

$$\frac{2}{a}\int_0^a \sin\frac{n\pi x}{a}\sin\frac{m\pi x}{a}\,dx = \frac{1}{a}\int_0^a \cos\frac{(n-m)\pi x}{a}\,dx - \frac{1}{a}\int_0^a \cos\frac{(n+m)\pi x}{a}\,dx$$

$$= 0 \text{ if } m \neq n$$

because $n - m$ and $n + m$ are both integers. The integrals are over complete cycles of $\cos\frac{(\text{integer})\pi x}{a}$ which are equal to zero.

11.22 Calculate the degeneracies of the first three levels for a particle in a cubical box.

SOLUTION

$$E = \frac{h^2}{8ma^2}(n_x^2 + n_y^2 + n_z^2) \qquad n_x, n_y, n_z \text{ are } 1, 2, 3 \ldots$$

n_x	n_y	n_x	$E/\frac{h^2}{8ma^2}$	
1	1	1	3	Degeneracy = 1
2	1	1		
1	2	1	6	Degeneracy = 3
1	1	2		
2	2	1		
2	1	2	9	Degeneracy = 3
1	2	2		

11.23 Later in Table 15.4 we will find that the following molecules have the indicated fundamental vibration frequencies.

$^{79}Br_2(325\ cm^{-1})$ $\quad$ $H^{81}Br(2649\ cm^{-1})$ $\quad$ $^1H_2(4401\ cm^{-1})$

What are the force constants for these molecules?

SOLUTION

$$k = \mu(2\pi c\quad)^2 = \mu(2\pi c\quad)^2$$

$$^{79}Br_2,\ k = \frac{39.46 \times 10^{-3}\ kg\ mol^{-1}}{6.022 \times 10^{23}\ mol^{-1}}[2\pi(2.998x10^8\ m\ s^{-1})(3.25x10^4\ m^{-1})]^2$$

$$= 245.5\ N\ m^{-1}$$

$$H^{81}Br,\ k = \frac{0.9954 \times 10^{-3}\ kg\ mol^{-1}}{6.022 \times 10^{23}\ mol^{-1}}[2\pi(2.998 \times 10^8)(2.649 \times 10^5\ cm^{-1})]^2$$

$$= 411.6\ N\ m^{-1}$$

$$^1H_2,\quad k = \frac{0.5039 \times 10^{-3}\ kg\ mol^{-1}}{6.022 \times 10^{23}\ mol^{-1}}[2\pi(2.998 \times 10^8)(4.40 \times 10^5\ m^{-1})]$$

$$= 575.1\ N\ m^{-1}$$

11.24 Check the normalization of ψ_0 and ψ_1 for the harmonic oscillator and show they are orthogonal.

<u>SOLUTION</u>

$$\int_{-\infty}^{\infty} \psi_0^{\,2}\, dx = \left(\frac{\alpha}{\pi}\right)^{1/2} \int_{-\infty}^{\infty} e^{-\alpha x^2}\, dx = \left(\frac{\alpha}{\pi}\right)^{1/2} \frac{\pi^{1/2}}{\alpha^{1/2}} = 1$$

$$\text{since} \int_{0}^{\infty} e^{-a^2 x^2}\, dx = \frac{1}{2a}\,\pi^{1/2}$$

$$\int_{-\infty}^{\infty} \psi_1^{\,2}\, dx = 2\left(\frac{\alpha^3}{\pi}\right)^{1/2} \int_{-\infty}^{\infty} x^2 e^{-\alpha x^2}\, dx = 4\left(\frac{\alpha^3}{\pi}\right)^{1/2} \int_{0}^{\infty} x^2 e^{-\alpha x^2}\, dx$$

$$= 4\left(\frac{\alpha^3}{\pi}\right)^{1/2} \frac{\pi^{1/2}}{2^2\,\alpha^{3/2}} = 1 \quad \text{since} \int_{0}^{\infty} x^2 e^{-ax^2}\, dx = \frac{\pi^{1/2}}{4a^{3/2}}$$

To check orthogonality

$$\int_{-\infty}^{\infty} \psi_0 \psi_1 dx = 2^{1/2}\left(\frac{\alpha}{\pi}\right)^{1/2} \int_{-\infty}^{\infty} x\, e^{-\alpha x^2}\, dx = 0$$

since x is an odd function and $e^{-\alpha x^2}$ is an even function.

11.25 For the harmonic oscillator substitute the wave function for the ground state into the Schrödinger equation and obtain the expression for the ground state energy.

SOLUTION

$$\psi_o = Ce^{-ax^2} \quad \text{where} \quad a = \frac{\pi}{h}\sqrt{km}$$

$$\frac{d\psi_o}{dx} = 2axCe^{-ax^2} \qquad \frac{d^2\psi_o}{dx^2} = 4a^2x^2Ce^{-ax^2} - 2aCe^{-ax^2}$$

Substituting in the Schrödinger equation

$$\left(-\frac{\hbar^2}{2m}\frac{d^2}{dx^2} + \frac{1}{2}kx^2\right)\psi = E\psi$$

yields $E = \frac{1}{2}h\left(\frac{1}{2\pi}\sqrt{\frac{k}{m}}\right) = \frac{1}{2}h\nu_o$

11.26 In the vibrational motion of HI, the iodine atom essentially remains stationary because of its large mass. Assuming that the hydrogen atom undergoes harmonic motion and that the force constant k is 317 N m^{-1}, what is the fundamental vibration frequency ν_o?

SOLUTION

$$m = \frac{1.0078 \times 10^{-3}\ \text{kg mol}^{-1}}{6.022 \times 10^{23}\ \text{mol}^{-1}} = 1.674 \times 10^{-27}\ \text{kg}$$

$$= \frac{1}{2\pi}\sqrt{\frac{k}{m}} = \frac{1}{2\pi}\sqrt{\frac{317\ \text{kg s}^{-2}}{1.674 \times 10^{-27}\ \text{kg}}} = 6.93 \times 10^{13}\ \text{s}^{-1}$$

11.27 What are the expectation values for $\langle x \rangle$ and $\langle x^2 \rangle$ for a quantum mechanical harmonic oscillator in the $v = 0$ state? What is the standard deviation Δx?

SOLUTION

As shown in Example 11.10, for $v = 0$, $\langle x \rangle = 0$ and

$$\langle x^2 \rangle = \frac{1}{2} \frac{\hbar}{(\mu k)^{1/2}}$$

$$\Delta x = \left(\langle x^2 \rangle - \langle x \rangle^2\right)^{1/2} = \left[\frac{1}{2} \frac{\hbar}{(\mu k)^{1/2}}\right]^{1/2}$$

11.28 What are the reduced mass and moment of inertia of $^{23}Na^{35}Cl$? The equilibrium internuclear distance R_e is 236 pm. What are the values of L, L_z, and E for the state with $J = 1$?

SOLUTION

$$\mu = \frac{1}{\frac{1}{m_1} + \frac{1}{m_2}} = \frac{m_1 m_2}{m_1 + m_2}$$

$$= \frac{(22.98977 \times 10^{-3} \text{ kg mol}^{-1})(34.96885 \times 10^{-3} \text{ kg mol}^{-1})}{[(22.98977 + 34.96885) \times 10^{-3} \text{ kg mol}^{-1}](6.022045 \times 10^{23} \text{ mol}^{-1})}$$

$$= 2.30332 \times 10^{-26} \text{ kg}$$

$$I = \mu R_e^{\,2} = (2.303 \times 10^{-26}\ \text{kg})(236 \times 10^{-12}\ \text{m})^2$$

$$= 1.283 \times 10^{-45}\ \text{kg m}^2$$

$$L = \sqrt{J(J+1)}\ \hbar = \frac{\sqrt{2}}{2\pi}\ (6.626 \times 10^{-34}\ \text{J s}) = 1.491 \times 10^{-34}\ \text{J s}$$

$$L_z = 0 \text{ or } \pm \frac{1}{2\pi}(6.626 \times 10^{-34}\ \text{J s}) = 0 \text{ or } \pm 1.054 \times 10^{-34}\ \text{J s}$$

$$E = \frac{\hbar^2}{2I} J(J + 1) = \frac{(6.626 \times 10^{-34}\ \text{J s})^2 (2)}{4\pi^2 (1.283 \times 10^{-45}\ \text{kg m}^2)}$$

$$= 1.734 \times 10^{-23}\ \text{J}$$

11.29 Show that the first two spherical harmonics in Table 11.2 give the same values of the energy and angular momentum as equations 11.123 and 11.135.

<u>SOLUTION</u>

$$Y_o^o = \frac{1}{(2\pi)^{1/2}}$$

Substituting in equation 11.121, $-\frac{\hbar^2}{2I}(0)\ Y_o^o = EY_o^o$

so that E = 0, in agreement with equation 11.123.

Substituting in equation 11.134, $-\hbar^2\ (0) Y_o^o = L^2\ Y_o^o$

so that $L^2 = 0$, in agreement with equation 11.135.

$$Y_1^o = \left(\frac{3}{4\pi}\right)^{1/2} \cos\theta$$

Substituting in equation 11.121,

$$-\frac{\hbar}{2I}\left[\frac{1}{\sin\theta}\ \frac{\partial}{\partial\theta}\left(-\sin^2\theta\left(\frac{3}{4\pi}\right)^{1/2}\right)\right] = EY_1{}^{o}$$

$$\frac{\partial}{\partial\theta}\left(\sin^2\theta\right) = 2\sin\theta\cos\theta$$

$$-\frac{\hbar^2}{2I}\left[\frac{1}{\sin\theta}\left(-\left(\frac{3}{4\pi}\right)^{1/2}\right) 2\sin\theta\cos\theta\right] = E_1^o\left(\frac{3}{4\pi}\right)^{1/2}\cos\theta$$

$$+\frac{\hbar^2}{I}\left(\frac{3}{4\pi}\right)^{1/2}\cos\theta = E_1^o\left(\frac{3}{4\pi}\right)^{1/2}\cos\theta$$

$$E_1^o = \frac{\hbar^2}{I} \qquad \text{as expected.}$$

Substituting in equation 11.134

$$-\hbar^2\left[\frac{1}{\sin\theta}\ \frac{\partial}{\partial\theta}\left(\sin\theta\,\frac{\partial}{\partial\theta}\right) + \frac{1}{\sin^2\theta}\ \frac{\partial^2}{\partial\phi^2}\right] Y_1^o = L^2\, Y_1^o$$

$$-\hbar^2\left[-\left(\frac{3}{4\pi}\right)^{1/2} 2\cos\theta\right] = L^2\left(\frac{3}{4\pi}\right)^{1/2}\cos\theta$$

$$L^2 = 2\hbar^2 \quad \text{as expected from } L^2 = 1(1+1)\hbar^2$$

11.30 A diatomic molecule is in the J = 2 rotational state. What is the value of the angular momentum? What are the values of the angular momentum in a particular direction?

SOLUTION

$$L = \sqrt{J(J+1)}\ \hbar = (\sqrt{6}\ /2\pi)(6.626 \times 10^{-34}\ \text{J s})$$

$$= 2.583 \times 10^{-34}\ \text{J s}$$

Possible values of L_Z are

-2	$(6.626 \times 10^{-34}$ J s$)/2\pi$			=	-2.109×10^{-34} J s
-1	"	"	"	=	-1.055×10^{-34} J s
0	"	"	"	=	0
1	"	"	"	=	1.055×10^{-34} J s
2	"	"	"	=	2.109×10^{-34} J s

11.31 857.7 watts

11.32 290 nm

11.33 1397 J min^{-1}

11.34 6.62×10^{-34} J s, 1.38×10^{-23} J K^{-1}

11.35 2.74×10^{14} s^{-1}, 1094 nm

11.36 4052, 2626 nm

11.37 1.87×10^{7} m s^{-1}

11.38 1.325×10^{-27} kg m s^{-1}, 0.0113 m s^{-1}

11.39 5.93×10^{5} m s^{-1}

11.40 0.130 nm

11.41 0.00549 nm

11.42 5.28×10^{-26} J 1.6×10^{-5} %

11.44 (a) - 6a (b) 1/k

11.45 $E = n^2h^2/8ml^2$

11.46 145.2, 582, 1305 kJ mol^{-1}

11.47

$n_x n_y n_z$	111	211	121	112	122	212	221	113	131	311
$E(8ma^2/h^2)$	3	6	6	6	9	9	9	11	11	11
Degeneracy	1		3			3			3	

11.49

11.50 $$E = (n_x + \frac{1}{2})h\nu_x + (n_y + \frac{1}{2})h\nu_y + (n_z + \frac{1}{2})h\nu_z$$

$$E_o = \frac{h\nu_x + h\nu_y + h\nu_z}{2}$$

11.51

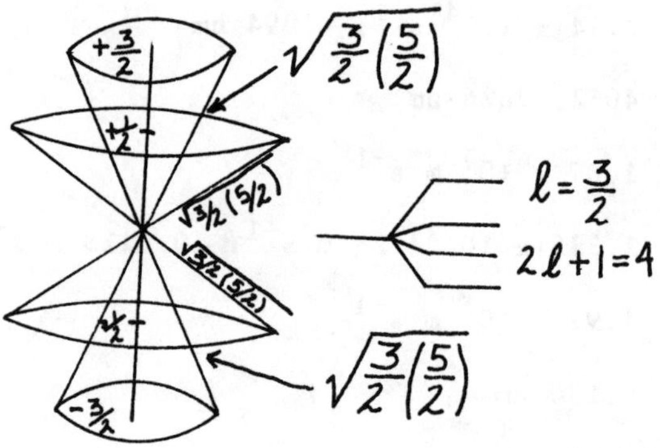

11.52

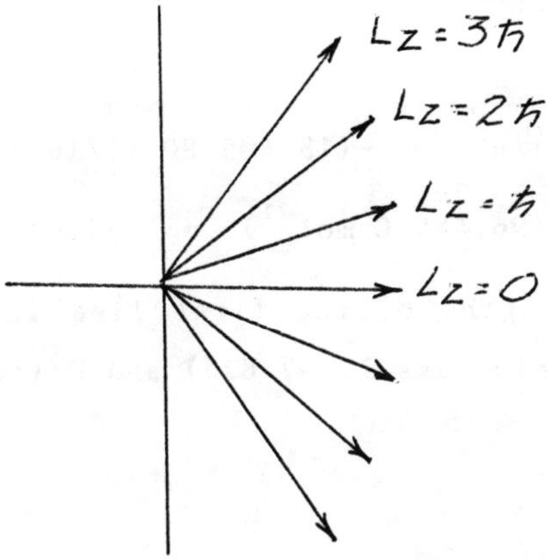

11.53 1.138×10^{-26} kg, 1.454×10^{-46} kg m^2, 2.583×10^{-34} J s, 1.442×10^{-21} J

CHAPTER 12: Atomic Structure

12.1 How much energy is required to remove a 4p electron from a hydrogen atom in eV and in kJ mol^{-1}?

SOLUTION

$$E = -(13.605\ 80\ V)/n^2 = -(13.605\ 80\ V)/16 = -0.850\ 36\ V$$

$$= (-0.850\ 36\ V)(96{,}485\ C\ mol^{-1}) = -82{,}047\ J\ mol^{-1}$$

12.2 What are the wavelengths of the first line in the Balmer series for H (relative atomic mass 1.007 825) and D (relative atomic mass 2.014 10)?

SOLUTION

The mass of the proton is given in the Appendix, but let us calculate it from the relative atomic mass since we are going to need to do this to obtain the mass of the deuteron.

$$m_p = (M_H/N_A) - m_e$$

$$= (1.007\ 825 \times 10^{-3}\ kg\ mol^{-1})/(16.027\ 045 \times 10^{23}\ mol^{-1}) - 9.110 \times 10^{-31}\ kg$$

$$= 1.627\ 648\ 5 \times 10^{-27}\ kg$$

$$R_H = R_\infty/(1 + m_e/m_p)$$

$$= \frac{1.097\ 373\ 177 \times 10^7\ m^{-1}}{1 + 9.109\ 53 \times 10^{-31}\ kg/1.672\ 649 \times 10^{-27}\ kg}$$

$$= 1.096\ 776 \times 10^7\ m^{-1}$$

$$\frac{1}{\lambda} = R_H\left(\frac{1}{2^2} - \frac{1}{3^2}\right) = \frac{R}{7.2}$$

$$\lambda = 7.2/R_H = 656.469\ 5\ nm$$

$$m_D = (M_D/N_A) - m_e = 3.343\ 634 \times 10^{-27}\ \text{kg}$$

$$R_D = \frac{1.097\ 373\ 177 \times 10^{7}\ \text{m}^{-1}}{(1 + 9.109\ 53 \times 10^{-31}\ \text{kg}/3.343\ 634 \times 10^{-27}\ \text{kg})}$$

$$= 1.097\ 707\ 4 \times 10^{7}\ \text{m}^{-1}$$

$$\lambda = 7.2/1.097\ 707\ 4 \times 10^{7}\ \text{m}^{-1} = 656.290\ 9\ \text{nm}$$

12.3 Calculate the ionization potentials for He^{+}, Li^{2+}, Be^{3+}, B^{4+}, and C^{5+}.

SOLUTION

$$E = (13.61\ \text{V})Z^2$$

	Z	E/V		Z	E/V
He^{+}	2	54.44	B^{4+}	5	340.25
Li^{2+}	3	122.49	C^{5+}	6	489.96
Be^{3+}	4	217.76			

12.4 Calculate the energies of ionization of H and D using the appropriate reduced masses.

SOLUTION

$$\mu_H = \frac{1}{1/m_e + 1/m_p} = \frac{m_e m_p}{m_p + m_e}$$

$$= \frac{(9.109\ 534 \times 10^{-31}\ \text{kg})(1.672\ 648\ 5 \times 10^{-27}\ \text{kg})}{(1.672\ 648\ 5 \times 10^{-27}\ \text{kg} + 9.109\ 534 \times 10^{-31}\ \text{kg})}$$

$$= 9.104\ 575 \times 10^{-31}\ \text{kg}$$

$$R_H = R_\infty\ \mu_H/m_e$$

$$= \frac{(1.097\ 373\ 177 \times 10^{7}\ \text{m}^{-1})(9.104\ 575 \times 10^{-31}\ \text{kg})}{9.109\ 534 \times 10^{-31}\ \text{kg}}$$

$$= 1.096\ 775\ 8 \times 10^{7}\ m^{-1}$$

Using the mass of the deuteron from problem 10.2

$$\mu_D = \frac{(9.109\ 534 \times 10^{-31}\ kg)(3.343\ 634 \times 10^{-27}\ kg)}{(3.343\ 634 \times 10^{-27}\ kg + 9.109\ 534 \times 10^{-31}\ kg)}$$

$$= 9.107\ 052 \times 10^{-31}\ kg$$

$$R_D = \frac{(1.097\ 373\ 177 \times 10^{7}\ m^{-1})(9.107\ 052 \times 10^{-31}\ kg)}{9.109\ 534 \times 10^{-31}\ kg}$$

$$= 1.097\ 074\ 2 \times 10^{7}\ m^{-1}$$

The ionization energy from the ground state is

$$E_\infty - E_1 = hcR\left(\frac{1}{n_1^2} - \frac{1}{n_2^2}\right) = hcR$$

The ionization energy of a hydrogen atom is

$$E_H = (6.626\ 176 \times 10^{-34}\ J\ s)(2.997\ 924\ 58 \times 10^{8}\ m\ s^{-1}) \times 1.096\ 775\ 8 \times 10^{7}\ m^{-1}$$

$$= 2.178\ 720 \times 10^{-18}\ J = 13.598\ 397\ eV$$

The ionization energy of a deuterium atom is

$$E_D = (6.626\ 176 \times 10^{-34}\ J\ s)(2.997\ 924\ 58 \times 10^{8}\ m\ s^{-1}) \times 1.097\ 074\ 2 \times 10^{7}\ m^{-1}$$

$$= 2.179\ 313 \times 10^{-18}\ J = 13.602\ 097\ eV$$

12.5 Using data from Table A.2 at 0 K what is the ionization energy of H(g)?

$$H(g) = H^+(g) + e^-$$

SOLUTION

For H(g) at 0 K $\Delta_f H^\circ = 216.037\ kJ\ mol^{-1}$

$H^+(g)$ at 0 K $\Delta_f H^\circ = 1528.085\ kJ\ mol^{-1}$

$e^-(g)$ at 0 K $\Delta_f H^\circ = 0$

$$\Delta H^\circ = 1312{,}048 \text{ J mol}^{-1}$$

$$= \frac{1312{,}048 \text{ J mol}^{-1}}{(1.602\ 189\ 2 \times 10^{-19} \text{ C})(6.022\ 045 \times 10^{23} \text{ mol}^{-1})}$$

$$= 13.598\ 53 \text{ eV}$$

12.6 Calculate the Bohr radius with the reduced mass of the hydrogen atom.

SOLUTION

$$\mu = \frac{1}{\dfrac{1}{m_e} + \dfrac{1}{m_p}}$$

$$= \frac{1}{\dfrac{1}{9.109\ 534 \times 10^{-31} \text{ kg}} + \dfrac{1}{1.672\ 648\ 5 \times 10^{-27} \text{ kg}}}$$

$$= 9.104\ 575 \times 10^{-31} \text{ kg}$$

$$a_o = \frac{h^2(4\pi\varepsilon_o)}{4\pi^2 \mu_e e^2} = \frac{h^2 \varepsilon_o}{\pi\mu_e e^2}$$

$$= \frac{(6.626\ 176 \times 10^{-34} \text{ J s})^2 (8.854\ 187\ 82 \times 10^{-12} \text{ C}^2 \text{ N}^{-1} \text{ m}^{-2})}{\pi(9.104\ 575 \times 10^{-31} \text{ kg})(1.602\ 189\ 2 \times 10^{-19} \text{ C})^2}$$

$$= 0.052\ 946\ 5 \text{ nm}$$

12.7 For a 1s level in a hydrogenlike atom the Schrödinger equation reduces to

$$\frac{1}{r^2}\frac{\partial}{\partial r}\left(r^2 \frac{\partial\varphi}{\partial r}\right) + \frac{8\pi^2 m_e}{h^2}\left(E + \frac{Ze^2}{4\pi\varepsilon_o r}\right)\varphi = 0$$

Using the functional form for ψ_{1s} obtain the expression for the energy of the ground state.

SOLUTION

$$\psi_{1s} = \frac{1}{\sqrt{\pi}}\left(\frac{Z}{a_0}\right)^{3/2} e^{-Zr/a_0} = Ce^{-Zr/a_0}$$

$$\frac{\partial\psi_{1s}}{\partial r} = -\frac{ZC}{a_0} e^{-Zr/a_0}$$

$$\frac{\partial}{\partial r}\left(-\frac{ZCr^2}{a_0} e^{-Zr/a_0}\right) = \left(\frac{Z^2Cr^2}{a_0^2} - \frac{2ZCr}{a_0}\right) e^{-Zr/a_0}$$

Substituting in the Schrodinger equation and eliminating a_0 by use of $a_0 = \dfrac{h^2(4\pi\varepsilon_0)}{4\pi^2 m_e e^2}$ yields $E = -\dfrac{2\pi^2 m_e e^4 Z^2}{h^2(4\pi\varepsilon_0)^2}$

12.8 What are the degeneracies of the following orbitals for hydrogenlike atoms? (a) n = 1, (b) n = 2, and (c) n = 3.

SOLUTION

(a) $n = 1$ $\ell = 0$ $m = 0$ $m_s = \pm\frac{1}{2}$
degeneracy = 2

(b) $n = 2$ $\ell = 0$ $m = 0$ $m_s = \pm\frac{1}{2}$

$\ell = 1$ $m = -1,0,+1$ $m_s = \pm\frac{1}{2}$
degeneracy = 8

(c) $n = 3$ $\ell = 0$ $m = 0$ $m_s = \pm\frac{1}{2}$

$\ell = 1$ $m = -1,0,+1$ $m_s = \pm\frac{1}{2}$

$\ell = 2$ $m = -2,-1,0,+1,+2$ $m_s = \pm\frac{1}{2}$
degeneracy = 18

[Note that the degeneracy is $2n^2$.]

12.9 Show that for a 1s orbital of a hydrogenlike atom the most probable distance from proton to electron is a_o/Z.

SOLUTION

$$\psi_{1s} \propto e^{-Zr/a_o}$$

Probability density of $p(r) \propto r^2 \psi_{1s}^{\ 2} = r^2 \times e^{-2Zr/a_o}$.

Setting the derivative of the probability density equal to zero,

$$r^2 e^{-2Zr/a_o}\left(\frac{-2Z}{a_o}\right) + 2re^{-2Zr/a_o} = 0 \qquad r = \frac{a_o}{Z}$$

12.10 What is the average distance from an orbital electron to the nucleus for a 2s and 2p electron in (a) H and (b) Li^{2+}

SOLUTION

(a) $$\langle r \rangle_{n,\ell} = \frac{n^2 a_o}{Z}\left\{1 + \frac{1}{2}\left[1 - \frac{\ell(\ell+1)}{n^2}\right]\right\}$$

For H $$\langle r \rangle_{2,0} = \frac{2^2(52.92\ \text{pm})}{1} \cdot \frac{3}{2} = 317.5\ \text{pm}$$

$$\langle r \rangle_{2,1} = \frac{2^2(52.92\ \text{pm})}{1}\left\{1 + \frac{1}{2}[1 - \frac{2}{4}]\right\} = 264.6\ \text{pm}$$

(b) For Li^{2+}

$$\langle r \rangle_{2,0} = \frac{2^2(52.92\ \text{pm})}{3} \cdot \frac{3}{2} = 105.8\ \text{pm}$$

$$\langle r \rangle_{2,1} = \frac{2^2(52.92\ \text{pm})}{3}\left\{1 + \frac{1}{2}[1 - \frac{2}{4}]\right\} = 88.2\ \text{pm}$$

12.11 Use equation 12.19 to derive the expression for the average distance between the electron and proton in a hydrogenlike atom in the 1s level.

SOLUTION

$$\langle r\rangle = \int_0^\infty r p(r)dr = \int_0^\infty r4\pi r^2 \frac{Z^3}{\pi a_o^3} e^{-2Zr/a_o} dr = \frac{3}{2}\frac{a_o}{Z}$$

12.12 The wave function for the ground state of the hydrogen atom may be represented by

$$\psi = Ce^{-r/a_o}$$

Where a_o is a constant. Determine the normalization constant c.

SOLUTION

$$\int \psi^* \psi \, dz = \int \psi^* \psi r^2 dr \sin\theta \, d\theta \, d\phi$$

$$= \int_0^\infty dr\, r^2 \int_0^\pi d\theta \sin\theta \int_0^{2\pi} d\phi [c^2 e^{-2r/a_o}] = c^2\pi a_o^3$$

Therefore $c = \left(1/\pi a_o^3\right)^{1/2}$

12.13 Calculate the expectation value of the distance between the nucleus and the electron of a hydrogenlike atom in the $2p_z$ state using equation 11.54. Show that the same result is obtained using equation 12.20.

SOLUTION

$$\langle r\rangle = \int \psi_{21}^* r \psi_{21} d\tau = \int \psi_{21}^2 r d\tau$$

$$\psi_{21} = \frac{1}{\sqrt{4}\;\; 2\pi}\left(\frac{Z}{a_o}\right)^{3/2}\left(\frac{Zr}{a_o}\right) \cos\theta \exp(-Zr/2a_o)$$

$$\langle r\rangle = \frac{1}{16(2\pi)}\left(\frac{Z}{a_0}\right)^5 \int_0^{2\pi}\int_0^{\pi}\int_0^{\infty} r^3 \exp(-Zr/a_0)\cos^2\theta\, r^2 \sin\theta\, d\theta d\phi dr$$

The element of volume in spherical polar coordinates is

$$d\tau = r^2 \sin\theta\, d\theta d\phi dr$$

and the ranges of the coordinates are

$$0 < \theta\ \pi \qquad 0 < \phi < 2\pi \qquad 0 < r < \infty$$

as may be seen in Figure 11.8.

When the integration is carried out using

$$\int_0^{\infty} x^n e^{-bx}\, dx = n!/b^{n+1} \quad \text{we obtain} \quad \langle r\rangle = 5a_0/Z$$

Equation 12.20 yields

$$\langle r\rangle = \frac{4a_0}{Z}\left[1 + \frac{1}{2} \quad 1 - \frac{1}{2}\right] = 5a/Z$$

12.14 It can be shown that a linear combination of two eigenfunctions belonging to the same degenerate level is also an eigenfunction of the hamiltonian with the same energy. In terms of mathematical formulas, if $H\psi_1 = E_1\psi_1$ and $H\psi_2 = E_1\psi_2$, then $H(c_1\psi_1 + c_2\psi_2) = E_1(c_1\psi_1 + c_2\psi_2)$. The wave function $c_1\psi_1 + c_2\psi_2$ still needs to be normalized. Equation 12.5 yields the following expressions for the 2p eigenfunctions

$$\psi_{2p_{+1}} = be^{-Zr/2a}\, r \sin\theta\, e^{i\theta}$$

$$\psi_{2p_{-1}} = be^{-Zr/2a}\, r \sin\theta\, e^{-\theta}$$

$$\psi_{2p_0} = ce^{-Zr/2a}\, r \cos\theta$$

Use this information to find the real functions for the 2p orbitals that are given in Table 12.1. Given:

$$e^{i\phi} = \cos\phi + i\sin\phi$$

$$e^{-i\phi} = \cos\phi - i\sin\phi$$

SOLUTION

As indicated in the footnote in Section 12.2, we define ψ_{2p_x} and ψ_{2p_y} as the following linear combinations:

$$\psi_{2p_x} = \left(\psi_{2p_{+1}} + \psi_{2p_{-1}}\right)/\sqrt{2}$$

$$\psi_{2p_z} = \left(\psi_{2p_{+1}} - \psi_{2p_{-1}}\right)/i\sqrt{2}$$

Substituting in these expressions we obtain

$$\psi_{2p_x} = be^{-Zr/2a} r \sin\theta(e^{i\theta} + e^{-i\theta})/\sqrt{2}$$

$$= 2be^{-Zr/2a} r \sin\theta \cos\theta/\sqrt{2}$$

$$\psi_{2p_y} = be^{-Zr/2a} r \sin\theta(e^{i\theta} - e^{-i\theta})/i\sqrt{2}$$

$$= 2be^{-Zr/2a} r \sin\theta \sin\phi/\sqrt{2}$$

12.15 For a 2p electron in a hydrogenlike atom what is the magnitude of the orbital angular momentum and what are the possible values of L_z?

SOLUTION

$n = 2$ $\ell = 1$ $m = -1, 0, +1$

$$|\underset{\sim}{L}| = [\ell(\ell + 1)]^{1/2}\hbar$$

$$= \sqrt{2}\ (6.626 \times 10^{-34}\ \text{J s})/2\pi = 1.491 \times 10^{-34}\ \text{J s}$$

$$L_z = m\hbar = m(6.626 \times 10^{-34}\ \text{J s})/2\pi$$

$$= \begin{cases} +\ 1.054 \times 10^{-34}\ \text{J s} \\ 0 \\ -\ 1.054 \times 10^{-34}\ \text{J s} \end{cases}$$

12.16 What is the magnitude of the angular momentum for electrons in 3s, 3p and 3d orbitals? How many radial and angular modes are there for each of these orbitals?

SOLUTION

	3s	3p	3d
$\lvert \underset{\sim}{L} \rvert = \sqrt{\ell(\ell + 1)}\ \hbar$	0	$\sqrt{2}\ \hbar$	$\sqrt{6}\ \hbar$
Radial modes	2	1	0
Angular modes	0	1	2

12.17 Derive $\hat{L}_z = m\hbar$ (equation 12.23) using $\hat{L}_z = -i\hbar\partial/\partial\phi$ (11.132c) and $\Phi = e^{im\phi}$.

SOLUTION

$$\hat{L}_z \Phi = -i\hbar \frac{\partial}{\mu\phi} e^{im\phi} = mhe^{im\phi} = m\hbar\Phi$$

12.18 For the wave function

$$\psi = \begin{vmatrix} \psi_A(1) & \psi_A(2) \\ \psi_B(1) & \psi_B(2) \end{vmatrix}$$

show that (a) the interchange of two columns changes the sign of the wave function, (b) the interchange of two rows changes the sign of the wave function, and (c) the two electrons cannot have the same spin orbital.

SOLUTION

(a) The interchange of two columns yields

$$\begin{vmatrix} \varphi_A(2) & \varphi_A(1) \\ \varphi_B(2) & \varphi_B(1) \end{vmatrix} = \varphi_A(2)\ \varphi_B(1) - \varphi_A(1)\ \varphi_B(2)$$

This is the wave function for the atom with the two electrons interchanged, and it is the negative of the function given in the problem, as required by the anti-symmetrization principle.

(b) The interchange of two rows yields

$$\begin{vmatrix} \varphi_B(1) & \varphi_B(2) \\ \varphi_A(1) & \varphi_A(2) \end{vmatrix} = \varphi_A(2)\ \varphi_B(1) - \varphi_A(1)\ \varphi_B(2)$$

which is the negative of the wave function given in the problem. Thus the interchange of two columns or two rows is equivalent to the interchange of two electrons between orbitals.

(c) If two electrons are represented by the same spin orbitals, the determinant

$$\begin{vmatrix} \varphi_A(1) & \varphi_A(2) \\ \varphi_A(1) & \varphi_A(2) \end{vmatrix} = 0$$

is in accord with the principle that two electrons in an atom cannot have four identical quantum numbers.

12.19 What are the electron configurations for H^-, Li^+, O^{2-}, F^-, Na^+, and Mg^{2+}?

SOLUTION

$1s^2$, $1s^2$, $1s^22s^22p^6$, $1s^22s^22p^6$, $1s^22s^22p^6$, $1s^22s^22p^6$.

12.20 How many electrons can enter the following sets of atomic orbitals: 1s, 2s, 2p, 3s, 3p, 3d?

SOLUTION

There are $2\ell + 1$ values of m for n given ℓ and each of these orbitals may be occupied by 2 electrons so $2(2\ell + 1)$ electrons can enter a set of orbitals.

1s:	$\ell = 0$,	$2(2\ell + 1)$	= 2
2s:	$\ell = 0$,	$2(2\ell + 1)$	= 2
2p:	$\ell = 1$,	$2(2\ell + 1)$	= 6
3s:	$\ell = 0$,	$2(2\ell + 1)$	= 2
3p:	$\ell = 1$,	$2(2\ell + 1)$	= 6
3d:	$\ell = 2$,	$2(2\ell + 1)$	= 10

12.21 Since the outer electron for Li is quite a bit further out than the two 1s electrons, this atom is something like a hydrogen atom in the 2s state. The first ionization potential of Li is 5.39 V. What ionization potential would be expected from this simple model of a Li atom? What effective nuclear charge Z′ seen by the outer electron would give the correct first ionization potential?

SOLUTION

$$E = \frac{13.61\ \text{V}}{n^2} = 13.61\ \text{V}/4 = 3.40\ \text{V}$$

$$5.39\ \text{V} = (Z')^2(13.61\ \text{V})/4$$

$$Z' = 1.259$$

12.22 The ionization potential of atomic hydrogen is 13.60 V. Calculate the wavelength of the light produced when a free electron without kinetic energy returns to the inner orbit.

SOLUTION

$$h\nu = hc/\lambda = Ee$$

$$\lambda = \frac{hc}{Ee} = \frac{(6.626 \times 10^{-34}\ \text{J s})(2.998 \times 10^{8}\ \text{m s}^{-1})}{(13.60\ \text{V})(1.602 \times 10^{-19}\ \text{C})} = 91.18\ \text{nm}$$

12.23 What is the electronic energy in hartrees of He, Li, and Be with respect to the nuclei and free electrons? Ionization energies are given in Table 12.4.

SOLUTION

The sums of the ionization energies are

He	79.003 V
Li	203.481 V
Be	399.139 V

The ionization energies expressed in eV are given by the same numbers. The energies of the neutral atoms with respect to the nucleus and free electrons are the ionization energies with a negative sign. The energies of the neutral atoms are expressed in hartrees (E_h) by dividing by 27.211 608 eV E_h^{-1}.

He	$-2.903\ E_h$
Li	$-7.478\ E_h$
Be	$-14.668\ E_h$

12.24 Why is the ionization energy of boron less than beryllium and the ionization energy of oxygen less than nitrogen?

SOLUTION

The first ionization energy of beryllium is due to the removal of one of the two 2s electrons, and the first ionization energy of boron is due to the removal of the electron from the singly occupied 2p subshell. The 2p electron is on the average further from the nucleus and is more easily removed.

The first ionization energy of nitrogen is due to the removal of one of the three 2p electrons, and the first ionization energy of oxygen is due to the removal of one of the four 2p electrons. Since the fourth electron in $2p^4$ in the oxygen atom has to join a half-filled orbital, it is not bound as strongly as the $2p^3$ electrons of nitrogen.

12.25 The enthalpy of formation of $H^-(g)$ at 0 K is given as 143.264 kJ mol^{-1} in Table A.2. What is the electron affinity of H(g)?

SOLUTION

The electron affinity is the energy released in the reaction

$$H(g) + e^- = H^-(g)$$

$$\Delta H^\circ = 143.264 - 216.037 = -72.773 \text{ kJ mol}^{-1}$$

$$= -\frac{(72{,}773 \text{ J mol}^{-1})}{(1.6022 \times 10^{-19} \text{ C})(6.022 \times 10^{23} \text{ mol}^{-1})} = -0.7542 \text{ eV}$$

Thus the electron affinity of H(g) is 0.7542 eV.

12.26 For a carbon atom in the configuration $[He]2s^22p^2$ the following term symbols are all possible, ${}^1S_0$, ${}^3P_0$, ${}^3P_1$, ${}^3P_2$, and ${}^1D_2$. According to Hund's rules, which is the most stable state?

SOLUTION

According to Hund's first rule the most stable states are ${}^3P_0$, ${}^3P_1$, and ${}^3P_2$. In using the third rule we have case (a). According to this rule the most stable is ${}^3P_0$.

12.27 Which of the following transitions are allowed in the electronic spectrum of a hydrogenlike atom: (a) 2s → 1s, (b) 2p → 1s, (c) 3d → 1s, (d) 3d → 3p?

SOLUTION

(a) $\Delta \ell$ = 0, not allowed (b) $\Delta \ell$ = −1, allowed

(c) $\Delta \ell$ = −2, not allowed (d) $\Delta \ell$ = −1, allowed

12.28 1.153×10^{-18} J 7.197 eV 694.3 kJ mol^{-1}

12.29 983 kJ mol^{-1} 265

12.30 656.172 12, 656.163 58 nm

12.31 0.198 μm.

12.32 $\langle r \rangle = 6a_0$

12.33 (a) 243 nm (b) 6.80 V

12.34 (0, 0, 106 pm) and (0, 0, −106 pm)

12.36

$$\left\langle \psi_{1s} / \psi_{2s} \right\rangle = (\text{const.}) \int_0^{\infty} r^2 \left(2 - \frac{r}{a_0}\right) e^{-\frac{r}{a_0}} \; e^{-\frac{r}{2a_0}} \, dr$$

$$= (\text{const.}) \int_0^{\infty} (2x^2 - x^3)\, e^{-\frac{3}{2}x} \, dx \quad \text{where } x = r/a_0$$

$$= (\text{const.}) \left(\frac{32}{27} - \frac{32}{27}\right) = 0$$

12.37

$$\left\langle \psi_{1s} / \psi_{1s} \right\rangle = \frac{4}{\pi a_0^3} \int_0^{\infty} r^2 \, e^{-2r/a_0} \, dr$$

12.38 17.639 pm

12.39 0.7144, 0.6615, 0.5557 nm

12.40 0, 0, $\sqrt{2}\ \hbar$, $\sqrt{6}\ \hbar$

12.41

	4s	4p	4d	4f
$\lvert \underset{\sim}{L} \rvert = \sqrt{\ell(\ell + 1)}\ \hbar$	0	$\sqrt{2}\ \hbar$	$\sqrt{6}\ \hbar$	$\sqrt{12}\ \hbar$
Radial modes	3	2	1	0
Angular modes	0	1	2	3

12.42 $\psi = 1s\alpha(1)\ 1s\beta(2) - 1s\alpha(2)\ 1s\beta(1)$

Interchanging coordinates gives

$1s\alpha(2)\ 1s\beta(1) - 1s\alpha(1)\ 1s\beta(2) = -\psi$

12.43 The inner electrons (1s) screen the nucleus, but since the probability density of the 2s orbital is greater near the nucleus than that of the 2p, the 2s electrons are not screened as much as the 2p electrons, and therefore are bound more tightly (have a lower energy).

12.44 Those with outer electrons in s orbitals are spherically symmetrical: H, He, Li, Be, Na, and Mg.

12.45

Sc	$1s^2 2s^2 2p^6 3s^2 3p^6$	$3d4s^2$
Sc^+	[argon core]	3d4s
Sc^{2+}	[argon core]	3d
Sc^{3+}	[argon core]	

12.46 230.1, 16.40 nm

12.47 54.4 V 164.0 nm 121.0 nm

CHAPTER 13: Symmetry

13.1 E, $C_2(z)$, $\sigma_v(xz)$, $\sigma_v'(yz)$

13.2 E, C_3, $3\sigma_v$

13.3 E, $C_4(z)$, $C_2(z)$, $S_4(z)$, $C_2'(x)$, $C_2'(y)$, $2C_2''$, i, σ_h, $\sigma_v(xz)$, $\sigma_v(yz)$, $2\sigma_d$

13.4 E, $4C_3$, $4S_6$, $C_4(x)$, $C_4(y)$, $C_4(z)$, $3S_4$, $C_2(x)$, $C_2(y)$, $C_2(z)$, $6C_2'$, i, $3\sigma_h$, $6\sigma_d$

13.5 E, C_2

13.6 E, $C_3(z)$, $S_3(z)$, $3C_2$, σ_h, $3\sigma_v$

13.7 E

13.8 E, C_4, C_2, $2\sigma_v$, $2\sigma_d$

13.9 E, C_3, $3C_2$, i, $3\sigma_d$

13.10 E, $C_2(z)$, $C_2(y)$, $C_2(x)$, i, $\sigma(xy)$, $\sigma(xz)$, $\sigma(yz)$

13.11 E, C_∞, S_∞, σ_h, i, $\infty\sigma_v$, ∞C_2

13.12 E, $C_2(z)$, $C_2(y)$, $C_2(x)$, i, $\sigma(xy)$, $\sigma(xz)$, $\sigma(yz)$

13.13 E, $S_4(z)$, $C_2(z)$, $C_2'(x)$, $C_2'(y)$, $2\sigma_d$

13.14 E, $C_2(z)$, $\sigma_v(xz)$, $\sigma_v'(yz)$

13.15 E, $C_2(z)$, $C_2(y)$, $C_2(x)$, i, $\sigma(xy)$, $\sigma(xz)$, $\sigma(yz)$

13.16 E, $4C_3$, $S_4(x)$, $S_4(y)$, $S_4(z)$, $C_2(x)$, $C_2(y)$, $C_2(z)$, $6\sigma_d$

13.17 E, $C_3(z)$, $S_3(z)$, $3C_2$, σ_h, $3\sigma_v$

13.18 E, i

13.19 E, C_3, $3C_2$, i, $3\sigma_d$

13.20 Construct the operator multiplication table for the point group C_{2h}.

SOLUTION

	E	C_2^1	σ_h	i
E	E	C_2^1	σ_h	i
C_2^1	C_2^1	E	i	σ_h
σ_h	σ_h	i	E	C_2^1
i	i	σ_h	C_2^1	R

13.21 List the operators associated with the S^6 elements and their equivalents, if any. How many distinct operations are produced.

SOLUTION

S_6^1	S_6^2	S_6^3	S_6^4	S_6^5	S_6^6
	C_3^1	i	C_3^2		E

Thus an S_6 axis implies the existence of both a C_3 axis coincident with the S_6 axis and a center of symmetry (i) Only the S_6^1 and S_6^5 operations are distinct operations characteristic of only the S_6 axis.

13.22 The C_i point group has only the symmetry elements E and i. Write the matrices for these two symmetry elements and show that they follow the same multiplication table as E and i.

SOLUTION

$$E = \begin{bmatrix} 1 & 0 & 0 \\ 0 & 1 & 0 \\ 0 & 0 & 1 \end{bmatrix} \qquad i = \begin{bmatrix} -1 & 0 & 0 \\ 0 & -1 & 0 \\ 0 & 0 & -1 \end{bmatrix}$$

$$EE = \begin{bmatrix} 1 & 0 & 0 \\ 0 & 1 & 0 \\ 0 & 0 & 1 \end{bmatrix} = E \qquad ii = \begin{bmatrix} 1 & 0 & 0 \\ 0 & 1 & 0 \\ 0 & 0 & 1 \end{bmatrix} = E$$

$$iE = \begin{bmatrix} -1 & 0 & 0 \\ 0 & -1 & 0 \\ 0 & 0 & -1 \end{bmatrix} = i \qquad Ei = \begin{bmatrix} -1 & 0 & 0 \\ 0 & -1 & 0 \\ 0 & 0 & -1 \end{bmatrix} = i$$

The multiplication table is

	E	i
E	E	i
i	i	E

13.23 O_h E, $4C_3$, $4S_6$, $3C_4$, $3S_4$, $3C_2$, $6C_2'$, i, $3\sigma_h$, $6\sigma_d$

13.24 D_{2d} E, S_4, C_2, $2C_2'$, $2\sigma_d$

13.25 C_s E, σ_h

13.26 D_{5h} E, C_5, S_5, $5C_2$, σ_h, $5\sigma_v$

13.27 C_{2v} E, C_2, σ_v, σ_v'

13.28 D_{6h} E, C_6, C_3, S_6, S_3, C_2, $3C_2'$, $3C_2''$, i, σ_h, $3\sigma_v$, $3\sigma_d$

13.29 D_{4d} E, S_8, C_4, C_2, $4C_2'$, $4\sigma_d$

13.30 O_h See 13.23

13.31 D_{3h} E, C_3, S_3, $3C_2$, σ_h, $3\sigma_v$

13.32 T_d E, $4C_3$, $3S_4$, $3C_2$, $6\sigma_d$

13.33 C_{2v} See 13.27

13.34 C_s See 13.25

13.35 D_{3h} See 13.31

13.36 C_3 E, C_3

13.37 C_{2v} See 13.27

13.38 C_{3v} E, C_3, $3\sigma_v$

13.39 T_d See 13.32

13.40 C_s See 13.25

13.41 D_{4h} E, C_4, C_2, S_4, $2C_2'$, $2C_2''$, i, σ_h, $2\sigma_v$, $2\sigma_d$

13.42 D_{2d} See 13.24

13.45
$$\begin{bmatrix} 1 & 0 & 0 \\ 0 & -1 & 0 \\ 0 & 0 & 1 \end{bmatrix} \quad \begin{bmatrix} 1 & 0 & 0 \\ 0 & -1 & 0 \\ 0 & 0 & 1 \end{bmatrix} = \begin{bmatrix} 1 & 0 & 0 \\ 0 & 1 & 0 \\ 0 & 0 & 1' \end{bmatrix}$$

$$\sigma_v \cdot \sigma_v = E$$

CHAPTER 14: Molecular Electronic Structure

14.1 The internuclear repulsion $Z_\alpha Z_\beta e^2/4\pi\varepsilon_o R_{\alpha\beta}$ is omitted from the hamiltonian operator $\hat{H}_{el}$ of a diatomic molecule. If ψ_{el} is an eigenfunction of $\hat{H}_{el}$ with eigenvalue E_{el}, what is the eigenvalue of the operator $\hat{H}_{el} + Z_\alpha Z_\beta e^2/4\pi\varepsilon_o R_{\alpha\beta}$?

SOLUTION

$$\left(\hat{H}_{el} + \frac{Z_\alpha Z_\beta e^2}{4\pi\varepsilon_o R_{\alpha\beta}}\right)\psi_{el} = E'\psi_{el}$$

$$\hat{H}_{el}\psi_{el} + \frac{Z_\alpha Z_\beta e^2}{4\pi\varepsilon_o R_{\alpha\beta}}\psi_{el} = \left(E_{el} + \frac{Z_\alpha Z_\beta e^2}{4\pi\varepsilon_o R_{\alpha\beta}}\right)\psi_{el} = E'\psi_{el}$$

$$E' = E_{el} + Z_\alpha Z_\beta e^2/4\pi\varepsilon_o R_{\alpha\beta}$$

14.2 Given the equilibrium dissociation energy D_e for N_2 in Table 14.2 and the fundamental vibration frequency 2331 cm^{-1}, calculate the spectroscopic dissociation energy D_o in kJ mol^{-1}.

SOLUTION

$$\frac{1}{2}h\tilde{\nu}_o = \frac{(6.626 \times 10^{-34}\ J\ s)(2.998 \times 10^8\ m\ s^{-1})(2.331 \times 10^5\ m^{-1})}{2(1000\ J\ kJ^{-1})} \times \frac{(6.022 \times 10^{23}\ mol^{-1})}{1}$$

$$= 13.93\ kJ\ mol^{-1}$$

$$D_o = D_e - \frac{1}{2}h\tilde{\nu}_o = 955.42 - 13.93$$

$$= 941.49\ kJ\ mol^{-1}$$

This is the thermodynamic dissociation energy at absolute zero. The standard change in enthalpy for dissociation at 0 K calculated from Table A.2 is 941.57 kJ mol^{-1}.

14.3 Derive the values of the normalization constants given in equations 14.27 and 14.28.

SOLUTION

$$\varphi_g = C(1s_A + 1s_B)$$

$$\int \varphi_g{}^2 \, d\tau = C^2 \int (1s_A{}^2 + 21s_A 1s_B + 1s_B{}^2) d\tau$$

$$1 = C^2 \left[1 + 2 \int 1s_A 1s_B d\tau + 1 \right]$$

$$C = \frac{1}{(2 + 2S)^{1/2}} \quad \text{where} \quad S = \int 1s_A 1s_B d\tau$$

$$\varphi_u = C'(1s_A - 1s_B)$$

$$\int \varphi_u d\tau = (C')^2 \int (1s_A{}^2 - 21s_A 1s_B + 1s_B{}^2) d\tau$$

$$= (C')^2 [2 - 2S]$$

$$C' = \frac{1}{(2 - 2S)^{1/2}}$$

14.4 Plot φ_g and φ_u versus distance along the internuclear axis for H_2^+ in the ground state without worrying about normalization (R_e = 106 pm).

SOLUTION

The values of these wavefunctions along the line through the two nuclei are given by

$$\varphi_g = C\left(e^{-|Z|/a_o} + e^{-(Z - R_e)/a_o}\right)$$

$$\varphi_u = C'\left(e^{-|Z|/a_o} + e^{-(Z - R_e)/a_o}\right)$$

where Z is the distance measured from nucleus A toward nucleus B, and C and C′ are the normalization constants that we are going to ignore. These plots are given in Fig. 14.5.

14.5 Plot φ_g^2 and φ_u^2 versus distance along the internuclear axis for H_2^+ in the ground state without worrying about normalization (R_e = 106 nm).

SOLUTION

See Fig. 14.5.

14.6 What are the normalization constants for φ_g and φ_u for the hydrogen molecule ion? It can be shown that the overlap integral S is given by $S = [1 + (R/a_o) + (R^2/3a_o^2)]\exp(-R/a_o)$.

SOLUTION

Since $R/a_o = 106 \text{ pm}/52.91 \text{ pm} = 2.003$

$$S = (1 + 2.003 + 2.003^2/3)\exp(-2.003) = 0.5856$$

According to Table 12.1 the normalization constant for a 1s wave function for H_2^+ is

$$\frac{1}{\pi^{1/2} a_o^{3/2}} = \frac{1}{\pi^{1/2}(52.41 \text{ pm})^{3/2}} = 1.466 \times 10^{-3} \text{ pm}^{-3/2}$$

For φ_g the normalization constant is

$$\frac{1.466 \times 10^{-3} \text{ pm}^{-3/2}}{[2(1 + S)]^{1/2}} = 8.232 \times 10^{-4} \text{ pm}^{-3/2}$$

For φ_u the normalization constant is

$$\frac{1.466 \times 10^{-3} \text{ pm}^{-3/2}}{[2(1 - S)]^{1/2}} = 1.610 \times 10^{-3} \text{ pm}^{-3/2}$$

14.7 Express the four VB wave functions for H_2 as Slater determinants.

SOLUTION

$$\varphi_1 = \frac{1}{2^{1/2}}\begin{vmatrix} 1s_A(1)\alpha(1) & 1s_A(2)\alpha(2) \\ 1s_B(1)\beta(1) & 1s_B(2)\beta(2) \end{vmatrix} - \frac{1}{2^{1/2}}\begin{vmatrix} 1s_A(1)\beta(1) & 1s_A(2)\beta(2) \\ 1s_B(1)\alpha(1) & 1s_B(2)\alpha(2) \end{vmatrix}$$

$$\psi_2 = \frac{1}{2^{1/2}} \begin{vmatrix} 1s_A(1)\alpha(1) & 1s_A(2)\alpha(2) \\ 1s_B(1)\alpha(1) & 1s_B(2)\alpha(2) \end{vmatrix}$$

ψ_3 is the same as ψ_2 with β replacing α. ψ_4 is obtained from ψ_1 by replacing the minus sign with a plus sign.

14.8 Show that the hybrid orbitals for tetravalent carbon given by equations 14.53 to 14.56 are orthogonal.

SOLUTION

$$\int \psi_{sp^3}(i)\, \psi_{sp^3}(ii)\, d\tau$$

$$= \frac{1}{4} \int (\psi_s + \psi_{px} + \psi_{py} + \psi_{pz})(\psi_s + \psi_{px} - \psi_{py} - \psi_{pz})\, d\tau$$

cross terms do not contribute since atomic orbitals are orthogonal

$$= \frac{1}{4} \int \left(\psi_s^2 + \psi_{px}^2 - \psi_{py}^2 - \psi_{pz}^2\right) d\tau$$

$$= \frac{1}{4}(1 + 1 - 1 - 1) = 0$$

since orbitals are normalized.

14.9 Show that the normalizing factor for an sp hybrid wave function is $1/\sqrt{2}$.

SOLUTION

$$\psi = N(\psi_s + \psi_p)$$

$$N^2 \int (\psi_s + \psi_p)^2 d\tau = 1 = N^2 \left[\int \psi_s^2\, d\tau + 2 \int \psi_s \psi_p\, d\tau + \int \psi_p^2\, d\tau\right]$$

$$1 = N^2 [1 + 2 \times 0 + 1] = N^2(2) \qquad N^2 = \frac{1}{2} \qquad N = \frac{1}{\sqrt{2}}$$

14.10 Discuss the structure of the methyl radical and the location of the unpaired electron.

SOLUTION

The CH_3 radical is planar with the 3-two-electron bonds due to the overlapping of the sp^2 orbitals of C with the s orbitals of H. The odd electron is in the remaining unhybridized p orbital which is perpendicular to the plane.

14.11 How many electrons are involved in sigma and pi bonding orbitals in the following substances: (a) ethylene, (b) ethane, (c) butadiene, (d) benzene?

SOLUTION

(a) 10σ, 2π (b) 14σ, 0π (c) 18σ, 4π (d) 24σ, 6π

[σ electron pairs not labeled]

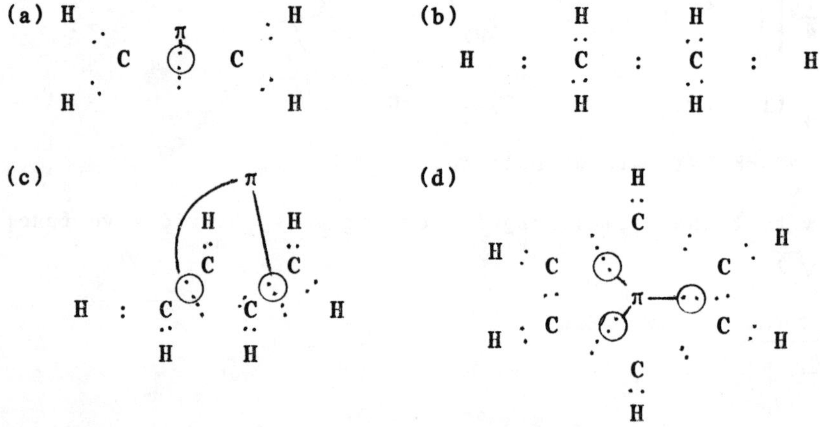

14.12 The heat of hydrogenation of cyclohexene is −121 kJ mol^{-1} and the heat of hydrogenation of benzene is −209 kJ mol^{-1}. What is the reduction of energy due to the formation of the π bond system in benzene?

SOLUTION

If the molecule hexatriene existed, its heat of hydrogenation would be expected to be $3(-121 \text{ kJ mol}^{-1}) = -362 \text{ kJ mol}^{-1}$.

hexatriene benzene

The resonance stabilization is equal to the difference in heats of hydrogenation.

$$-362 \text{ kJ mol}^{-1} - (-209 \text{ kJ mol}^{-1}) = -153 \text{ kJ mol}^{-1}$$

14.13 Using data in Table A.1, calculate the electronegativities of Cl, Br, and I. The electronegativity of H is taken to be 2.1.

SOLUTION

$$|x_A - x_B| = 0.0497 \left\{ E(A-B) - \frac{1}{2} [E(A-A) + E(B-B)] \right\}^{1/2}$$

For HCl

$$|x_{Cl} - x_H| = 0.0497 \left\{ 92.307 + 217.965 + 121.679 - \frac{1}{2} [2(217.965) + 2(121.679)] \right\}^{1/2}$$

$$x_{Cl} - 2.1 = 1.0 \qquad x_{Cl} = 3.1$$

For HBr

$$|x_{Br} - x_H| = 0.0497 \left\{ 36.401 + 217.965 + 111.884 - \frac{1}{2} [2(217.965) + 2 \times (111.884) - 30.907] \right\}^{1/2}$$

$$x_{Br} - 2.1 = 0.7 \qquad x_{Br} = 2.8$$

For HI

$$|x_I - x_H| = 0.0497 \left\{-26.49 + 217.965 + 106.838 - \frac{1}{2}[2(217.965) + 2(106.838) - 62.438]\right\}^{1/2}$$

$$x_I - 2.1 = 0.2 \qquad X_I = 2.3$$

14.14 Explain the relative enthalpies of formation (Table A.1) of HCl(g), HBr(g), and HI(g) in terms of electronegativity concept.

SOLUTION

	Cl	Br	I
Electronegativity of halogen atom	3.0	2.8	2.5
$\Delta_f H^\circ$ (298 K) of hydride in kJ mol^{-1}	-92	-36	+26

According to the electronegativity concept, the bond in HCl should be stronger than in HBr, and the bond in HBr should be stronger than in HI. This is in accord with the fact that more heat is evolved when HCl is formed from its elements than HBr, and more heat is evolved in forming HBr than HI.

14.15 For KF(g) the dissociation constant D_e is 5.18 eV and the dipole moment is 28.7 x 10^{-30} C m. Estimate these values assuming that the bonding is entirely ionic. The ionization potential K(g) is 4.34 eV, and the electron affinity of F(g) is 3.40 eV. The equilibrium internuclear distance in KF(g) is 0.217 nm.

SOLUTION

The work required to separate $K^+ F^-$ from the equilibrium internuclear distance to infinity is

$$\phi = \frac{Q_1 Q_2}{4\pi\varepsilon_0 R_e} = \frac{(1.602 \times 10^{-19}\ C)^2 (0.8988 \times 10^{10}\ N\ m^2\ C^{-2})}{217 \times 10^{-12}\ m}$$

$$= 1.063 \times 10^{-18}\ J = \frac{1.063 \times 10^{-18}\ J}{1.602 \times 10^{-19}\ J\ eV^{-1}} = 6.64\ eV$$

Since KF actually dissociates into neutral atoms, the ionization potential of the metal has to be subtracted and the electron affinity of the nonmetal has to be added to calculate D_e:

$$D_e = 6.64 - 4.34 + 3.40 = 5.70 \text{ eV}$$

This is higher than the experimental value (5.18 eV) because there is some repulsion of K^+ and F^- due to inner electron clouds.

The dipole moment expected for the oversimplified structure $K^+ F^-$ is

$$(1.602 \times 10^{-19} \text{ C})(217 \times 10^{-12} \text{ m}) = 34.7 \times 10^{-30} \text{ Cm}$$

14.16 The equilibrium internuclear distance for NaCl(g) is 236.1 pm. What dipole moment is expected? The actual value is 3.003×10^{-29} C m. How do you explain the difference?

SOLUTION

$$\mu = (1.602 \times 10^{-19} \text{ C})(0.2361 \times 10^{-9} \text{ m})$$

$$= 3.783 \times 10^{-29} \text{ C m}$$

The actual value is less because the ions polarize each other. The positive ion attracts the electrons of the negative ion toward itself, and the negative ion repels the electrons of the positive ion. The more polarizable the ions the larger the effect is.

14.17 Show that the SI units are the same on the two sides of equation 14.84.

SOLUTION

$$\frac{\kappa - 1}{\kappa + 2}\frac{M}{\rho} = \frac{N_A}{3\varepsilon_0}\left(\frac{\mu^2}{3kT} + \alpha_e + \alpha_v\right) \qquad \mu = \text{C m}$$

$$\alpha = \frac{\text{dipole moment}}{\text{electric field strength}} = \frac{\text{C m}}{\text{V m}^{-1}} = \frac{\text{C m}^2}{\text{J C}^{-1}} = \frac{\text{C}^2\text{ m}^2}{\text{kg m}^2\text{ s}^{-2}}$$

$$\varepsilon_0 = \text{C}^2\text{ N}^{-1}\text{ m}^{-2}$$

Left side: $\dfrac{\text{kg mol}^{-1}}{\text{kg m}^{-3}} = \text{m}^3\ \text{mol}^{-1}$

Right side: $\dfrac{\text{mol}^{-1}}{\text{C}^2\ \text{N}^{-1}\ \text{m}^{-2}}\left(\dfrac{\text{C}^2\ \text{m}^2}{\text{kg m}^2\ \text{s}^{-2}} + \dfrac{\text{C}^2\ \text{m}^2}{\text{kg m}^2\ \text{s}^{-2}}\right) = \text{m}^3\ \text{mol}^{-1}$

14.18 The molar polarization $\mathcal{P}$ of ammonia varies with temperature as follows:

$t/^\circ\text{C}$	19.1	35.9	59.9	113.9	139.9	172.9
$\mathcal{P}/\text{cm}^3\ \text{mol}^{-1}$	57.57	55.01	51.22	44.99	42.51	39.59

Calculate the dipole moment of ammonia.

<u>SOLUTION</u>

$$\mathcal{P} = \frac{N_A}{3\varepsilon}\left(\frac{\mu^2}{3kT} + \alpha_e + \alpha_v\right)$$

The slope of a plot of $\mathcal{P}$ versus 1/T is

$$\frac{N_A\mu^2}{9k\varepsilon_o} = \frac{(66 - 5)\ \text{cm}^3\ \text{mol}^{-1}}{4 \times 10^{-3}\ \text{K}^{-1}} = 1.525 \times 10^4\ \text{cm}^3\ \text{mol}^{-1}\ \text{K}$$

$$= 1.525 \times 10^{-2}\ \text{m}^3\ \text{mol}^{-1}\ \text{K}$$

$$\mu = \left[\frac{(1.525 \times 10^{-2}\ \text{m}^3\ \text{mol}^{-1}\ \text{K})9k\varepsilon_o}{N_A}\right]$$

$$= \left[\frac{(1.525\text{x}10^{-2}\text{m}^3\text{mol}^{-1}\text{K})(9)(1.381\text{x}10^{-23}\text{J K}^{-1})(8.854\text{x}10^{-12}\text{C}^2\text{N}^{-1}\text{m}^{-2})}{6.022 \times 10^{23}\ \text{mol}^{-1}}\right]$$

$$= 5.28 \times 10^{-30}\ \text{C m}$$

(Solution to 14.18 continued top of page 217)

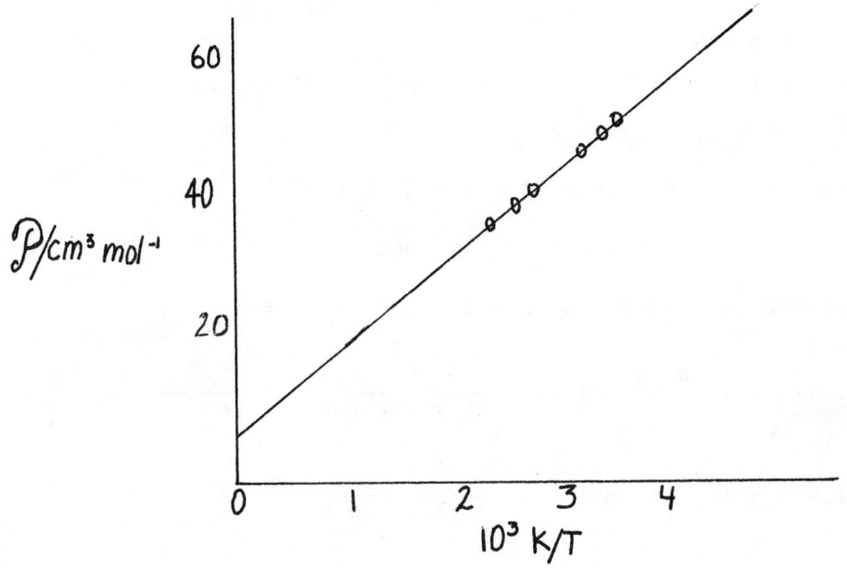

14.19 At 1 bar the relative permittivity of NH_3 gas is 1.007 20 at 292.2 K and 1.003 24 at 446.0 K. Calculate the dipole moment μ and the polarizability α.

<u>SOLUTION</u>

$$(\kappa - 1)\, M/\rho = \frac{N_A}{\varepsilon_o}\left(\frac{\mu^2}{3kT} + \alpha\right)$$

Assuming NH_3 is a perfect gas $P = \rho RT/M$

At T = 292.2 K,

$$M/\rho = RT/P = (8.324\ \mathrm{J\ K^{-1}\ mol^{-1}})(292.2\ \mathrm{K})/(101{,}325\ \mathrm{Pa})$$

$$= 23.98 \times 10^{-3}\ \mathrm{m^3\ mol^{-1}}$$

$$(\kappa - 1)\,M/\rho = (1.007\ 20 - 1)(23.98 \times 10^{-3}\ \text{m}^3\ \text{mol}^{-1})$$

$$= 1.727 \times 10^{-4}\ \text{m}^3\ \text{mol}^{-1}$$

At $T = 446.0$ K

$$M/\rho = (8.314\ \text{J K}^{-1}\ \text{mol}^{-1})(446.0\ \text{K})/(101{,}325\ \text{Pa})$$

$$= 36.60 \times 10^{-3}\ \text{m}^3\ \text{mol}^{-1}$$

$$(\kappa - 1)\,M/\rho = (1.003\ 24 - 1)(36.60 \times 10^{-3}\ \text{m}^3\ \text{mol}^{-1})$$

$$= 1.186 \times 10^{-4}\ \text{m}^3\ \text{mol}^{-1}$$

We have the following two simultaneous equations

$$1.727 \times 10^{-4}\ \text{m}^3\ \text{mol}^{-1} = \frac{N_A\mu^2}{3\varepsilon_o k(292.2\ \text{K})} + \frac{N_A\alpha}{\varepsilon_o}$$

$$1.186 \times 10^{-4}\ \text{m}^3\ \text{mol}^{-1} = \frac{N_A\mu^2}{3\varepsilon_o k(446.0\ \text{k})} + \frac{N_A\alpha}{\varepsilon_o}$$

Taking the difference

$$0.541 \times 10^{-4}\ \text{m}^3\ \text{mol}^{-1} = \frac{N_A\mu^2}{3\varepsilon_o K}\left(\frac{1}{292.2\ \text{K}} - \frac{1}{446.0\ \text{K}}\right)$$

$$\frac{N_A\mu^2}{3\varepsilon_o k} = 0.0458\ \text{m}^3\ \text{mol}^{-1}\ \text{K}$$

$$\mu = \left[\frac{(0.0458\text{m}^3\text{mol}^{-1}\text{K})(3)(8.854\text{x}10^{-12}\text{C}^2\text{N}^{-1}\text{m}^{-2})(1.3807\text{x}10^{-23}\text{J K}^{-1})}{6.022 \times 10^{23}\ \text{mol}^{-1}}\right]^{1/2}$$

$$= 5.28 \times 10^{-30}\ \text{C m}$$

Solving the first equation for α

$$\alpha = \frac{\varepsilon_o}{N_A}\left[1.727 \times 10^{-4}\ \text{m}^3\ \text{mol}^{-1} - \frac{N_A\mu^2}{3\varepsilon_o k(292.2\ \text{K})}\right]$$

$$= \frac{8.854 \times 10^{-12}\ C^2\ N^{-1}\ m^{-1}}{6.022 \times 10^{23}\ mol^{-1}} \left[1.727 \times 10^{-4}\ m^3\ mol^{-1} - \frac{(6.022 \times 10^{23}\ mol^{-1})(5.28 \times 10^{-30}\ C\ m)^2}{3(8.854 \times 10^{-12}\ C^2\ N^{-1}\ m^{-2})(1.3807 \times 10^{-23}\ J\ K^{-1})(292.2\ K)} \right]$$

$$= 2.36 \times 10^{-40}\ C\ m/V\ m^{-1}$$

14.20 The dipole moments of HCl, HBr, and HI are given in Table 14.4. Explain the relative magnitudes of these moments in terms of electronegativity.

SOLUTION

	$\mu/10^{-30}$ C m	Electronegativity of halogen atom
HCl	3.44	3.0
HBr	2.64	2.8
HI	1.00	2.4

The electronegativities increase in going from I to Cl and so the negative charge on the halogen is greater in HCl than in HI and the dipole moment is larger.

14.21 Calculate the dipole moment HCl would have if it consisted of a proton and a chloride ion (considered to be a point charge) separated by 127 pm (the internuclear distance obtained from the infrared spectrum). The experimental value is 3.44×10^{-30} C m. How do you explain the difference?

SOLUTION

$$\mu = QR = (1.602 \times 10^{-19}\ C)(0.127 \times 10^{-9}\ m)$$

$$= 20.4 \times 10^{-30}\ C\ m$$

$$\mu_{expt} = 3.44 \times 10^{-30}\ C\ m$$

The charges are not completely separated in HCl.

14.22 A beam of sodium D light (589 nm) is passed through 100 cm of an aqueous solution of sucrose containing 10 g sucrose per 100 cm^3. Calculate I/I_o, where I_o is the intensity that would have been obtained with pure water, given that M = 342.30 g mol^{-1} and $dn/dc = 0.15\ g^{-1}\ cm^3$ for sucrose. The refractive index of water at 20 °C is 1.333 for the sodium D line.

SOLUTION

$$\tau = \frac{32\pi^3 n_o^2 (dn/dc)^2 Mc}{3N_A\lambda^4}$$

$$= \frac{32\pi^3(1.333)^2(0.15\ g^{-1}\ cm^3)^2(342.30\ g\ mol^{-1})(0.1\ g\ cm^{-3})}{3(6.02\times 10^{23}\ mol^{-1})(5.89\ 10^{-5}\ cm)^4}$$

$$= 6.25\times 10^{-5}\ cm^{-1}$$

$$\frac{I}{I_o} = e^{-\tau x} = e^{-(6.25\times 10^{-5}\ cm^{-1})(100\ cm)} = 0.9938$$

14.23 The Lennard-Jones parameters for nitrogen are ε/k = 95.1 K and σ = 0.37 nm. Plot the potential energy (expressed as V/k in K) for the interaction of two molecules of nitrogen.

SOLUTION

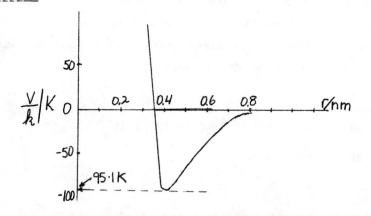

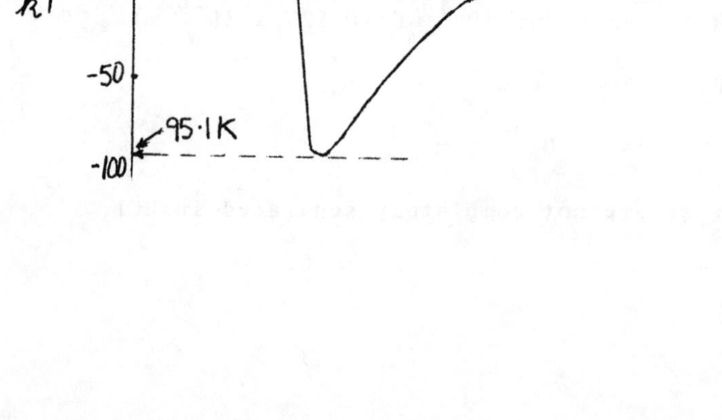

14.24 For Ne the parameters of the Lennard-Jones 6-12 potential are $\varepsilon/k = 35.6$ K and $\sigma = 275$ pm. Plot V in J mol^{-1} versus r and calculate the distance r_m where $dV/dr = 0$.

SOLUTION

$$r_m = 2^{1/6}\sigma = 2^{1/6}(0.275 \text{ nm}) = 0.309 \text{ nm}$$

$$\varepsilon = (8.314 \text{ J K}^{-1} \text{ mol}^{-1})(35.6 \text{ K}) = 296.0 \text{ J mol}^{-1}$$

$$V = 4\varepsilon\left[\left(\frac{\sigma}{r}\right)^{12} - \left(\frac{\sigma}{r}\right)^{6}\right]$$

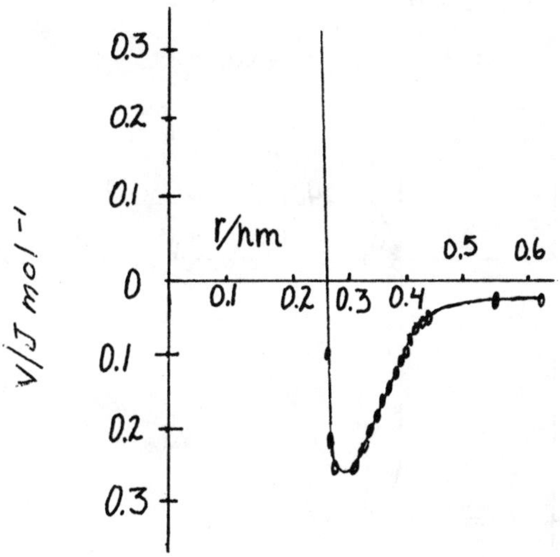

$$V = 4(296.0 \text{ J mol}^{-1})\left[\left(\frac{2.75}{r}\right)^{12} - \left(\frac{2.75}{r}\right)^{6}\right]$$

14.25 494 kJ mol^{-1} 493.570 kJ mol^{-1} from Table A.2

14.28

14.29 0.899 976 E_h

14.30 0.375 E_h

14.31 $E = 4.7483 + 2(13.5984) = 31.9451$ eV

14.35

14.36 787.38 kJ mol^{-1}

14.37 6.408×10^{-29} C m

14.39 Because of the higher electronegativity of Cl, the Cl atom in CH_3Cl is more negative than the Br atom in CH_3Br.

14.40 4.0×10^{-30} C m 0.88×10^{-40} C m/V m^{-1}

14.41 5.29×10^{-30} C m

14.42 19,500 g mol^{-1} 0.9980

14.43

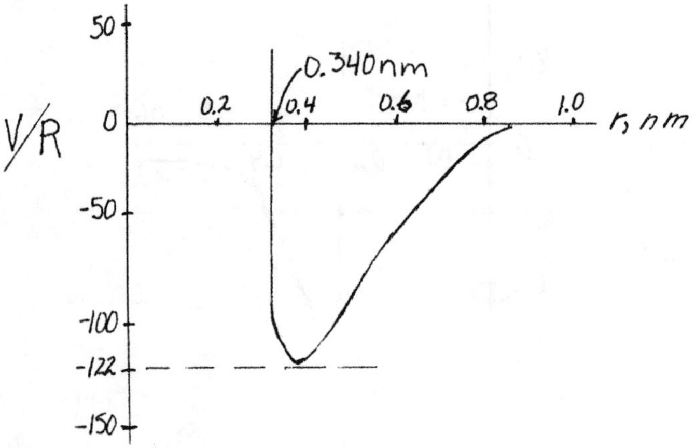

14.44 0.429 nm

CHAPTER 15: Molecular Spectroscopy

15.1 Since the energy of a molecular quantum state is divided by kT in the Boltzmann distribution, it is of interest to calculate the temperature at which kT is equal to the energy of photons of different wavelength. Calculate the temperature at which kT is equal to the energy of photons of wavelength 10^3 cm, 10^{-1} cm, 10^{-3} cm, 10^{-5} cm.

SOLUTION

$$kT = hc/\lambda$$

$$T = \frac{hc}{k\lambda} = \frac{(6.626 \times 10^{-34}\ \text{J s})(2.998 \times 10^{8}\ \text{m s}^{-1})}{(1.381 \times 10^{-23}\ \text{J K}^{-1})(10\ \text{m})}$$

$$= 1.44 \times 10^{-4}\ \text{K} \quad \text{for } \lambda = 10^{3}\ \text{cm}$$

For $\lambda = 10^{-1}$ cm, $T = 14.4$ K

For $\lambda = 10^{-3}$ cm, $T = 1440$ K

For $\lambda = 10^{-5}$ cm, $T = 144{,}000$ K

Note that conversion factors between temperature and various forms energy are given in the back cover.

15.2 Most chemical reactions require activation energies ranging between 40 and 400 kJ mol^{-1}. What are the equivalents of 40 and 400 kJ mol^{-1} in terms of (a) nm, (b) wave numbers, and (c) electron volts?

SOLUTION

For 40 kJ mol^{-1}

(a) $\lambda = \dfrac{hc}{E}$

$$= \frac{(6.626 \times 10^{-34}\ \text{J s})(2.998 \times 10^{8}\ \text{m s}^{-1})(6.022 \times 10^{23}\ \text{mol}^{-1})}{(4 \times 10^{4}\ \text{J mol}^{-1})}$$

$$= 2.991 \times 10^{-6}\ \text{m} = 2991\ \text{nm}$$

(b) $\tilde{\nu} = \frac{1}{\lambda} = \frac{1}{2.991 \times 10^{-4}\ \text{cm}} = 3343\ \text{cm}^{-1}$

(c) $\frac{(4 \times 10^{4}\ \text{J mol}^{-1})}{(6.022 \times 10^{23}\ \text{mol}^{-1})(1.602 \times 10^{-19}\ \text{J eV}^{-1})} = 0.415\ \text{eV}$

For 400 kJ mol^{-1}

(a) $\lambda = (2991\ \text{nm}) \frac{4 \times 10^{4}\ \text{J mol}^{-1}}{4 \times 10^{5}\ \text{J mol}^{-1}} = 299.1\ \text{nm}$

(b) $\tilde{\nu} = \frac{1}{\lambda} = \frac{1}{2.991 \times 10^{-5}\ \text{cm}} = 33{,}430\ \text{cm}^{-1}$

(c) $(0.415\ \text{eV}) \frac{4 \times 10^{5}\ \text{J mol}^{-1}}{4 \times 10^{4}\ \text{J mol}^{-1}} = 4.15\ \text{eV}$

15.3 (a) What vibrational frequency in wave numbers corresponds to a thermal energy of kT at 25 °C? (b) What is the wavelength of this radiation?

SOLUTION

$$hc\tilde{\nu} = kT$$

$$\tilde{\nu} = \frac{kT}{hc} = \frac{(1.381 \times 10^{-23}\ \text{J K}^{-1})(298.15\ \text{K})(10^{-2}\ \text{m cm}^{-1})}{(6.626 \times 10^{-34}\ \text{J K}^{-1})(2.998 \times 10^{8}\ \text{ms}^{-1})}$$

$$= 207\ \text{cm}^{-1}$$

$$\lambda = \frac{1}{\tilde{\nu}} = 4.83 \times 10^{-3}\ \text{cm}$$

15.4 What are the SI units for the Einstein coefficients when the energy densities are expressed in terms of frequencies?

SOLUTION

$$A_{mn} = -\frac{dN_n/dt}{N_n} = \frac{\text{m}^{-3}\ \text{s}^{-1}}{\text{m}^{-3}} = \text{s}^{-1}$$

$$B_{mn} = -\frac{dN_n/dt}{\rho_{\tilde{\nu}} N_n} = \frac{m^{-3}\ s^{-1}\ m^3}{kg\ m^2\ s^{-2}\ s^{-1}\ m^{-3}} = \frac{m}{kg\ s^{-2}} = \frac{m^3}{J}$$

$$\rho_{\tilde{\nu}}\ d\tilde{\nu} = \frac{J}{m^3} = \rho_{\tilde{\nu}}\ s^{-1} \qquad \rho_{\tilde{\nu}} = \frac{J\ s^{-1}}{m^3}$$

15.5 Calculate the reduced mass and the moment of inertia of $D^{35}Cl$, given that R_e = 127.5 pm.

SOLUTION

$$\mu = \frac{m_D m_{Cl}}{m_D + m_{Cl}}$$

$$= \frac{(2.014\ 10 \times 10^{-3}\ kg\ mol^{-1})(34.968\ 85 \times 10^{-3}\ kg\ mol^{-1})}{[(2.014\ 10 + 34.968\ 85) \times 10^{-3}\ kg\ mol^{-1}](6.022\ 045 \times 10^{23}\ mol^{-1})}$$

$$= 3.1624 \times 10^{-27}\ kg$$

$$I = \mu r^2 = (3.1624 \times 10^{-27}\ kg)(1.275 \times 10^{-10}\ m)^2$$

$$= 5.141 \times 10^{-47}\ kg\ m^2$$

15.6 Calculate the frequency in wave numbers and the wavelength in cm of the first rotational transition ($J = 0 \longrightarrow 1$) for $D^{35}Cl$.

SOLUTION

$$\tilde{\nu} = 2BJ = \frac{h}{4\pi^2 cI} \quad \text{since } J = 1$$

The moment of inertia was calculated in the preceding problem,

$$\tilde{\nu} = \frac{(6.626 \times 10^{-34}\ J\ s)}{4\pi^2(2.998 \times 10^8\ m\ s^{-1})(5.141 \times 10^{-47}\ kg\ m^2)} = 1089\ m^{-1}$$

$$= (1089\ m^{-1})(10^{-2}\ m\ cm^{-1}) = 10.89\ cm^{-1}$$

$$\lambda = 1/\tilde{\nu} = 1/(10.89\ cm^{-1}) = 0.09183\ cm$$

15.7 The internuclear distance of $^{12}C^{16}O$ is 112.83 pm. Calculate the frequencies in wave numbers for the first four lines in the pure rotational spectrum.

SOLUTION

$$\mu = \frac{1/6.022\ 045 \times 10^{23}\ \text{mol}^{-1}}{\dfrac{1}{12 \times 10^{-3}\ \text{kg}} + \dfrac{1}{15.99491 \times 10^{-3}\ \text{kg}}}$$

$$= 1.138\ 52 \times 10^{-26}\ \text{kg}$$

$$I = (1.138\ 52 \times 10^{-26}\ \text{kg})(112.83 \times 10^{-12}\ \text{m})^2$$

$$= 1.4994 \times 10^{-46}\text{kg m}^2$$

$$\tilde{\nu} = \frac{2hJ}{8\pi^2 Ic} = \frac{2(6.626 \times 10^{-34}\text{J s})(10^{-2}\ \text{m cm}^{-1})J}{8\pi^2(1.4494 \times 10^{-46}\ \text{kg m}^2)(2.998 \times 10^{8}\ \text{m s}^{-1})}$$

$$= (3.863\ \text{cm}^{-1})J$$

J		$\tilde{\nu}$
0	1	3.863 cm^{-1}
1	2	7.725 cm^{-1}
2	3	11.588 cm^{-1}
3	4	15.450 cm^{-1}

15.8 The far infrared spectrum of HI consists of a series of equally spaced lines with $\Delta\tilde{\nu} = 12.8\ \text{cm}^{-1}$. What is (a) the moment of inertia, and (b) the internuclear distance?

SOLUTION

(a) $$\Delta\tilde{\nu} = 2B = \frac{2h}{8\pi^2 cI}$$

$$I = \frac{h}{8\pi^2 cB} = \frac{(6.63 \times 10^{-34}\ \text{J s})}{8\pi^2(2.998 \times 10^{8}\ \text{m s}^{-1})(6.4\ \text{cm}^{-1})(10\ \text{cm m}^{-1})}$$

$$= 4.37 \times 10^{-47}\ \text{kg m}^2$$

(b) $I = \frac{m_A m_B}{m_A + m_B} R^2$

$$R_e = \left[\frac{(4.37 \times 10^{-47} \text{ kg m}^2)(128 \times 10^{-3} \text{ kg})(6.02 \times 10^{23} \text{ mol}^{-1})}{(127 \times 10^{-3} \text{ kg})(1 \times 10^{-3} \text{ kg})}\right]^{1/2}$$

$$= 163 \text{ pm}$$

15.9 Show that for a rotational spectrum of a diatomic molecule the rotational quantum numer (to the nearest integer value) for the maximum populated level is given by

$$J_{max} = \sqrt{\frac{kT}{2hcB}} - \frac{1}{2}$$

SOLUTION

$$\frac{dN_J}{dJ} = 2e^{\frac{-BhcJ(J+1)}{kT}} - (2J+1)\frac{Bhc}{kT} e^{\frac{-BhcJ(J+1)}{kT}}$$

At the maximum value of N_J this derivative is equal to zero so that the above expression is obtained. For $H^{35}Cl$ at 300 K

$$J_{max} = \sqrt{\frac{(1.3806 \times 10^{-23})(300)}{2(6.626 \times 10^{-34})(2.998 \times 10^{8})(1044)}} - \frac{1}{2}$$

$$= 2.7, \text{ which rounds to } 3$$

15.10 What are the relative populations of the first five rotational levels for $H^{35}Cl$ and $^{12}C^{16}O$ at room temperature?

SOLUTION

$$N_J = K(2J + 1)\ e^{-[hcJ(J + 1)B]/kT}$$

For $H^{35}Cl$, $B = 10.59\ cm^{-1}$

$$hcB = (6.626 \times 10^{-34}\ J\ s)(2.9979 \times 10^{8}\ m\ s^{-1})(1059\ m^{-1})$$
$$= 2.104\ c\ 10^{-22}\ J$$

$$N_J/K = (2J + 1)e^{-(2.104 \times 10^{-22})J(J + 1)/(1.3806 \times 10^{-23})(300)}$$
$$= (2J + 1)e^{-0.05080\ J(J + 1)}$$

For $^{12}C^{16}O$, $B = 1.9313\ cm^{-1}$

$$\frac{hcB}{kT} = 9.26 \times 10^{-3}$$

$$N_J/K = (2J + 1)e^{-(9.26 \times 10^{-3})J(J + 1)}$$

J	$H^{35}Cl$	$^{12}C^{16}O$
0	1	1
1	2.71	2.95
2	3.69	4.73
3	3.81	6.26
4	3.26	7.48

15.11 The moment of inertia of $^{16}O^{12}C^{16}O$ is $7.167 \times 10^{-46}\ kg\ m^2$. (a) Calculate the CO bond length, R_{CO}, in CO_2. (b) Assuming that isotopic substitution does not alter R_{CO}, calculate the moments of inertia of (1) $^{18}O^{12}C^{18}O$ and (2) $^{16}O^{13}C^{16}O$.

SOLUTION

(a) Since the CO_2 molecule is symmetrical, the carbon atom is on the axis of rotation and does not contribute to the moment of inertia.

$$I = 2mR_{CO}^2$$

$$R_{CO} = (I/2m)^{1/2}$$

$$= \left[\frac{(71.67 \times 10^{-47}\ \text{kg m}^2)(6.022\ 045 \times 10^{23}\ \text{mol}^{-1})}{2(15.994\ 91 \times 10^{-3}\ \text{kg mol}^{-1})}\right]^{1/2}$$

$$= 0.1162\ \text{nm}$$

(b) For $^{18}O^{12}C^{18}O$

$$I = 2mR_{CO}^2$$

$$= \frac{2(17.999\ 16 \times 10^{-3}\ \text{kg mol}^{-1})(0.1162 \times 10^{-9}\ \text{m})^2}{6.022\ 045 \times 10^{23}\ \text{mol}^{-1}}$$

$$= 8.071 \times 10^{-46}\ \text{kg m}^2$$

For $^{16}O^{13}C^{16}O$ the moment of inertia is the same as for $^{16}O^{12}C^{16}O$.

15.12 Derive the expression for the moment of inertia of a symmetrical tetrahedral molecule like CH_4 in terms of the bond length R and the masses of the four tetrahedral atoms. The easiest way to derive the expression is to consider an axis along one CH bond. Show that the same result is obtained if the axis is taken perpendicular to the plane defined by one group of three atoms HCH.

SOLUTION

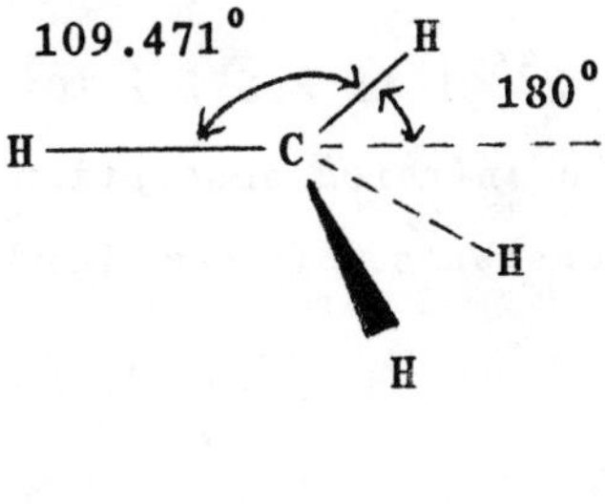

$180° - 109.471° = 70.529°$

$\sin 70.529° = h/R$ where h is the perpendicular distance to the axis

$$I = \Sigma m_i h_i^2 = 3mR^2 \sin^2(70.529°) = \frac{8}{3} mR^2$$

If the axis is taken perpendicular to the page through C

$$I = 2mR^2 + 2[R^2 \sin^2(90° - \frac{109.471°}{2})] = \frac{8}{3} mR^2$$

15.13 What are the rotational energies for the symmetric top NH_3 for $J = 1$ and $J = 2$? Given: $I_{\parallel} = 4.41 \times 10^{-47}$ kg m^2 and $I_{\perp} = 2.81 \times 10^{-47}$ kg m^2.

SOLUTION

$$B = \frac{\hbar}{4\pi c I_{\perp}} = \frac{1.054 \times 10^{-34} \text{ J s}}{4\pi(2.997 \times 10^{10} \text{ cm s}^{-1})(2.81 \times 10^{-47} \text{ kg m}^2)}$$

$$= 9.96 \text{ cm}^{-1}$$

$$A = \frac{\hbar}{4\pi c I_{\parallel}} = \frac{1.054 \times 10^{-34} \text{ J s}}{4\pi(2.997 \times 10^{10} \text{ cm s}^{-1})(4.41 \times 10^{-47} \text{ kg m}^2)}$$

$$= 6.35 \text{ cm}^{-1}$$

$$E/\text{cm}^{-1} = (9.96 \text{ cm}^{-1})(J(J+1)) = (-3.61 \text{ cm}^{-1}) K^2$$

If $J = 1$, $K = 0$, $E = (9.96 \text{ cm}^{-1})(2) = 19.92 \text{ cm}^{-1}$

$K = \pm 1$, $E = 19.92 - 3.61 = 16.31 \text{ cm}^{-1}$

If $J = 2$, $K = 0$, $E = (9.96 \text{ cm}^{-1})(6) = 59.76 \text{ cm}^{-1}$

$K = \pm 1$, $E = 59.76 \text{ cm}^{-1} - 3.61 \text{ cm}^{-1} = 56.15 \text{ cm}^{-1}$

$K = \pm 2$, $E = 59.76 \text{ cm}^{-1} - (3.61 \text{ cm}^{-1})(4) = 45.32 \text{ cm}^{-1}$

15.14 The fundamental vibration frequency of $H^{35}Cl$ is 8.667×10^{13} s^{-1}. What would be the separation between the infrared absorption lines for $H^{35}Cl$ and $H^{37}Cl$ if the force constants of the bonds are assumed to be the same?

SOLUTION

$$\mu = \frac{m_H m_{Cl}}{m_H + m_{Cl}}$$

For $H^{35}Cl$

$$\mu_1 = \frac{(1.007\ 825 \times 10^{-3}\ kg\ mol^{-1})(34.968\ 85 \times 10^{-3}\ kg\ mol^{-1})}{(35.976\ 68 \times 10^{-3}\ kg\ mol^{-1})(6.022\ 045 \times 10^{23}\ mol^{-1})}$$

$$= 1.62668 \times 10^{-27}\ kg$$

For $H^{37}Cl$

$$\mu_2 = \frac{(1.007\ 825 \times 10^{-3}\ kg\ mol^{-1})(36.947 \times 10^{-3}\ kg\ mol^{-1})}{(37.955 \times 10^{-3}\ kg\ mol^{-1})(6.022\ 045 \times 10^{23}\ mol^{-1})}$$

$$= 1.62911 \times 10^{-27}\ kg$$

$$\nu = \frac{1}{2\pi}\left(\frac{k}{\mu}\right)^{1/2} \qquad \frac{\nu_1}{\nu_2} = \left(\frac{\mu_2}{\mu_1}\right)^{1/2} = \frac{\lambda_2}{\lambda_1}$$

$$\lambda_1 = \frac{(2.9979 \times 10^{8}\ m\ s^{-1})}{8.667 \times 10^{13}\ s^{-1}} = 3459.0 \times 10^{-9}\ nm$$

$$\lambda_2 = \lambda_1\left(\frac{\mu_2}{\mu_1}\right)^{1/2} = (3459.0\ nm)\left(\frac{1.629\ 11 \times 10^{-27}\ kg}{1.626\ 68 \times 10^{-27}\ kg}\right)^{1/2} = 3461.6\ nm$$

$$\lambda_2 - \lambda_1 = 2.6\ nm$$

15.15 Calculate the force constants k for $^{14}N_2$ in the ground state and in the $3\pi_g$ state. (See Table 15.4)

SOLUTION

$$\mu = \frac{m}{2} = \frac{14.0067 \times 10^{-3}\ kg\ mol^{-1}}{2(6.022\ 045 \times 10^{23}\ mol^{-1})} = 1.162\ 952 \times 10^{-26}\ kg$$

$$\nu = \frac{1}{2\pi}\left(\frac{k}{\mu}\right)^{1/2}$$

$$k = (2\pi\nu)^2\mu = [2\pi(235\ 857\ m^{-1})(2.997\ 924 \times 10^{8}\ m\ s^{-1})]^2\mu$$

$$= 2295.41 \text{ N m}^{-1} \quad \text{for} \quad {}^{1}\Sigma_{g}^{+}$$

$$k = [2\pi(173\ 339 \text{ m}^{-1})(2.997\ 924 \times 10^{8} \text{ m s}^{-1})]^{2}\mu$$

$$= 1239.81 \text{ N m}^{-1} \quad \text{for} \quad {}^{3}\pi_{g}$$

15.16 (a) What fractions of $H^{35}Cl$ molecules are in the v = 1, 2, and 3 states at 1000 K and (b) 2000 K?

SOLUTION

$$\frac{hc\tilde{\nu}}{k} = \frac{(6.626 \times 10^{-34} \text{ J s})(2.998 \times 10^{10} \text{ cm s}^{-1})(2990.95 \text{ cm}^{-1})}{1.381 \times 10^{-23} \text{ J K}^{-1}}$$

$$= 4302 \text{ K}$$

(a) At 1000 K, equation 15.70 yields

$$f_1 = (1 - e^{-4.302})e^{-4.302} = 0.0133$$

$$f_2 = (1 - e^{-4.302})e^{-(2)(4.302)} = 0.0018$$

$$f_3 = (1 - e^{-4.302})e^{-(3)(4.302)} = 0.000002$$

(b) At 2000 K

$$f_1 = (1 - e^{-2.151})e^{-2.151} = 0.1028$$

$$f_2 = (1 - e^{-2.151})e^{(2)(2.151)} = 0.0120$$

$$f_3 = (1 - e^{-2.151})e^{-(3)(2.151)} = 0.0014$$

15.17 Given the following fundamental vibration frequencies:

$H^{35}Cl$	2989 cm^{-1}	$H^{2}D$	3817 cm^{-1}
$^{2}D^{35}Cl$	2144 cm^{-1}	$^{2}D^{2}D$	2990 cm^{-1}

Calculate ΔH^{o} for the reaction

$$H^{35}Cl\ (\nu = 0) + {}^{2}D^{2}D(\nu = 0) = {}^{2}D^{35}Cl(\nu = 0) + H^{2}D(\nu = 0)$$

SOLUTION

$\Delta H^\circ = E_{DCl} + E_{HD} - E_{HCl} - E_{D_2}$ $\quad E = hc\tilde{\nu}/2$ for $v = 0$

$\Delta\tilde{\nu} = 2144 + 3817 - 2989 - 2990 = -18\ \text{cm}^{-1}$

$\Delta H^\circ = -\frac{1}{2} hc\Delta\tilde{\nu}$

$= -\frac{1}{2}(6.63 \times 10^{-34}\ \text{J s})(3 \times 10^{8}\ \text{m s}^{-1})(-18\ \text{cm}^{-1})(100\ \text{cm m}^{-1})$

$= 1.79 \times 10^{-22}\ \text{J}$

$= (1.79 \times 10^{-22}\ \text{J})(6.022 \times 10^{23}\ \text{mol}^{-1}) = 108\ \text{J mol}^{-1}$

15.18 Calculate D_e for H_2^+ using the information that $D_0 = 2.650\ 78$ eV, $\omega_e = 2321.7\ \text{cm}^{-1}$, $\omega_e x_e = 66.2\ \text{cm}^{-1}$, and $\omega_e y_e = 0.6\ \text{cm}^{-1}$.

SOLUTION

$D_e = D_o + \omega_e/2 - w_e x_e/4 + w_e y_e/8$

$= (2.650\ 78\ \text{eV})(8065.47\ \text{cm}^{-1}) + (2321.7\ \text{cm}^{-1})/2$

$- (66.2\ \text{cm}^{-1})/4 + (0.6\ \text{cm}^{-1})/8$

$= 22{,}540\ \text{cm}^{-1}$

The value in Chapter 14 is 22,527 cm^{-1}.

15.19 Using the values for ω_e and $\omega_e x_e$ in Table 15.4 for $^1H^{35}Cl$ estimate the dissociation energy assuming the Morse potential is applicable.

SOLUTION

$\omega_e = 2990.95\ \text{cm}^{-1}$ $\quad \omega_e x_e = 52.819\ \text{cm}^{-1}$ $\quad x_e = 0.01766$

The zero point energy lies at

$G(0) = \frac{1}{2}\omega_e - \frac{1}{4}\omega_e x_e = \frac{1}{2}(2991\ \text{cm}^{-1}) - \frac{1}{4}(52.8\ \text{cm}^{-1})$

$= 1482\ \text{cm}^{-1}$

$$D_e = \frac{\omega_e}{4x_e} = \frac{2990.95\ \text{cm}^{-1}}{4(0.017\ 66)} = 42{,}341\ \text{cm}^{-1}$$

$$D_o = D_e - G(0) = 42{,}341 - 1482 = 40{,}859\ \text{cm}^{-1}$$

$$= (40{,}859\ \text{cm}^{-1})(1.2399 \times 10^{-4}\ \text{eV cm}) = 5.07\ \text{eV}$$

The actual value is 4.434 eV

15.20 (a) What fraction of $H_2(g)$ molecules are in the v = 1 state at room temperature? (b) What fractions of $Br_2(g)$ molecules are in the v = 1, 2, and 3 states at room temperature?

SOLUTION

(a) Using equation $f_1 = (1 - e^{-h\nu/kT})e^{-h\nu/kT}$

$$\frac{h\nu}{kT} = \frac{hc\tilde{\nu}}{kT} = \frac{(6.626 \times 10^{-34}\ \text{J s})(2.998 \times 10^{10}\ \text{cm s}^{-1})(4401\ \text{cm}^{-1})}{(1.381 \times 10^{-23}\ \text{J K}^{-1})(298\ \text{K})}$$

$$= 21.2 \qquad f_1 = (1 - e^{-21.2})e^{-21.2} = 6.2 \times 10^{-10}$$

(b) For Br_2, $\omega_e = 325.321\ \text{cm}^{-1}$

$$\frac{h\nu}{kT} = \frac{hc\tilde{\nu}}{kT} = \frac{(6.626 \times 10^{-34}\ \text{J s})(2.998 \times 10^{10}\ \text{cm s}^{-1})(325.3\ \text{cm}^{-1})}{(1.381 \times 10^{-23}\ \text{J K}^{-1})(298\ \text{K})}$$

$$= 1.570$$

$$f_1 = (1 - e^{-1.570})e^{-1.570} = 0.165$$

$$f_2 = (1 - e^{-1.570})e^{-(2)(1.570)} = 0.034$$

$$f_3 = (1 - e^{-1.570})e^{-(3)(1.570)} = 0.007$$

15.21 The first three lines in the R branch of the fundamental vibration-rotation band of $H^{35}Cl$ have the following frequencies in cm^{-1}: 2906.25(0), 2925.78(1), 2944.89(2), where the numbers in parentheses are the J values for the initial level. What are the values of ω_0, B_v', B_v'', B_e, and α?

SOLUTION

According to equation 15.75

$$\tilde{\nu} = \omega_o(B_v' + B_v'')m + (B_v' - B_v'')m^2$$

$$= a + bm + cm^2 \quad \text{where} \quad m = J'' + 1$$

$$2906.25 = a + b + c$$

$$2925.78 = a + 2b + 4c$$

$$2944.89 = a + 3b + 9c$$

$$a = 2886.30\ \text{cm}^{-1} = \omega_o$$

$$b = 20.16 = B_v' + B_v''$$

$$c = -0.21 = B_v' - B_v''$$

$$B_v' = 9.98\ \text{cm}^{-1} = B_e - \frac{3}{2}\alpha$$

$$B_v'' = 10.19\ \text{cm}^{-1} = B_e - \frac{1}{2}\alpha$$

$$B_e = 10.30\ \text{cm}^{-1}$$

$$\alpha = 0.21\ \text{cm}^{-1}$$

15.22 In Table 14.1 D_e for H_2 is given as 4.7483 eV or 458.135 kJ mol^{-1}. Given the vibrational parameters for H_2 in Table 15.4 calculate the value you would expect for $\Delta_f H^o$ for H(g) at 0 K.

SOLUTION

$$D_o = D_e - \frac{1}{2}\omega_e + \frac{1}{4}\omega_e x_e$$

$$= (4.7483\ \text{eV})(8065.478\ \text{cm}^{-1}\ \text{eV}^{-1}) - \frac{1}{2}(4400.39) + \frac{1}{4}(120.815)$$

$$= 36{,}127\ \text{cm}^{-1} \quad \text{or} \quad 432.175\ \text{kJ mol}^{-1}$$

$$H_2(g) = 2H(g)$$

$$\Delta H_o^o = 2\Delta_f H^o(H) = 432.175\ \text{kJ mol}^{-1}$$

$\Delta_f H^\circ(H) = 216.088 \text{ kJ mol}^{-1}$

Table A.2 gives 216.037 kJ mol^{-1}.

15.23 Calculate the wavelengths in (a) wave numbers, and (b) micrometers of the center two lines in the vibration-rotation spectrum of HBr for the fundamental vibration. The necessary data are to be found in Table 15.4.

SOLUTION

The reduced mass of $H^{80}Br$ is given in Table 15.4 as

$$\mu = \frac{0.995\ 58 \times 10^{-3}\ \text{kg mol}^{-1}}{6.022\ 045 \times 10^{-23}\ \text{mol}^{-1}} = 1.653\ 23 \times 10^{-27}\ \text{kg}$$

$$I = \mu R^2 = (1.653\ 23 \times 10^{-27}\ \text{kg})(1.4138 \times 10^{-10}\ \text{m})^2$$

$$= 3.304\ 52 \times 10^{-47}\ \text{kg m}^2$$

$$B = \frac{h}{8\pi^2 cI}$$

$$= \frac{(6.626 \times 10^{-34}\ \text{J s})(10^{-2}\ \text{m cm}^{-1})}{8\pi^2(2.998 \times 10^8\ \text{m s}^{-1})(3.3045 \times 10^{-47}\ \text{kg m}^2)} = 8.47\ \text{cm}^{-1}$$

(a) $\tilde{\nu}_P = \tilde{\nu}_o - 2BJ''$ where $J'' = 1,2,3,\ldots.$

$$= 2649.67\ \text{cm}^{-1} - 2(8.47\ \text{cm}^{-1}) = 2632.72\ \text{cm}^{-1}$$

$\tilde{\nu}_R = \tilde{\nu}_o + 2B + 2BJ''$ where $J'' = 0,1,2,\ldots.$

$$= 2649.67\ \text{cm}^{-1} + 2(8.47\ \text{cm}^{-1}) = 2666.61\ \text{cm}^{-1}$$

(b) $\lambda_P = 1/\tilde{\nu}_P = 3.798\ 34 \times 10^{-4}\ \text{cm} = 3.798\ 34\ \mu\text{m}$

$\lambda_R = 1/\tilde{\nu}_R = 3.750\ 08 \times 10^{-4}\ \text{cm} = 3.750\ 08\ \mu\text{m}$

15.24 How many normal modes of vibration are there for (a) SO_2(bent), (b) H_2O_2(bent), (c) HC≡CH(linear), and (d) C_6H_6?

SOLUTION

The number of normal modes of vibration is 3N-6 for a non-linear molecule and 3N-5 for a linear molecule.

(a) 3N-6 = 9-6 = 3 (c) 3N-5 = 12-5 = 7
(b) 3N-6 = 12-6 = 6 (d) 3N-6 = 36-6 = 30

15.25 List the numbers of translational, rotational, and vibrational degrees of freedom for (a) Ne, (b) N_2, (c) CO_2, and (d) CH_2O.

SOLUTION

	Molecule	Trans.	Rot.	Vib.	Total = 3N
(a)	Ne	3	0	0	3
(b)	N_2	3	2	1	6
(c)	CO_2	3	2	4*	9
(d)	CH_2O	3	3	6**	12

*For linear molecules 3N-5 **For nonlinear molecules 3N-6

15.26 Acetylene is a symmetrical linear molecule. It has seven normal modes of vibration, two of which are doubly degenerate. These normal modes may be represented as follows:

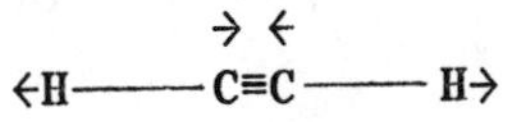

$\tilde{\nu}_1 = 3374\ cm^{-1}$

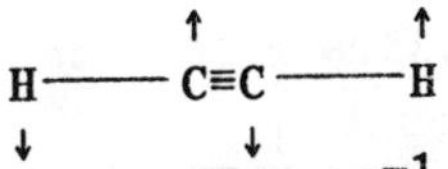

$\tilde{\nu}_4 = 612\ cm^{-1}$

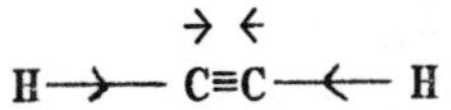

$\tilde{\nu}_2 = 1974\ cm^{-1}$

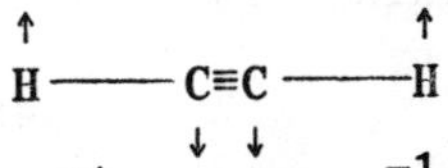

$\tilde{\nu}_5 = 729\ cm^{-1}$

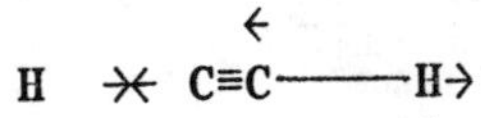

$\tilde{\nu}_3 = 3287\ cm^{-1}$

(a) Which are the doubly degenerate vibrations?
(b) Which vibrations are infrared active?
(c) Which vibrations are Raman active?

SOLUTION

(a) $\tilde{\nu}_4$ and $\tilde{\nu}_5$ since they can also take place in a plane perpendicular to the plane of the paper

(b) $\tilde{\nu}_3$ and $\tilde{\nu}_5$ because the dipole moment changes during the vibration

(c) $\tilde{\nu}_1$, $\tilde{\nu}_2$, and $\tilde{\nu}_4$ because the polarizability of the molecule changes during these vibrations

15.27 When CCl_4 is irradiated with the 435.8 nm mercury line, Raman lines are obtained at 439.9, 444.6, and 450.7 nm. Calculate the Raman frequencies of CCl_4 (expressed in wave numbers). Also calculate the wavelengths (expressed in μm) in the infrared at which absorption might be expected.

SOLUTION

$$\tilde{\nu}_{Raman} = \frac{1}{\lambda_{incid}} - \frac{1}{\lambda_{scatt}}$$

$$= \frac{1}{435.8 \times 10^{-7}\ \text{cm}} - \frac{1}{439.9 \times 10^{-7}\ \text{cm}} = 214\ \text{cm}^{-1}$$

$$\lambda = \frac{1}{\tilde{\nu}} = \frac{1}{214\ \text{cm}^{-1}} = 46.7 \times 10^{-6}\ \text{m}$$

For the remaining lines

$\tilde{\nu}_{Raman}/\text{cm}^{-1}$	312	454	759
$\lambda/\mu\text{m}$	32.0	22.0	13.2

15.28 The first several Raman frequencies of $^{14}N_2$ are 19.908, 27.857, 35.812, 43.762, 51.721, and 59.662 cm^{-1}. These lines are due to pure rotational transitions with $J = 1,2,3,4,5$, and 6. The spacing between the lines is $4B_e$. What is the internuclear distance?

SOLUTION

$$\mu = \frac{(14.003\ 07 \times 10^{-3}\ \text{kg})^2}{(28.006\ 14 \times 10^{-3}\ \text{kg})(6.022\ 045 \times 10^{23}\ \text{mol}^{-1})}$$

$$= 1.162\ 651 \times 10^{-26}\ \text{kg}$$

The average spacing between lines is 7.951 cm^{-1}, and so

$B_e = (7.951\ \text{cm}^{-1})(10^2\ \text{cm m}^{-1})/4 = 198.78\ \text{m}^{-1}$ since $\Delta J = 2$.

$$B_e = \frac{h}{8\pi^2 c\mu R_e^2}$$

$$R_e = \left(\frac{h}{8\pi^2 c\mu B_e}\right)^{1/2}$$

$$= \left[\frac{6.626\ 176 \times 10^{-34}\ \text{J s}}{8\pi^2(2.997\ 924\ 58 \times 10^8\ \text{m s}^{-1})(1.162\ 651 \times 10^{-26}\ \text{kg})(198.78\ \text{m}^{-1})}\right]$$

$$= 110\ \text{pm}$$

15.29 What Raman shifts are expected for the first four Stokes lines for CO_2?

SOLUTION

$I = 7.167 \times 10^{-46}\ \text{kg m}^2$ (Problem 15.11)

$$B_e = \frac{h}{8\pi^2 Ic} = \frac{(6.626 \times 10^{-34}\ \text{J s})(10^{-2}\ \text{m cm}^{-1})}{8\pi^2(7.167 \times 10^{-46}\ \text{kg m}^2)(2.9979 \times 10^8\ \text{m s}^{-1})}$$

$= 0.3906\ \text{cm}^{-1}$ $\qquad \Delta\tilde{\nu}_R = 2B_e(2J + 3)$

J''	$\Delta\tilde{\nu}_R$
0	2.3436 cm^{-1}
1	3.9060
2	5.4684
3	7.0308

15.30 (a) 3.10 nm (b) 414 nm

15.31 11,604 K

15.32 (a) $1.138\ 52 \times 10^{-26}$ kg (b) 1.4492×10^{-46} kg m^2

15.33 21.1, 42.2, 63.3, 189.9 cm^{-1}

474, 237, 158, 53 μm

15.34 (a) 4.61×10^{-48} (b) 6.15×10^{-48}
(c) 6.92×10^{-48} (d) 9.22×10^{-48} kg m^2

15.35 0.4515 kJ 0.0265 cm

15.36 113 pm

15.37 $\nu = Jh/4\pi^2 I$

15.38

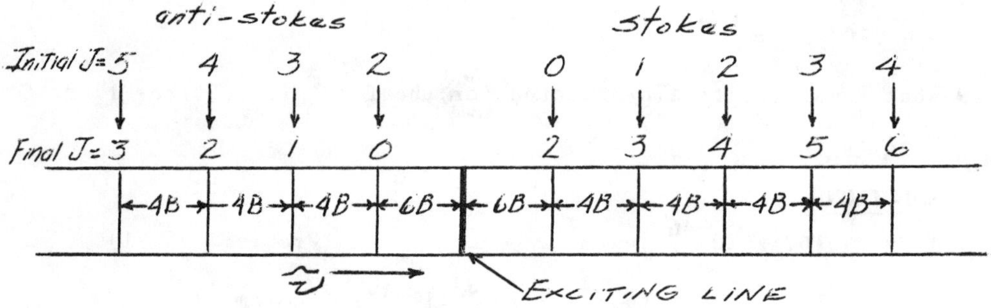

15.40 12,168.5, 24,337.0, 36,505.6 Mc

15.41 9.10, 14.39 K

15.42 0.1855, 0.1737 eV

15.43 8.731×10^{-20} J 52.59 kJ 0.5449 eV

15.44 (a) 26.288 (b) 3.381 kJ mol^{-1}

15.45 $6.216 \times 10^{13}\ s^{-1}$ 4.83 μm

15.46 $4.09 \times 10^{8}\ N\ m^{-1}$

15.47 8.9×10^{-7} 0.358

15.48 3.025 eV

15.49 87,510 cm^{-1} 10.85 eV

15.50 4592.8 4605.9 4632.1 4645.2 cm^{-1}

2.1773 2.1154 2.1588 2.1528 μm

15.51 1870 cm^{-1} 8.45 cm^{-1}

15.52 $B_v' = 5.29$ $B_v'' = 5.42$ $B_e = 5.49$ $\alpha = 0.13\ cm^{-1}$

R_e = 127.0 pm compared with 127.455 pm in Table 15.4.

15.53

Molecule	Translational	Rotational	Vibrational
Cl_2	3	2	3N − 5 = 1
H_2O	3	3	3N − 6 = 3
HC≡CH	3	3	3N − 5 = 7

15.54

Molecule	Translational	Rotational	Vibrational
NNO	3	2	3N − 5 = 4
NH_3	3	3	3N − 6 = 6

15.55 0.075 19 nm

15.56 109.7 pm

15.57 655 J mol^{-1}

CHAPTER 16: Electronic Spectroscopy of Molecules

16.1 The spectroscopic dissociation energy of $H_2(g)$ is 4.4763 eV, and the fundamental vibrational frequency is 4395.24 cm^{-1}. What is the spectroscopic dissociation energy of $D_2(g)$ if it has the same force constant?

SOLUTION

Since deuterium atoms are more massive, the fundamental vibration frequency will be lower in D_2 than H_2. Therefore the zero point energy for D_2 is lower, and its dissociation energy is higher.

For H_2

$$\mu = \frac{(1.007\ 825 \times 10^{-3}\ \text{kg mol}^{-1})^2}{2(1.007\ 825 \times 10^{-3}\ \text{kg mol}^{-1})(6.022\ 045 \times 10^{23}\ \text{mol}^{-1})}$$

$$= 8.367\ 797 \times 10^{-28}\ \text{kg}$$

For D_2

$$\mu = \frac{(2.014\ 10 \times 10^{-3}\ \text{kg mol}^{-1})^2}{2(2.014\ 10 \times 10^{-3}\ \text{kg mol}^{-1})(6.022\ 045 \times 10^{23}\ \text{mol}^{-1})}$$

$$= 16.722\ 725 \times 10^{-28}\ \text{kg}$$

According to equation 11.95

$$\frac{\omega_{D_2}}{\omega_{H_2}} = \left[\frac{8.367\ 797 \times 10^{-28}\ \text{kg}}{16.722\ 725 \times 10^{-28}\ \text{kg}}\right]^{1/2}$$

$$\omega_{H_2} = 4395.24\ \text{cm}^{-1} \quad \text{so} \quad \omega_{D_2} = 3109.10\ \text{cm}^{-1}$$

$$\Delta\omega = 1286.1\ \text{cm}^{-1}$$

$$D^{o}_{D_2} = 4.4763\ \text{eV} + hc\Delta\omega/2$$

$$= 4.4763\ \text{eV} + \frac{(6.626\text{x}10^{-34}\ \text{J s})(2.9979\text{x}10^{8}\ \text{m s}^{-1})(1.286\text{x}10^{5}\ \text{m}^{-1})}{2(1.602 \times 10^{-19}\ \text{J eV}^{-1})}$$

$= 4.4763 \text{ eV} + 0.0797 \text{ eV} = 4.5560 \text{ eV}$

16.2 According to the hypothesis of Franck, the molecules of the halogens dissociate into one normal atom and one excited atom. The wavelength of the convergence limit in the spectrum of iodine is 499.5 nm. (a) What is the energy of dissociation of iodine into one normal and one excited atom? (b) The lowest excitation energy of the iodine atom is 0.94 eV. What is the energy corresponding to this excitation? (c) Compute the heat of dissociation of the iodine molecule into two normal atoms, and compare it with the value obtained from thermochemical data, 144.4 kJ mol^{-1}.

<u>SOLUTION</u>

(a) $E = hcN_A/\lambda$

$$= \frac{(6.626 \times 10^{-3} \text{ J s})(2.998 \times 10^{8} \text{ m s}^{-1})(6.022 \times 10^{23} \text{ mol}^{-1})}{(4995 \times 10^{-10} \text{ m})(10^{3} \text{ J kJ}^{-1})}$$

$= 239.5 \text{ kJ mol}^{-1}$

(b) $(0.94 \text{ eV})(96{,}485 \text{ C mol}^{-1}) = 90.7 \text{ kJ mol}^{-1}$

(c)

$I_2 = I + I^*$		$239.5 \text{ kJ mol}^{-1}$
$I^* = I$		$-90.7 \text{ kJ mol}^{-1}$
$I = 2I$		$148.8 \text{ kJ mol}^{-1}$

16.3 The ultraviolet absorption of O_2 includes a series of lines (the Schumann-Runge bands) due to transitions from the ${}^3\Sigma_u^-$ ground state to the excited electronic state ${}^3\Sigma_u^-$, which are shown in Fig. 16.1. These lines converge to 175.9 nm, which corresponds to dissociation to one O atom in its ground state ${}^3P$ and one O atom in an excited state ${}^1D$. What is D_0 for O_2? How does this compare with the enthalpy of formation at 0 K? Given: The ${}^1D$ state of O is 1.970 eV above the ground state ${}^3P$.

SOLUTION

The dissociation energy D_0 for $O_2(^3\Sigma_g^-) = O(^3P) + O('D)$ is

$$\Delta E = \frac{hcN_A}{\lambda} = \frac{(6.626\times10^{-34}\ J\ s)(2.998\times10^{8}\ m\ s\ \ )(6.022\ x\ 10^{23}\ mol\ \ )}{175.9\ x\ 10^{-9}\ m}$$

$$= 680.2\ kJ\ mol^{-1} = (680.2\ kJ\ mol^{-1})/(96.485\ kJ^{-1}\ eV^{-1})$$

$$= 7.050\ eV$$

Since $O(^3P) = O('D)$ $\Delta E = 1.970$ eV

The dissociation energy of $O_2(^3\Sigma_g^-)$ into $2O(^3P)$ is

$D_0 = 7.050 - 1.970 = 5{,}080$ eV or 490.14 kJ mol^{-1}.

From Table A.2 $\Delta_f H^o$ (0 K) = 2(246.785) = 493.57 kJ mol^{-1}.

16.4 The spectroscopic dissociation energy of $^{127}I_2$ is 1.542 38 eV according to Table 15.4. What wavelength of light would you use to dissociate ground state molecules to ground state atoms if you wanted the atoms to fly away with velocities of 10^3 m s^{-1}?

SOLUTION

$$E = (1.542\ 38\ eV)(96{,}485\ J\ mol^{-1}\ eV^{-1}) + 2(\tfrac{1}{2}\ 126.904\ x\ 10^{-3}\ kg\ mol^{-1})(10^{3}\ m\ s^{-1})^2$$

$$= (1.488\ x\ 10^{5} + 1.269\ x\ 10^{5})\ J\ mol^{-1}$$

$$= 2.757\ x\ 10^{5}\ J\ mol^{-1} = N_A hc/\lambda$$

$$\lambda = \frac{(6.022\ x\ 10^{23}\ mol^{-1})(6.626\ x\ 10^{-34}\ J\ s)(2.998\ x\ 10^{8}\ m\ s^{-1})}{2.757\ x\ 10^{5}\ J\ mol^{-1}}$$

$$= 4.339\ x\ 10^{-7}\ m = 433.9\ nm$$

16.5 A solution of a dye containing 0.1 mol L^{-1} transmits 80% of the light at 435.6 nm in a glass cell 1 cm thick. (a) What percent of light will be absorbed by a solution containing 2 mol L^{-1} in a cell 1 cm thick? (b) What concentration will be required to absorb 50% of the light? (c) What percent of the light will be transmitted by a solution of the dye containing 0.1 mol L^{-1} in a cell 5 cm thick? (d) What thickness should the cell be in order to absorb 90% of the light with solution of this concentration?

SOLUTION

$$\log (I_o/I) = \varepsilon c l$$

$$\log (100/80) = \varepsilon(0.1 \text{ mol L}^{-1})(1 \text{ cm})$$

$$\varepsilon = 0.969 \text{ L mol}^{-1} \text{ cm}^{-1}$$

(a) $I = I_o e^{-2.303\, \varepsilon c l}$

$$= (100)\exp [-(2.303)(0.969 \text{ L mol}^{-1} \text{ cm}^{-1})(2 \text{ mol L}^{-1})(1 \text{ cm})]$$

$$= 1.2\%$$

(b) $\log (100/50) = (0.969 \text{ L mol}^{-1} \text{ cm}^{-1})(1 \text{ cm})c$

$$c = 0.311 \text{ mol L}^{-1}$$

(c) $I = (100)\exp[-(2.303)(0.969 \text{ L mol}^{-1}\text{cm}^{-1})(0.1 \text{ mol L}^{-1})(5 \text{ cm})]$

$$= 32.8\%$$

(d) $\log (100/10) = (0.969 \text{ L mol}^{-1} \text{ cm}^{-1})(0.1 \text{ mol L}^{-1})l$

$$l = 10.3 \text{ cm}$$

16.6 The absorption coefficient α for a solid is defined by $I = I_o e^{-\alpha x}$, where x is the thickness of the sample. The absorption coefficients for NaCl and KBr at a wavelength of 28 μm are 14 and 0.25 cm^{-1}. Calculate the percentage of this infrared radiation transmitted by 0.5 cm thicknesses of these crystals.

SOLUTION

For NaCl $\quad I = I_o e^{-\alpha x} = 100\, e^{-(14 \text{ cm}^{-1})(0.5 \text{ cm})} = 0.091\%$

For KBr $\quad I = 100\ e^{-(0.25\ \text{cm}^{-1})(0.5\ \text{cm})} = 88.2\%$

16.7 The following absorption data are obtained for solutions of oxyhemoglobin in pH 7 buffer at 575 nm in a 1-cm cell:

g/cm^3	**Transmission, %**
3×10^{-4}	53.5
5×10^{-4}	35.1
10×10^{-4}	12.3

The molar mass of hemoglobin is 64.0 kg mol^{-1}. (a) Is Beer's law obeyed? What is the molar absorption coefficient? (b) Calculate the percent transmission for a solution containing 10^{-4} g/cm^3.

<u>SOLUTION</u>

(a)

grams per cm^3	**mol L^{-1}**	**I/I_o**	**$\log(I/I_o)$**	**ε/L mol^{-1} cm^{-1}**
3×10^{-4}	4.69×10^{-6}	0.535	−0.272	5.80×10^4
5×10^{-4}	7.82×10^{-6}	0.351	−0.455	5.82×10^4
10×10^{-4}	15.64×10^{-6}	0.123	−0.910	5.82×10^4

Beer's law is obeyed and the molar absorption coefficient is 5.81×10^4 L mol^{-1} cm^{-1}.

(b) $\log(I/I_o) = -\varepsilon c l$

$$= -[5.81 \times 10^4\ \text{L mol}^{-1}\ \text{cm}^{-1}](1.564 \times 10^{-6}\ \text{mol L}^{-1})(1\ \text{cm})$$

$$= -0.091$$

$I/I_o = 0.81$ or 81% transmission

16.8 The protein metmyoglobin and imidazole form a complex in solution. The molar absorption coefficients in L mol^{-1} cm^{-1} of the metmyoglobin (Mb) and the complex (C) are as follows:

λ	ε_{Mb} / 10^3 L mol^{-1} cm^{-1}	ε_C / 10^3 L mol^{-1} cm^{-1}
500 nm	9.42	6.88
630 nm	3.58	1.30

An equilibrium mixture in a cell of 1-cm path length has an absorbance of 0.435 at 500 nm and 0.121 at 630 nm. What are the concentrations of metmyoglobin and complex?

SOLUTION

$$\log (I_o/I) = A = (\varepsilon_1 c_1 + \varepsilon_2 c_2)l$$

At 500 nm $\quad 0.435 = 9.42\times10^3 c_1 + 6.88\times10^3 c_2$

At 630 nm $\quad 0.121 = 3.58\times10^3 c_1 + 1.30\times10^3 c_2$

Solving these equations simultaneously

$c_1 = 2.17 \times 10^{-5}$ mol L^{-1} metmyoglobin

$c_2 = 3.37 \times 10^{-5}$ mol L^{-1} complex

16.9 The absorption spectrum for benzene in Fig. 16.6 shows maxima at about 180, 200, and 250 nm. Estimate the integrated absorption coefficients using $\varepsilon_{max}\Delta\tilde{\nu}_{1/2}$ and assuming that the width at half maximum is 5000 cm^{-1} in each case. What are the three oscillator strengths?

SOLUTION

At 180 nm

$$\varepsilon_{max}\Delta\tilde{\nu}_{1/2} = (50.1 \times 10^3 \text{ L mol}^{-1} \text{ cm}^{-1})(5000 \text{ cm}^{-1})$$
$$= 2.50 \times 10^8 \text{ L mol}^{-1} \text{ cm}^{-2}$$

$$f = (4.33 \times 10^{-9} \text{ mol L}^{-1} \text{ cm}^2)(2.50 \times 10^8 \text{ L mol}^{-1} \text{ cm}^{-2}) = 1.08$$

At 200 nm

$$\varepsilon_{max}\Delta\tilde{\nu}_{1/2} = (7000 \text{ L mol}^{-1} \text{ cm}^{-1})(5000 \text{ cm}^{-1})$$

$$= 3.5 \times 10^{7} \text{ L mol}^{-1} \text{ cm}^{-2}$$

$$f = (4.33 \times 10^{-9} \text{ mol L}^{-1} \text{ cm}^{2})(3.5 \times 10^{7} \text{ L mol}^{-1} \text{ cm}^{-2}) = 0.152$$

At 250 nm

$$\varepsilon_{max}\Delta\tilde{\nu}_{1/2} = (100 \text{ L mol}^{-1} \text{ cm}^{-1})(5000 \text{ cm}^{-1})$$

$$= 5 \times 10^{5} \text{ L mol}^{-1} \text{ cm}^{-2}$$

$$f = (4.33 \times 10^{-9} \text{ mol L}^{-1} \text{ cm}^{2})(5 \times 10^{5} \text{ L mol}^{-1} \text{ cm}^{-2}) = 0.0022$$

16.10 An absorption band has a maximum molar absorbancy index of 2×10^{4} L mol^{-1} cm^{-1} and a width at half-height of 4000 cm^{-1}. Assuming the band has a triangular shape when plotted against wavenumber, what is the integrated absorption coefficient for the band?

SOLUTION

$$\int \varepsilon d\tilde{\nu} = 1.06\, \varepsilon_{max}\Delta\tilde{\nu}_{1/2}$$

$$= (1.06)(2 \times 10^{4} \text{ L mol}^{-1} \text{ cm}^{-1})(4000 \text{ cm}^{-1})$$

$$= 8.48 \times 10^{7} \text{ L mol}^{-1} \text{ cm}^{-2}$$

16.11 What is the oscillator strength f for the transition in the preceding problem?

SOLUTION

$$f = (4.32 \times 10^{-9} \text{ L}^{-1} \text{ mol cm}^{2}) \int \varepsilon d\tilde{\nu}$$

$$= (4.32 \times 10^{-9} \text{ L}^{-1} \text{ mol cm}^{2})(8.48 \times 10^{7} \text{ L mol}^{-1} \text{ cm}^{-2}) = 0.366$$

16.12 (a) Calculate the energy levels for n = 1 and n = 2 for an electron in a potential well of width 0.5 nm with infinite barriers on either side. The energies should be expressed in J and in kJ mol^{-1}. (b) If an electron makes a transition from n = 2 to n = 1, what will be the wavelength of the radiation emitted?

SOLUTION

$$\text{(a) } E = \frac{n^2h^2}{8m_e a^2} = \frac{n^2(6.626 \times 10^{-34}\ \text{J s})^2}{8(9.110 \times 10^{-31}\ \text{kg})(5 \times 10^{-10}\ \text{m})^2}$$

$$= n^2\ 2.410 \times 10^{-19}\ \text{J} \qquad 2.410 \times 10^{-19}\ \text{J for } n = 1$$

$$9.640 \times 10^{-19}\ \text{J for } n = 2$$

$$= \frac{n^2(2.410 \times 10^{-19}\ \text{J})(6.022 \times 10^{23}\ \text{mol}^{-1})}{(10^3\ \text{J kJ}^{-1})}$$

$$= n^2\ 145.14\ \text{kJ mol}^{-1} \qquad 145.14\ \text{kJ mol}^{-1}\ \text{for } n = 1$$

$$580.57\ \text{kJ mol}^{-1}\ \text{for } n = 2$$

$$\text{(b) } E_2 - E_1 = \frac{hc}{\lambda}$$

$$\lambda = \frac{hc}{E_2 - E_1}$$

$$= \frac{(6.626 \times 10^{-34}\ \text{J s})(2.998 \times 10^8\ \text{m s}^{-1})}{(9.640 - 2.410) \times 10^{-19}\ \text{J}} = 275\ \text{nm}$$

16.13 The electronic spectrum of a molecule

$$R\text{---}(CH{=}CH)_k\text{---}R'$$

can be described by considering the transitions of $2k\pi$ electrons among the levels in a one-dimensional box of length kl, where l is the length of the unit —CH=CH—. (a) Ignore electron-electron interactions and assign electrons to orbitals in accord with the Pauli principle. Obtain an expression for the lowest energy transition as a function of k. (b) If $l = 0.28$ nm, what is the wavelength of the lowest energy transition for $k = 2$, 5, and 10?

SOLUTION

$$\text{(a) } E = \frac{n^2h^2}{8mk^2l^2}$$

According to the Pauli principle assign 2 electrons per energy level. We assign 2k electrons to the k lowest levels. The transition of lowest energy is $k \rightarrow k + 1$.

$$\Delta E = \frac{(2k + 1)h^2}{8mk^2 l^2} = \frac{hc}{\lambda}$$

$$\lambda = (2.59 \times 10^{-7} \text{ m})k^2/(2k + 1)$$

(b) For $l = 0.28$ nm, $k = 2$, $\lambda = 207$ nm
$k = 5$, $\lambda = 589$ nm
$k = 10$, $\lambda = 1230$ nm

16.14 The life times of vibrationally excited states of molecules of a liquid are limited by the collision rates in the liquid. If one in ten collisions deactivates a vibrationally excited state, what is the broadening of vibrational lines if a molecule undergoes 10^{13} collisions per second?

SOLUTION

$$\Delta E = hc\Delta\tilde{\nu} \geqslant h/2\pi\Delta t$$

$$\Delta\tilde{\nu} \geqslant (2\pi c\Delta t)^{-1} = [2\pi(3 \times 10^{10} \text{ cm s}^{-1})(10^{-12} \text{ s})]^{-1}$$

$$\Delta\tilde{\nu} \geqslant 5.3 \text{ cm}^{-1}$$

16.15 A sample of oxygen gas is irradiated with $MgK_{\alpha_1\alpha_2}$ radiation of 0.99 nm (1253.6 eV). A strong emission of electrons with velocities of 1.57×10^7 m s^{-1} is found. What is the binding energy of these electrons?

SOLUTION

$$\frac{1}{2}mv^2 = h\nu - I$$

$$I = h\nu - \frac{1}{2}mv^2$$

$$= 1253.6 \text{ eV} - \frac{1}{2}\frac{(9.109 \times 10^{-31} \text{ kg})(1.57 \times 10^7 \text{ m s}^{-1})^2}{(1.602 \times 10^{-19} \text{ J eV}^{-1})}$$

$$= 553 \text{ eV}$$

16.16 The ionization potential of an atom may be determined by exposing it to high energy monochromatic radiation and measuring the speed of ejected electrons. When krypton is radiated with 584 pm light from a helium discharge lamp, ejected electrons have a velocity of 1.59×10^{6} m s^{-1}. What is the ionization potential?

SOLUTION

$$h\nu = \frac{1}{2} m_e \nu^2 + E_i$$

$$E_i = \frac{(6.626 \times 10^{-34} \text{ J s})(2.998 \times 10^{8} \text{ m s}^{-1})}{(58.4 \times 10^{-9} \text{ m})(1.602 \times 10^{-19} \text{ J eV}^{-1})} - \frac{1}{2} \frac{(9.110 \times 10^{-31} \text{ kg})(1.59 \times 10^{6} \text{ m s}^{-1})^2}{(1.602 \times 10^{-19} \text{ J eV}^{-1})}$$

$$E_i = 21.23 \text{ eV} - 7.19 \text{ eV} = 14.04 \text{ eV}$$

$$1 \text{ eV} = 1.602 \times 10^{-19} \text{ J}$$

16.17 When α-D-mannose ($[\alpha]_D^{20}$ = +29.3°) is dissolved in water, the optical rotation decreases as β-D-mannose is formed until at equilibrium $[\alpha]_D^{20}$ = +14.2°). This process is referred to mutarotation. As expected, when β-D-mannose ($[\alpha]_D^{20}$ = −17.2°) is dissolved in water, the optical rotation increases until $[\alpha]_D^{20}$ = +14.2° is obtained. Calculate the percentage of α form in the equilibrium mixture.

SOLUTION

$$[\alpha]_\alpha = 29.3^\circ \qquad [\alpha]_\beta = -17.0^\circ$$

$$[\alpha]_{mixture} = +14.2^\circ = 29.3^\circ f_\alpha - 17.0^\circ f_\beta$$

$$= 29.3^\circ f_\alpha - 17.0^\circ (1 - f_\alpha)$$

$$= -17.0^\circ + 36.3 f_\alpha$$

$$f_\alpha = \frac{31.2^\circ}{46.3^\circ} = 0.67 = \text{fraction of } \alpha \text{ form}$$

16.18 What is the difference in refractive index for left (n_1) and right (n_r) circularly polarized radiation if the measured rotation ϕ for a one decimeter path length is 10^{-3} degrees at a wavelength of 600 nm.

<u>SOLUTION</u>

$$n_1 - n_r = \frac{\phi\lambda}{1800} = \frac{(10^{-3} \text{ deg dm}^{-1})(6 \times 10^{-6} \text{ dm})}{1800 \text{ deg}}$$

$$= 0.3 \times 10^{-11}$$

16.19 -77.8 kJ mol^{-1}

16.20   (a) 15,885 cm^{-1} (b) 1.9695 eV

16.21   112.7 nm

16.22   25.2%

16.23 $(7.7 \pm 0.7) \times 10^{-5}$ g cm^{-3}

16.24   2.6 μg cm^{-3}

16.25   (a) 3.24×10^{-3} mol L^{-1} 5.12×10^{-3} mol L^{-1}

         (b) 2.95×10^4 (mol L^{-1})$^{-1}$

16.26 $\varepsilon_I = \varepsilon_{In}/(10^{-pH+pK} + 1)$

16.27 16×10^4 L mol^{-1} cm^{-2} 7×10^{-4}

16.28   0.522 nm

16.29   0.0053 cm^{-1}

16.30   5.20 eV

16.31   0.418

16.32   0.636

CHAPTER 17: Magnetic Resonance Spectroscopy

17.1 Calculate the magnetic flux density to give a precessional frequency for fluorine of 60 MHz.

SOLUTION

$$B = \frac{h\nu}{|g_N|\mu_N}$$

$$= \frac{(6.6262 \times 10^{-34}\ \mathrm{J\ s})(6 \times 10^{7}\ \mathrm{s^{-1}})}{(5.257)(5.0508 \times 10^{-27}\ \mathrm{J\ T^{-1}})} = 1.4973\ \mathrm{T}$$

17.2 Frequencies used in nuclear magnetic resonance spectrometers are of the order of 60 to 300 MHz. Calculate the corresponding energies in kilojoules per mole.

SOLUTION

$$E = h = \frac{(6.626 \times 10^{-34}\ \mathrm{J\ s})(6 \times 10^{7}\ \mathrm{s^{-1}})(6.022 \times 10^{23}\ \mathrm{mol^{-1}})}{(1000\ \mathrm{J\ kJ^{-1}})}$$

$$= 2.394 \times 10^{-5}\ \mathrm{kJ\ mol^{-1}}$$

$$E = (2.394 \times 10^{-5}\ \mathrm{kJ\ mol^{-1}})(300\ \mathrm{MHz})/(60\ \mathrm{MHz})$$

$$= 11.970 \times 10^{-5}\ \mathrm{kJ\ mol^{-1}}$$

17.3 What magnetic flux density is required for proton magnetic resonance at 220 MHz?

SOLUTION

$$\frac{220\ \mathrm{MHz}}{60\ \mathrm{MHz}} \times 1.4902\ \mathrm{T} = 5.1670\ \mathrm{T}$$

17.4 (a) What are the energy levels for a ^{23}Na nucleus in a magnetic field of 2 T? (b) What is the absorption frequency?

SOLUTION

$$E = |g_N|\mu_N m_I B = (1.478)(5.0508 \times 10^{-27}\ \mathrm{J\ T^{-1}})(2\ \mathrm{T})m_I$$

$= (1.493 \times 10^{-26}\ \text{J})m_I$

Since $I = \frac{3}{2}$, $m_I = -\frac{3}{2}, -\frac{1}{2}, +\frac{1}{2}, +\frac{3}{2}$ and

$E = -2.240 \times 10^{-26}$, -0.7465×10^{-26}, 0.7465×10^{-26}, and 2.240×10^{-26}

$$\nu = \frac{\Delta E}{h} = \frac{1.493 \times 10^{-26}\ \text{J}}{6.6262 \times 10^{-34}\ \text{J s}} = 22.53\ \text{MHz}$$

17.5 The magnetogyric ratio γ_N for a nucleus is defined by $\mu_N = \gamma_N \hbar I$ What is the value of γ_N for H?

SOLUTION

$\mu_N = g_N \mu_N I$ Therefore, $\gamma_N = g_N \mu_N / \hbar$

$$= \frac{2\pi(5.585)(5.0508 \times 10^{-27}\ \text{J T}^{-1})}{(6.6262 \times 10^{-34}\ \text{J s})} = 2.675 \times 10^{8}\ \text{s}^{-1}\ \text{T}^{-1}$$

17.6 What is the ratio of the number of ^{31}P spins in the lower state to the number in the upper state in a magnetic field of 1.724 T at room temperature? (Given: $g_N = 2.263$)

SOLUTION

$$\frac{N_\alpha}{N_\beta} = 1 + \frac{g_N \mu_N B}{kT}$$

$$= 1 + \frac{(2.263)(5.0508 \times 10^{-27}\ \text{J T}^{-1})(1.724\ \text{T})}{(1.3807 \times 10^{-23}\ \text{J K}^{-1})(298.15\ \text{K})}$$

$$= 1 + 4.79 \times 10^{-6}$$

17.7 In a magnetic field of 2 T, what fraction of the protons have their spin lined up with the field at room temperature?

<u>SOLUTION</u>

$$\frac{N_1}{N_h} = 1 + \frac{g_N \mu_N B}{kT}$$

$$= 1 + \frac{(5.585)(5.05 \times 10^{-27}\ J\ T^{-1})(2\ T)}{(1.38 \times 10^{-23}\ J\ K^{-1})(298\ K)} = 1.000\ 013\ 72$$

$$\frac{N_1}{N_1 + N_h} = \frac{1}{1 + N_h/N_1} = \frac{1}{1 + 1/(N_h/N_1)}$$

$$= \frac{1}{1 + 1/1.000\ 013\ 72} = 0.500\ 003\ 43$$

17.8 Using information from Tables 17.2 and 17.3, sketch the spectrum you would expect for ethyl acetate ($CH_3CO_2CH_2CH_3$).

<u>SOLUTION</u>

CH_3 – C(=O) – O – CH_2 – CH_3

CH_3 (acetyl): $\delta \approx 2$; CH_2: $\delta \approx 3.6$; CH_3 (ethyl): $\delta \approx 1$

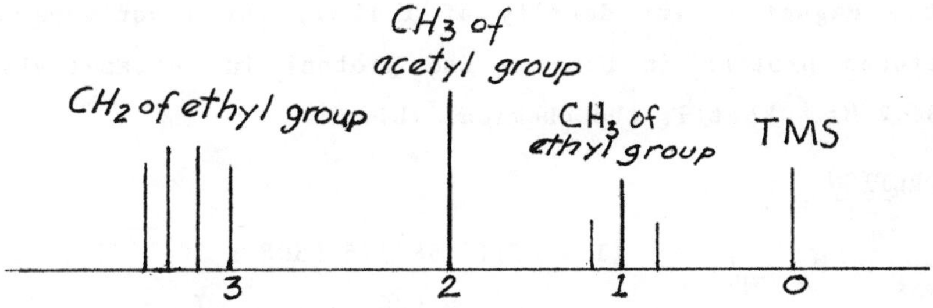

17.9 Chemical shifts δ are generally expressed in ppm, but it is also convenient to express them in Hz because coupling constants are expressed in Hz. What shifts in frequency correspond with a chemical shift of δ = 1 for protons in a spectrometer with (a) B = 1.41 T and (b) 7.00 T?

SOLUTION

(a) $\nu_i = |g_i|\mu_N(1 - \sigma_i)B/h$

$\Delta\nu_i = |g_i|\mu_N\sigma_i B/h$

$= \dfrac{(5.585)(5.051 \times 10^{-27}\ \mathrm{J\ T^{-1}})(10^{-6})(1.41\ \mathrm{T})}{6.626 \times 10^{-34}\ \mathrm{J\ s}} = 60\ \mathrm{Hz}$

(b) $\Delta\nu_i = (60\ \mathrm{Hz})(7.00\ \mathrm{T})/(1.41\ \mathrm{T}) = 298\ \mathrm{Hz}$

17.10 (a) What is the separation of the methyl and methylene proton resonances in ethanol at 60 MHz? (b) At 300 MHz? (See Table 17.2.)

SOLUTION

$\delta_{CH_3} = 1.17 \times 10^{-6}$ $\qquad\qquad \delta_{CH_2} = 3.59 \times 10^{-6}$

(a) $\nu_0\delta = (60 \times 10^6\ \mathrm{Hz})(2.42 \times 10^{-6}) = 14.52\ \mathrm{Hz}$

(b) $\nu_0\delta = (300 \times 10^6\ \mathrm{Hz})(2.42 \times 10^{-6}) = 726\ \mathrm{Hz}$

17.11 At a magnetic flux density of 1.41 T, the frequency separation between protons in benzene and protons in tetramethylsilane is 436.2 Hz. What is the chemical shift?

SOLUTION

$\nu_0 = Bg_N\mu_N/h = \dfrac{(1.41\ \mathrm{T})(5.585)(5.0508 \times 10^{-27}\ \mathrm{J\ T^{-1}})}{6.6262 \times 10^{-34}\ \mathrm{J\ s}}$

$= 60.0\ \mathrm{MHz}$

$\delta = 436.2\ \mathrm{Hz}/60\ \mathrm{MHz} = 7.270 \times 10^{-6}$

which is usually given as 7.270.

17.12 Calculate the factor for converting a chemical shift in terms of magnetic flux density to frequency, for proton magnetic resonance.

SOLUTION

$$\nu = g_N \mu_N B/h$$

$$\frac{g_N \mu_N}{h} = \frac{5.585(5.0508 \times 10^{-27}\ \mathrm{J\ T^{-1}})}{6.6262 \times 10^{-34}\ \mathrm{J\ s}} = 4257.2 \times 10^4\ \mathrm{Hz\ T^{-1}}$$

17.13 Are protons a and b magnetically equivalent in

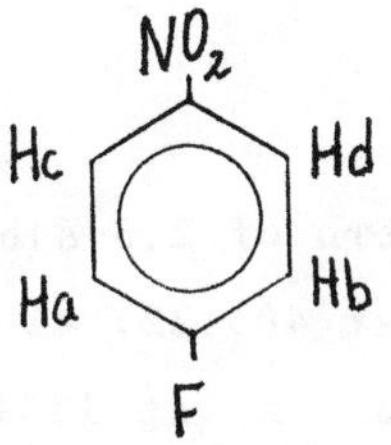

SOLUTION

They are not magnetically equivalent (although they are chemically equivalent) because the coupling $H_a - H_c$ is different from the coupling $H_a - H_d$.

17.14 Sketch the proton resonance spectrum of $D_2CHCOCD_3$ (deuteroacetone containing a little hydrogen). Indicate the relative intensities of the lines.

SOLUTION

The deuteron has a spin of I = 1. Some of the methyl groups will be CD_2H- groups, so that the proton sees two equivalent deuterons. The first deuteron will split the proton resonance into a triplet, and the second deuteron will further split these lines with the same spin-spin coupling constant.

(Please see figure, p. 258)

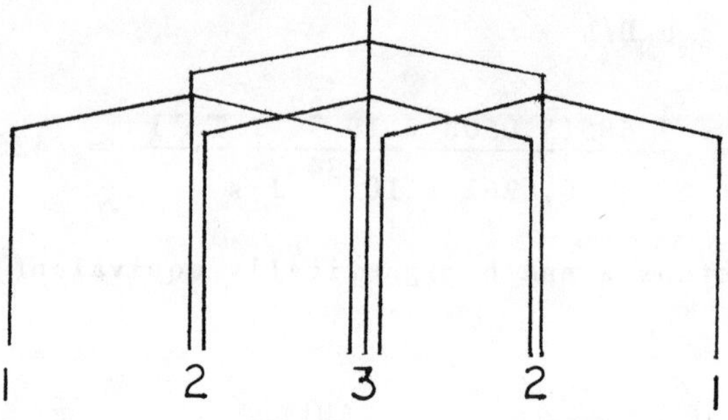

17.15 The proton resonance pattern of 2,3-dibromothiophene shows an AB-type spectrum with lines at 405.22, 410.85, 425.07, and 430.84 Hz measured from tetramethylsilane at 1.41 T [K. F. Kuhlmann and C. L. Braun, J. Chem. Ed., 46, 750 (1969)]. (a) What is the coupling constant J? (b) What is the difference in the chemical shifts of the A and B hydrogens? (c) At what frequencies would the lines be found at 2 T?

SOLUTION

(a) The average spacing of the two doublets is $J = 5.70 \pm 0.07$ Hz.

(b) $$\nu_0\delta = \left[(a - d)(b - c)\right]^{1/2} = \left[(25.62)(14.22)\right]^{1/2}$$

$$= 19.09 \text{ Hz}$$

$$\delta = (19.09 \text{ Hz})/60 \times 10^6 \text{ Hz}) = 0.318 \times 10^{-6} = 0.318 \text{ ppm}$$

(c) At 2 T $$\nu_0' = \frac{2}{1.41} 60 \times 10^6 \text{ Hz} = 85.1 \times 10^6 \text{ Hz}$$

The center of the spectrum shifts by ν_0' / ν_0

$$418.00 \frac{85.1 \times 10^6}{60 \times 10^6} = 592.86$$

$$\text{Distance of b and c from center} = \frac{[(\nu_0\delta)^2 + J^2]^{1/2} - J}{2}$$

$$= \frac{1}{2}\left\{[(85.1 \times 0.318)^2 + 5.70^2]^{1/2} - 5.40\right\} = 10.98$$

$$\begin{aligned} a &= 592.86 - 10.98 - 5.70 = 576.18 \text{ Hz} \\ b &= 592.86 - 10.98 = 581.88 \text{ Hz} \\ c &= 592.86 + 10.98 = 603.84 \text{ Hz} \\ d &= 592.86 + 10.98 + 5.70 = 609.54 \text{ Hz} \end{aligned}$$

17.16 At room temperature the chemical shift of cyclohexane protons is an average of the chemical shifts of the axial and equatorial protons. Explain.

SOLUTION

At room temperature the rate of conversion of cyclohexane from boat to chair forms is so fast that the protons are at the average local magnetic field.

17.17 As shown in Fig. 17.14, the two lines in the proton magnetic resonance spectrum for the two methyl groups connected to nitrogen in N,N-dimethylacetamide coalesce when the temperature is raised. What is the rate constant for the cis-trans isomerization when the multiplet structure is just lost at 331 K? The difference in chemical shifts between the two peaks is 10.85 Hz.

SOLUTION

$$k = \frac{\pi \Delta \nu_0}{\sqrt{2}} = \frac{\pi(10.85 \text{ Hz})}{\sqrt{2}} = 24.1 \text{ s}^{-1}$$

17.18 Calculate the precessional frequency of electrons in a 1.5-T field.

SOLUTION

$$\nu = \frac{B g_e \mu_B}{h} = \frac{(1.5 \text{ T})(2.0023)(9.2742 \times 10^{-24} \text{ J T}^{-1})}{6.6262 \times 10^{-34} \text{ J s}}$$

$$= 4.2037 \times 10^{10} \text{ s}^{-1} = 42{,}037 \text{ MHz}$$

17.19 Line separations in ESR may be expressed in G or MHz. Show how the conversion factor $1\ T = 2.80 \times 10^4$ MHz is obtained.

SOLUTION

The resonance frequency for electrons in a 1 gauss field is given by

$$\nu = \frac{g_e B \mu_B}{h} = \frac{(2.00)(9.274 \times 10^{-24}\ J\ T^{-1})(1\ T)}{6.626 \times 10^{-34}\ J\ s}$$

$$= 2.80 \times 10^{10}\ s^{-1} = 2.80 \times 10^4\ MHz$$

17.20 Sketch the ESR spectrum expected for p-benzosemiquinone radical ion.

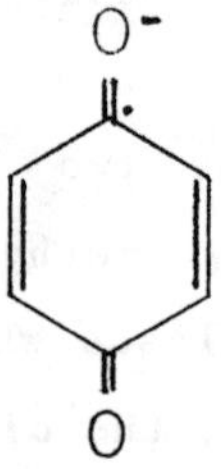

The four hydrogens are magnetically equivalent.

SOLUTION

		1		1					1H
	1		2		1				2H
1		3		3		1			3H
1	4		6		4		1		4H

Thus there will be five equally spaced lines with relative intensities of 1, 4, 6, 4, 1.

17.21 An unpaired electron in the presence of two protons gives the following four-line ESR spectrum: $\Delta B/10^{-4}\ T = 0, 1, 3, 4$. What are the two coupling constants in T and in MHz?

SOLUTION

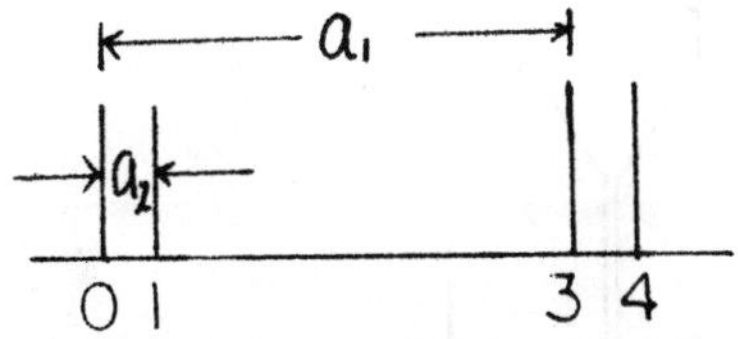

$$a_1 = 3 \times 10^{-4}\ T$$

$$a_2 = 1 \times 10^{-4}\ T$$

Multiplying by the factor in problem 17.19

$$a_1 = (3 \times 10^{-4}\ T)(2.80 \times 10^4\ MHz\ T^{-1}) = 8.40\ MHz$$

$$a_2 = (1 \times 10^{-4}\ T)(2.80 \times 10^4\ MHz\ T^{-1}) = 2.80\ MHz$$

17.22 Sketch the ESR spectrum for an unpaired electron in the presence of three protons for the following cases: (a) the protons are not equivalent, (b) the protons are equivalent, and (c) two protons are equivalent and the third is different.

SOLUTION

(a)

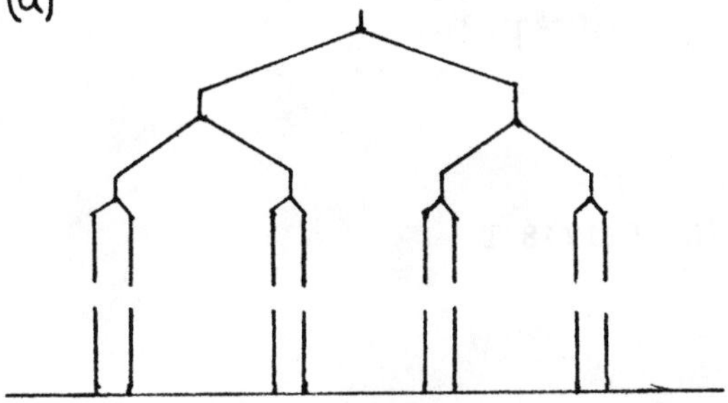

(b)

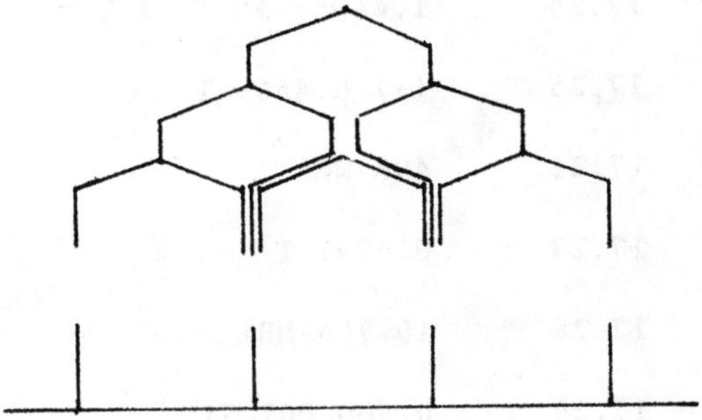

(c)

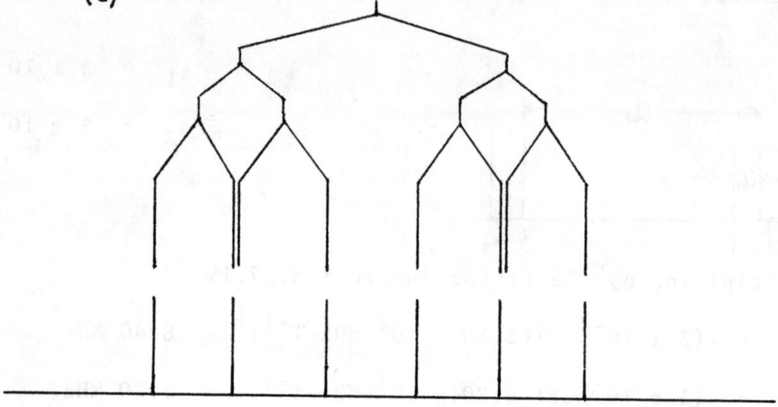

17.23 Sketch the ESR spectrum of CH_3.

<u>SOLUTION</u>

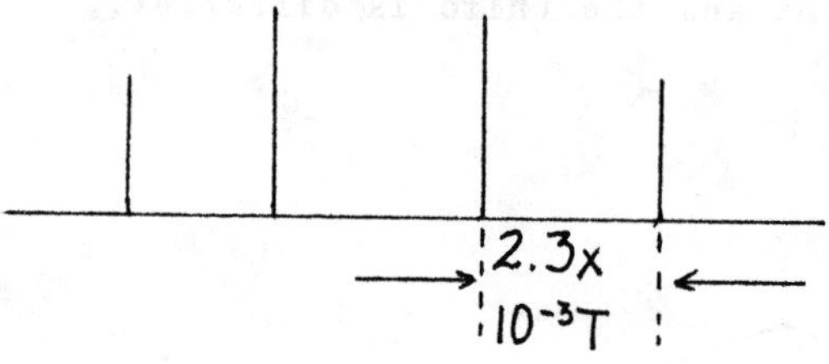

17.24 1.410×10^{-27} J T^{-1}

17.25 (a) 0.4669 T (b) 0.1248 T

17.26 498 MHz

17.27 0.4991 T

17.28 10.706 MHz

17.29 0.500 001 71

17.30

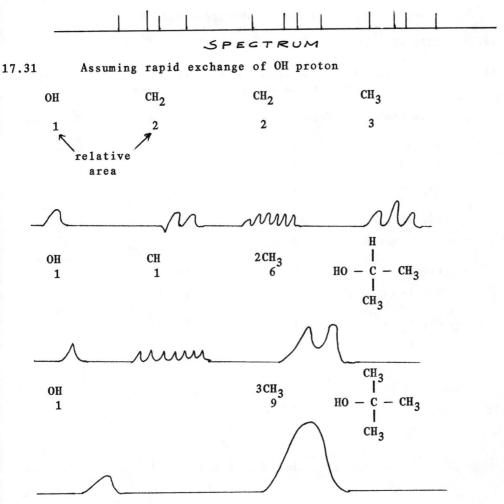

17.31 Assuming rapid exchange of OH proton

OH	CH_2	CH_2	CH_3
1	2	2	3

relative area

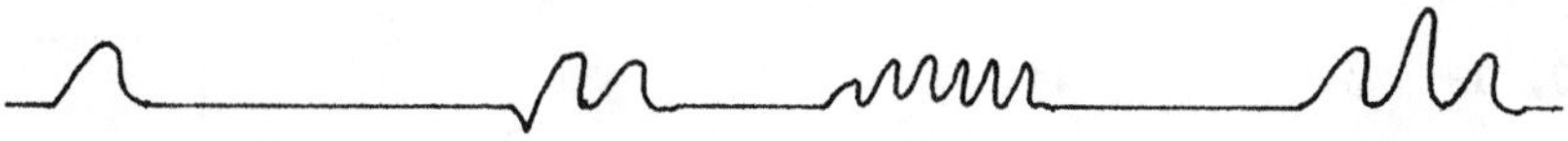

OH	CH	$2CH_3$
1	1	6

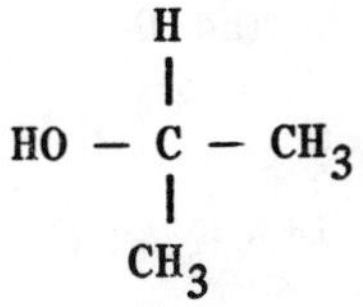

OH	$3CH_3$
1	9

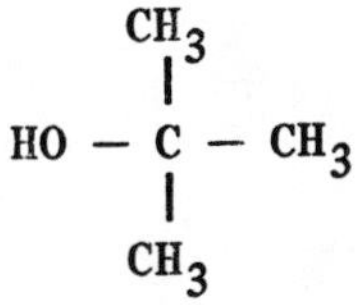

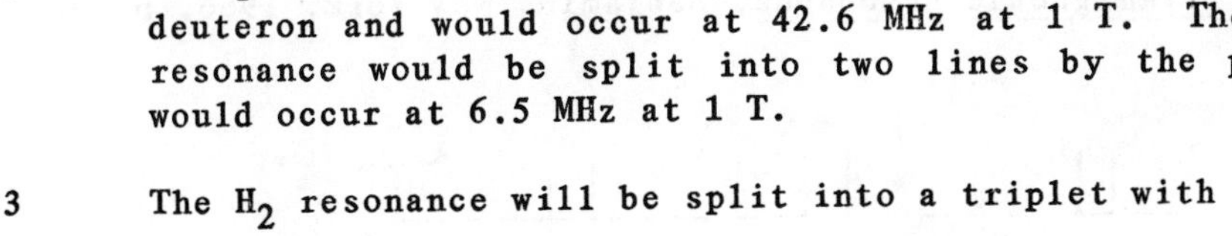

17.32 The proton resonance would be split into three lines by the deuteron and would occur at 42.6 MHz at 1 T. The deuteron resonance would be split into two lines by the proton and would occur at 6.5 MHz at 1 T.

17.33 The H_2 resonance will be split into a triplet with a coupling constant of 0.5 - 4 Hz. The H_5 resonance will be split into a triplet with a coupling constant of 6 - 9 Hz. The $H_4 - H_6$ resonance will produce an AB-type spectrum.

17.34 9230 MHz

17.35 The deuteron has a spin of +1, 0, or −1. Therefore the following combinations of spins are possible:

Spin comb.	Total spin	No. of ways	
all +	+3	1	
2+, 10	+2	3	
2+, 1−	+1	3	6
20, 1+	+1	3	
+, −, and 0	0	3! = 6	7
all 0	0	1	
2−, 1+	−1	3	6
20, 1−	−1	3	
2−, 10	−2	3	
all −	−3	1	

Spectrum

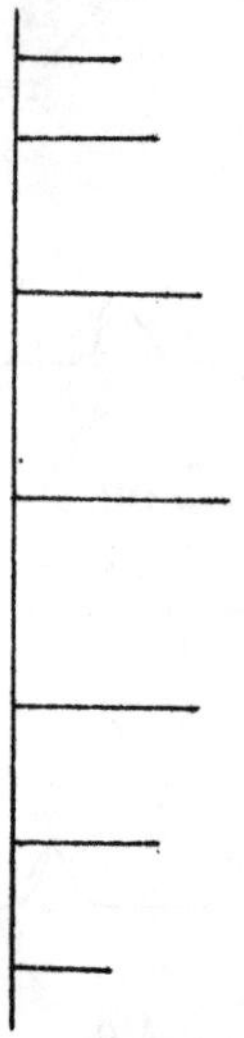

17.36

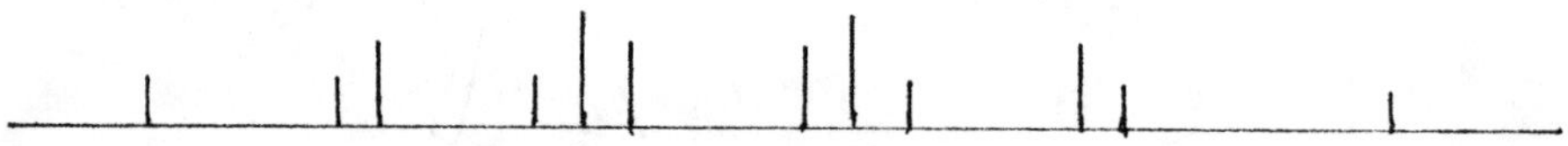

See M. Bersohn and J. C. Baird, An Introduction to Electron Paramagnetic Resonance, Benjamin, New York, 1966, p. 93.

17.37

CHAPTER 18: Statistical Mechanics

18.1 Using the Boltzmann distribution calculate the ratio of populations at 25 °C of energy levels separated by (a) 1000 cm^{-1}, and (b) 10 kJ mol^{-1}.

SOLUTION

(a) $$\exp\left[-\frac{hc\tilde{\nu}}{kT}\right] = \exp\left[-\frac{(6.626\text{x}10^{-34}\text{ J s})(2.998\text{x}10^{8}\text{ m s}^{-1})(10^{5}\text{ m}^{-1})}{(1.380\text{ x }10^{-23}\text{ J K}^{-1})(298.15\text{ K})}\right]$$

$$= 0.0080$$

(b) $$\exp\left[-\frac{10^4\text{ J mol}^{-1}}{(8.314\text{ J K}^{-1}\text{ mol}^{-1})(298\text{ K})}\right] = 0.0177$$

18.2 Calculate the ratio of populations at 25 °C of energy levels separated by (a) 1 eV and (b) 10 eV. (c) Calculate the ratios at 1000 °C.

SOLUTION

(a) $$\frac{N_A}{N_B} = \frac{e^{-E_A/RT}}{e^{-E_B/RT}} = e^{-(E_A - E_B)/RT}$$

$$1\text{ eV} = (1.602\text{ x }10^{-19}\text{ J})(6.022\text{ x }10^{23}\text{ mol}^{-1})$$

$$= 9.647\text{ x }10^{4}\text{ J mol}^{-1}$$

$$N_A/N_B = \exp(-\ 9.647\text{ x }10^{4}/8.314\text{ x }298.15)$$

$$= 1.253\text{ x }10^{-17}$$

(b) $$N_A/N_B = \exp(-\ 9.647\text{ x }10^{5}/8.314\text{ x }298.15)$$

$$= 1.147\text{ x }10^{-169}$$

(c) At 1000 K $\quad N_A/N_B = 1.100\text{ x }10^{-4}$

$N_A/N_B = 2.596\text{ x }10^{-40}$

18.3 Calculate the translational partition function for $H_2(g)$ at 1000 K and 1 bar.

SOLUTION

$$V = \frac{RT}{P} = \frac{(1\ \text{mol})(8.314\ \text{J K}^{-1}\ \text{mol}^{-1})(1000\ \text{K})}{10^5\ \text{N m}^{-2}} = 0.083\ 14\ \text{m}^3$$

$$m = 2(1.0078 \times 10^{-3}\ \text{kg mol}^{-1})/(6.022 \times 10^{23}\ \text{mol}^{-1})$$
$$= 3.347 \times 10^{-27}\ \text{kg}$$

$$q_{tr} = \left(\frac{2\pi mkT}{h^2}\right)^{3/2} V$$

$$= \left[\frac{2\pi(3.347 \times 10^{-27}\ \text{kg})(1.3806 \times 10^{-23}\ \text{J K}^{-1})(1000\ \text{K})}{(6.626 \times 10^{-34}\ \text{J s})^2}\right] \times (0.083\ 14\ \text{m}^3)$$

$$= 1.414 \times 10^{30}$$

18.4 A helium atom is in a volume of 10^{-9} m^3. What are the values of its translational partition function at 298 K, 1000 K, and 5000 K?

SOLUTION

At 298 K

$$q_t = \left(2\pi mkT/h^2\right)^{3/2} V$$

$$= \left[\frac{2\pi(4.0026 \times 10^{-3}\ \text{kg mol}^{-1})(1.38 \times 10^{-23}\ \text{J K}^{-1})(298\ \text{K})}{(6.022 \times 10^{23}\ \text{mol}^{-1})(6.626 \times 10^{-34}\ \text{J s})^2}\right]^{3/2} (10^{-9}\ \text{m}^3)$$

$$= 7.74 \times 10^{21}$$

At 1000 K

$$q_t = 7.74 \times 10^{21}\ (1000\ \text{K}/298\ \text{K})^{3/2} = 4.76 \times 10^{22}$$

At 5000 K

$$q_t = 7.74 \times 10^{21}\ (5000\ \text{K}/298\ \text{K})^{3/2} = 5.32 \times 10^{23}$$

18.5 The thermal de Broglie wavelength defined in connection with equation 18.70 is a little different from the de Broglie wavelength. (a) What is the de Broglie wavelength for hydrogen atoms at 3000 K? (b) How does it compare with the thermal de Broglie wavelength calculated in Example 18.1? (c) How does this thermal de Broglie wavelength compare with the mean distance between hydrogen atoms in a gas of hydrogen atoms at 3000 K and 1 bar?

SOLUTION

(a) $$\varepsilon_t = \frac{3}{2}kT(\text{equ. } 15.73) = \frac{1}{2}mv^2 = \frac{p^2}{2m}$$

$$m = \frac{1.0079 \times 10^{-3} \text{ kg mol}^{-1}}{6.022 \times 10^{23} \text{ mol}^{-1}} = 1.674 \times 10^{-27} \text{ kg}$$

$$p = (3mkT)^{1/2}$$

$$= \left[(3)(1.674 \times 10^{-27} \text{ kg})(1.38 \times 10^{-23} \text{ J s})(3000 \text{ K})\right]^{1/2}$$

$$= 1.442 \times 10^{-23} \text{ kg m s}^{-1}$$

$$\lambda = \frac{h}{p} = \frac{6.626 \times 10^{-34} \text{ J s}}{1.442 \times 10^{-23} \text{ kg m s}^{-1}} = 4.596 \times 10^{-11} \text{ m}$$

(b) This is a little larger than the thermal de Broglie wavelength of 3.175×10^{-11} m.

(c) $V = N_A l^3$

$$0.2494 \text{ m}^3 = (6.022 \times 10^{23} \text{ mol}^{-1})l^3$$

$$l = 7.45 \times 10^{-9} \text{ m}$$

Thus both the de Broglie wavelength and the thermal de Broglie wavelength are small compared with the mean distance between atoms.

18.6 Calculate the entropy of one mole of H-atom gas at 1000 K and (a) 1 bar, and (b) 1000 bar.

SOLUTION

$$S^{\circ} = R \ln \left[\frac{(2\pi mkT)^{3/2} V e^{5/2}}{h^3 N_A} \right] + R \ln 2$$

$$m = \frac{1.0079 \times 10^{-3} \text{ kg mol}^{-1}}{6.022\ 045 \times 10^{23} \text{ mol}^{-1}} = 1.673\ 85 \times 10^{-27} \text{ kg}$$

$$V = \frac{RT}{P} = \frac{(8.314 \text{ J K}^{-1} \text{ mol}^{-1})(1000 \text{ K})}{10^5 \text{ N m}^{-2}}$$

$$= 0.083\ 14 \text{ m}^3 \text{ mol}^{-1} \qquad \text{at 1 bar}$$

$$V = 8.314 \times 10^{-5} \text{ m}^3 \text{ mol}^{-1} \qquad \text{at 1000 bar}$$

(a) At 1000 K and 1 bar $\quad S^{\circ} = 139.871 \text{ J K}^{-1} \text{ mol}^{-1}$

(b) At 1000 K and 1000 bar $\quad S^{\circ} = 82.437 \text{ J K}^{-1} \text{ mol}^{-1}$

18.7 Calculate the entropy of neon at 25 °C and 1 bar.

SOLUTION

$$m = \frac{20.179 \times 10^{-3} \text{ kg mol}^{-1}}{6.022\ 045 \times 10^{23} \text{ mol}^{-1}} = 3.350\ 86 \times 10^{-26} \text{ kg}$$

$$V = \frac{RT}{P} = \frac{(8.314 \text{ J K}^{-1} \text{ mol}^{-1})(298.15 \text{ K})}{10^5 \text{ N m}^{-2}}$$

$$= 2.479 \times 10^{-2} \text{ m}^3 \text{ mol}^{-1}$$

$$S^{\circ} = R \left\{ \frac{5}{2} + \ln \left[\left(\frac{2\pi mkT}{h^2} \right)^{3/2} \frac{V}{N_A} \right] \right\} = 146.328 \text{ J K}^{-1} \text{ mol}^{-1}$$

18.8 Calculate S° and C_P° for argon ($M = 39.948 \text{ g mol}^{-1}$) at 25 °C and 1 bar.

SOLUTION

$$m = \frac{39.948 \times 10^{-3}\ \text{kg mol}^{-1}}{6.022\ 045 \times 10^{23}\ \text{mol}^{-1}} = 6.633\ 63 \times 10^{26}\ \text{kg}$$

$$V = \frac{RT}{P} = \frac{(8.314\ 41\ \text{J K}^{-1}\ \text{mol}^{-1})(298.15\ \text{K})}{10^5\ \text{N m}^{-2}}$$

$$= 2.478\ 94 \times 10^{-2}\ \text{m}^3\ \text{mol}^{-1}$$

$$S^\circ = R\left\{\frac{5}{2} + \ln\left[\left(\frac{2\pi mkT}{h^2}\right)^{3/2}\frac{V}{N_A}\right]\right\}$$

$$= 154.844\ \text{J K}^{-1}\ \text{mol}^{-1}$$

$$C_P^\circ = \frac{5}{2}R = 20.786\ \text{J K}^{-1}\ \text{mol}^{-1}$$

18.9 What are the translational partition functions of hydrogen atoms and hydrogen molecules at 500 K in a volume of $4.157 \times 10^{-2}\ \text{m}^3$? (This is the molar volume of a perfect gas at this temperature and a pressure of 1 bar.)

SOLUTION

$$q_t = \left(\frac{2\pi mkT}{h^2}\right)^{3/2} V$$

For H

$$q_t = \left[\frac{2\pi(1.008 \times 10^{-3}\ \text{kg mol}^{-1})(1.3806 \times 10^{-23}\ \text{J K}^{-1})(500\ \text{K})}{(6.022\ 045 \times 10^{23}\ \text{mol}^{-1})(6.626 \times 10^{-34}\ \text{J s})^2}\right] \times (4.157 \times 10^{-2}\ \text{m}^3)$$

$$= 8.84 \times 10^{28}$$

For H_2, $m_{H_2} = 2m_H$

$$q_{tH_2} = 2^{3/2}\ q_{tH} = 2.50 \times 10^{29}$$

18.10 Calculate the translational partition functions for H, H_2, and H_3 at 1000 K and 1 bar. What are the rotational partition functions of H_2 and H_3 (linear) at 1000 K? The internuclear distances in H_3 are 94 pm.

SOLUTION

$$V = \frac{RT}{P} = \frac{(8.314\ 41\ \text{J K}^{-1}\ \text{mol}^{-1})(1000\ \text{K})}{10^5\ \text{N m}^{-2}}$$

$$= 0.083\ 1441\ \text{m}^3\ \text{mol}^{-1}$$

$$q_t = \frac{(2\pi mkT)^{3/2}}{h^3} V$$

$$= \frac{[2\pi(1.0079\text{x}10^{-3}\ \text{kg}/6.022\text{x}10^{23}\ \text{mol}^{-1})(1.38\text{x}10^{-23}\ \text{J K}^{-1})(10^3\ \text{K})]^{3/2}\ V}{6.62\ \text{x}\ 10^{-34}\ \text{J s})^3}$$

$$= 6.026\ \text{x}\ 10^{30}\ V = 5.00\ \text{x}\ 10^{29}$$

For H_2 $\quad q_t = 6.026 \text{ x } 10^{30}\ V\ 2^{3/2} = 1.42 \text{ x } 10^{30}$

For H_3 $\quad q_t = 6.026 \text{ x } 10^{30}\ V\ 3^{3/2} = 2.60 \text{ x } 10^{30}$

$$q_r = \frac{8\pi^2 IkT}{2h^2}$$

For H_2

$$q_r = \frac{8\pi^2(4.6054\text{x}10^{-48}\ \text{kg m}^2)(1.38066\text{x}10^{-23}\ \text{J K}^{-1})(1000\ \text{K})}{2(6.62618\ \text{x}\ 10^{-34}\ \text{J s})^2} = 5.72$$

For H_3

$$I = \frac{m_1 m_3}{m_1 + m_3} R^2$$

$$= \frac{(1.0079\ \text{x}\ 10^{-3}\ \text{kg mol}^{-1})^2(1.88\ \text{x}\ 10^{-10}\ \text{m})^2}{2(1.0078\ \text{x}\ 10^{-3}\ \text{kg mol}^{-1})(6.02205\ \text{x}\ 10^{23}\ \text{mol}^{-1})}$$

$$= 2.96\ \text{x}\ 10^{-47}\ \text{kg m}^2$$

(Note that m_2 is on the axis of rotation, and does not

contribute to the moment of inertia.)

$$q_t = 5.72 \frac{29.6 \times 10^{-48} \text{ kg m}^2}{4.60 \times 10^{-48} \text{ kg m}^2} = 36.8$$

18.11 What is the rotational contribution to C_P^0 and S^o of CH_4 at 298.15 K?

SOLUTION

$$(C_P^0)_r = \frac{3}{2} R = \frac{3}{2} (8.314\ 41) = 12.472 \text{ J K}^{-1} \text{ mol}^{-1}$$

$$S_r^o = R \ln\left[\frac{\pi^{1/2}}{\sigma}\left(\frac{T^3 e^3}{\theta_a \theta_b \theta_c}\right)^{1/2}\right]$$

$$= 8.314\ 41 \ln\left[\frac{\pi^{1/2}}{12}\left(\frac{298.15^3 \times 2.7183^3}{435.6}\right)^{1/2}\right]$$

$$= 42.366 \text{ J K}^{-1} \text{ mol}^{-1}$$

18.12 What are the symmetry numbers of the following organic molecules, assuming free rotation of methyl groups? (a) ethane, (b) propane, (c) 2-methylpropane, (d) 2,2-dimethylpropane.

SOLUTION

(a) CH_3CH_3 $\sigma = 3^2 \times 2 = 18$ (b) $CH_3CH_2CH_3$ $\sigma = 3^2 = 9$

(c) $CH_3-CH(CH_3)-CH_3$ (central C bearing H and CH_3) $\sigma = 3^3 \times 3 = 81$

(d) $CH_3-C(CH_3)_2-CH_3$ (central C bearing two CH_3) $\sigma = 12 \times 3^4 = 972$

18.13 Derive the expression for the vibrational contribution to the internal energy $U = \dfrac{RTx}{e^x - 1}$ where $x = h\nu/kT$.

SOLUTION

$$q_v = \frac{1}{1 - e^{-h\nu/kT}}$$

$$U = RT^2 \frac{\partial \ln q}{\partial T} \qquad \ln q_v = -\ln(1 - e^{-h\nu/kT})$$

$$\frac{\partial \ln q_v}{\partial T} = \frac{e^{-h\nu/kT}}{1 - e^{-h\nu/kT}}\left(\frac{h\nu}{kT^2}\right) = \frac{x}{T(e^x - 1)} \qquad \text{where } x = h\nu/kT.$$

$$U = \frac{RTx}{e^x - 1}$$

18.14 Using $\left(\dfrac{\partial G}{\partial T}\right)_P = -S$ and the contribution of vibration to the Gibbs energy for a diatomic molecule in a perfect gas $G = RT \ln(1 - e^{-x})$ derive the expression for the corresponding contribution to the entropy.

SOLUTION

$$x = \frac{h\nu}{kT} \qquad \frac{dx}{dT} = -\frac{h\nu}{kT^2}$$

$$\frac{\partial G}{\partial T} = R\ln(1 - e^{-x}) + \frac{RT}{(1 - e^{-x})}\frac{d}{dT}(1 - e^{-x})$$

$$= R\ln(1 - e^{-x}) + \frac{RT}{(1 - e^{-x})}\left[-e^{-x}\frac{d(-x)}{dT}\right]$$

$$= R\ln(1 - e^{-x}) - \frac{Rxe^{-x}}{(1 - e^{-x})}$$

$$S = -R\ln(1 - e^{-x}) + \frac{Rx}{e^x - 1} = R\left[\frac{x}{e^x - 1} - \ln(1 - e^{-x})\right]$$

18.15 By use of series expansions show the vibrational contribution to C_V^o for a diatomic molecule approaches R as T ⟶ ∞.

SOLUTION

$$\left(C_V^o\right)_v = R\left(\frac{\theta_v}{T}\right)^2 \frac{e^{\theta_v/T}}{(e^{\theta_v/T} - 1)^2}$$

$$e^x = 1 + x + \frac{x^2}{2!} + \ldots \qquad e^{\theta_v/T} = 1 + \frac{\theta_v}{T} + \ldots$$

$$\left(C_V^o\right)_v = R\left(\frac{\theta_v}{T}\right) \frac{\left(1 + \frac{\theta_v}{T} + \ldots\right)}{\left(\frac{\theta_v}{T}\right)}$$

As T ⟶ ∞ $\left(C_V^o\right)_v$ ⟶ R

18.16 Calculate the entropy of nitrogen gas at 25 °C and 1 bar pressure. The equilibrium separation of atoms is 109.5 pm and the vibrational wave number is 2330.7 cm^{-1}.

SOLUTION

$$m = \frac{2(14.0067 \times 10^{-3} \text{ kg mol}^{-1})}{6.022\ 045 \times 10^{23} \text{ mol}^{-1}} = 4.651\ 81 \times 10^{-26} \text{ kg}$$

$$\frac{kT}{P^0} = \frac{(1.380\ 662 \times 10^{-23} \text{ J K}^{-1})(298.15 \text{ K})}{10^5 \text{ N m}^{-2}} = 4.116\ 44 \times 10^{-26} \text{ m}^3$$

$$S_t^o = R\left\{\frac{5}{2} + \ln\left[\left(\frac{2\pi mkT}{h^2}\right)^{3/2} \frac{kT}{P^0}\right]\right\}$$

$$\ln\left[\left(\frac{2\pi(4.65181\text{x}10^{-26} \text{ kg})(1.380662\text{x}10^{-23} \text{ J K}^{-1})(298.15 \text{ K})}{(6.626\ 176 \times 10^{-34} \text{ J s})^2}\right)^{3/2} \times (4.116\ 44 \times 10^{-26} \text{ m}^3)\right] = 15.591\ 36$$

$$S_t^\circ = (8.314\ 41\ \text{J K}^{-1}\ \text{mol}^{-1})(\tfrac{5}{2} + 15.591\ 36)$$

$$= 150.419\ \text{J K}^{-1}\ \text{mol}^{-1}$$

$$\theta_r = \frac{h^2}{8\pi^2 IK}$$

$$\mu = \frac{m_N^2}{2m_N} = \frac{m_N}{2} = \frac{14.0067 \times 10^{-3}\ \text{kg mol}^{-1}}{2(6.022\ 045 \times 10^{23}\ \text{mol}^{-1})}$$

$$= 1.162\ 95 \times 10^{-26}\ \text{kg}$$

$$I = \mu R^2 = (1.162\ 95 \times 10^{-26}\ \text{kg})(1.095 \times 10^{-10}\text{m})^2$$

$$= 1.394\ 41 \times 10^{-46}\ \text{kg m}^2$$

$$\theta_r = \frac{(6.626\ 176 \times 10^{-34}\ \text{J s})^2}{8\pi^2(1.394\ 41 \times 10^{-46}\ \text{kg m}^2)(1.380\ 662 \times 10^{-23}\ \text{J K}^{-1})}$$

$$= 2.888\ 41\ \text{K}$$

$$S_r^\circ = R\ \ln\left(\frac{eT}{\sigma\theta_r}\right)$$

$$= (8.314\ 41\ \text{J K}^{-1}\ \text{mol}^{-1})\ \ln\left[\frac{(2.718\ 28)(298.15\ \text{K})}{(2)(2.888\ 41\ \text{K})}\right]$$

$$= 41.104\ \text{J K}^{-1}\ \text{mol}^{-1}$$

$$x = \frac{h\tilde{\nu}}{kT} = \frac{hc\tilde{\nu}}{kT}$$

$$= \frac{(6.626176\text{x}10^{-34}\ \text{J s})(2.997925\text{x}10^{8}\ \text{m s}^{-1})(2.3307\text{x}10^{5}\ \text{m}^{-1})}{(1.380662 \times 10^{-23}\ \text{J K}^{-1})(298.15\ \text{K})}$$

$$= 11.247$$

$$S_v^\circ = R\left[\frac{x}{e^x - 1} - \ln(1 - e^{-x})\right]$$

$$= (8.314\ 41\ \text{J K}^{-1}\ \text{mol}^{-1})\left[\frac{11.247}{e^{11.247} - 1} - \ln(1 - e^{-11.247})\right]$$

$$= 1.21 \times 10^{-3}\ \text{J K}^{-1}\ \text{mol}^{-1}$$

$$S^\circ = S_t^\circ + S_r^\circ + S_v^\circ = 150.419 + 41.104 + 0 = 191.524\ \text{J K}^{-1}\ \text{mol}^{-1}$$

18.17 Calculate C_P for CO_2 at 1000 K. Compare the actual contributions to C_P from the various normal modes with the classical expectations.

SOLUTION

$$C_P^\circ = \frac{5}{2}R + R + \sum_{i=1}^{4} \frac{Rx_i^2 e^{x_i}}{(e^{x_i} - 1)^2}$$

$$x_i = \frac{hc\tilde{\nu}_i}{kT} = (1.438 \times 10^{-5}\ \text{m})\tilde{\nu}$$

$\tilde{\nu}_i/\text{m}^{-1}$	1.3512×10^5	6.722×10^4	6.722×10^4	2.3964×10^5
x_i	1.943	0.967	0.967	3.446

$$C_P^\circ = \frac{7}{2}(8.314\ 41) + 6.127 + 2(7.695) + 3.357$$

$$= 53.974\ \text{J K}^{-1}\ \text{mol}^{-1}$$

$$\text{Classically } C_P^\circ = C_{Pt}^\circ + C_{Pr}^\circ + C_{Pv}^\circ = \frac{5}{2}R + R + 4R$$

$$= \frac{15}{2}R = 62.355\ \text{J K}^{-1}\ \text{mol}^{-1}$$

18.18 Calculate the ratio of the number of HBr molecules in state v = 2, J = 5 to the number in state v = 1, J = 2 at 1000 K. Assume that all of the molecules are in their electronic ground states. (θ_v = 3700 K, θ_r = 12.1 K)

SOLUTION

$$\frac{N(v=2,J=5)}{N(v=1,J=2)} = \frac{g_{2vib}e^{\frac{-\varepsilon_{2vib}}{kT}} \quad g_{5rot}e^{\frac{-\varepsilon_{5rot}}{kT}}}{g_{1vib}e^{\frac{-\varepsilon_{1vib}}{kT}} \quad g_{2rot}e^{\frac{-\varepsilon_{2rot}}{kT}}}$$

$$g_{vib} = 1 \text{ for all } v$$

$$g_{rot} = 2J + 1 = 5 \text{ for } g_{2rot}$$
$$= 11 \text{ for } g_{5rot}$$

$$\varepsilon_{vib} = vh\tilde{\nu}$$

Given $\theta_v = h\tilde{\nu}/k = 3700$ K for HBr

$$\varepsilon_{vib} = vk\theta_v \qquad \frac{\varepsilon_{vib}}{kT} = \frac{v\theta_v}{T} \qquad \varepsilon_{rot} = \frac{J(J+1)h^2}{8\pi^2 I}$$

Given $\theta_r = \dfrac{h}{8\pi^2 Ik} = 12.1$ K for HBr

$$\varepsilon_{rot} = J(J+1)\theta_r k \qquad \frac{\varepsilon_{rot}}{kT} = \frac{J(J+1)\theta_r}{T}$$

$$\frac{N(v=2,J=5)}{N(v=1,J=2)} = \frac{e^{-(2)(3.7)} \times 11 \times e^{-(5)(5+1)(12.1)/10^3}}{e^{-3.7} \times 5 \times e^{-2(2+1)(12.1)/10^3}} = 0.0407$$

18.19 The ground state of Cl(g) is twofold degenerate. The first excited state is 875.4 cm^{-1} higher in energy and is twofold degenerate. What is the value of the electronic partition function at 25 °C? At 1000 K?

<u>SOLUTION</u>

$$q = g_0 e^{-0} + g_1 e^{-\varepsilon_1/kT} = 2 + 2e^{-hc\tilde{\nu}/kT}$$

$$= 2 + 2\exp\left[\frac{-(6.626 \times 10^{-34}\ \text{J s})(2.9979 \times 10^8\ \text{m s}^{-1})(8.754 \times 10^4\ \text{m}^{-1})}{(1.3806 \times 10^{-23}\ \text{J K}^{-1})(298.15\ \text{K})}\right]$$

$$= 2 + 2e^{-4.224} = 2.029$$

At 1000 K $\qquad q = 2 + 2e^{-1.259} = 2.568$

18.20 What is the partition function for oxygen atoms at 1000 K according to the date in Table 18.1? What are the relative populations of these levels at equilibrium?

SOLUTION

$$q_e = 5 + 3e^{-228.1/1000} + e^{-325.9/1000} = 8.11$$

$$P_1 = 5/q_e = 0.617$$

$$P_2 = 3e^{-228.1/1000}\Big/ q_e = 0.294$$

$$P_3 = e^{-325.9/1000}\Big/ q_e = 0.089$$

18.21 Derive the expression for the electronic internal energy of an atom or molecule. What is the electronic energy per mole for a chlorine atom at 298 K and 1000 K? (See Problem 18.19)

SOLUTION

$$U = N \frac{\sum_i \varepsilon_i e^{-\varepsilon_i/kT}}{\sum_i e^{-\varepsilon_i/kT}}$$

$$= N \frac{2\varepsilon_1 e^{-\varepsilon_1/kT}}{2 + 2e^{-\varepsilon_1/kT}}$$

where ε_0 is taken as zero and $\varepsilon_1 = 875.4 \text{ cm}^{-1}$

$$\varepsilon_1 = (875.4 \text{ cm}^{-1})(1.439 \text{ K cm}^{-1}) = 1{,}260 \text{ K}$$

$$N\varepsilon_1 = Nhc_1 = (6.022 \times 10^{23} \text{ mol}^{-1})(6.626 \times 10^{-34} \text{J s})(2.998 \times 10^{8} \text{ m s}^{-1})(875.4 \text{ cm}^{-1})(100 \text{ cm m}^{-1})$$

$$= 10.47 \text{ kJ mol}^{-1}$$

$$U_{298} = \frac{(10.47 \text{ kJ mol}^{-1})e^{-1260/298}}{1 + e^{-1260/298}} = 0.150 \text{ kJ mol}^{-1}$$

$$U_{1000} = \frac{(10.47 \text{ kJ mol}^{-1})e^{-1260/1000}}{1 + e^{-1260/1000}} = 2.31 \text{ kJ mol}^{-1}$$

18.22 Calculate the temperature at which 10% of the molecules in a system will be in the first excited electronic state if this state is 400 kJ mol^{-1} above the ground state.

SOLUTION

$$\frac{N_i}{N} = \frac{e^{-E_i/RT}}{\Sigma e^{-E_i/RT}} \qquad \frac{N_1}{N} = \frac{e^{-E_1/RT}}{e^{0/RT} + e^{-E_1/RT}}$$

$$0.1 = \frac{1}{e^{E_1/RT} + 1}$$

$$e^{E_1/RT} = \frac{0.9}{0.1} = 9 \qquad \frac{E_1}{RT} = \ln 9$$

$$T = \frac{E_1}{R \ln 9} = \frac{(400{,}000 \text{ J mol}^{-1})}{(8.314 \text{ J K}^{-1} \text{ mol}^{-1}) \ln 9} = 22{,}000 \text{ K}$$

18.23 Calculate the fraction of hydrogen atoms that at equilibrium at 1000 °C would have n = 2.

SOLUTION

Since the fraction will be very small, it is given by the ratio of the number with N = 2 to the number N = 1.

$$\text{Fraction} = \frac{e^{-E_1/kT}}{e^{-E_1/kT}} = e^{-(E_2 - E_1)/kT}$$

From Example 12.1 $\quad E = \dfrac{-2.179\ 907 \times 10^{-18} \text{ J}}{n^2}$

$$\text{Fraction} = \exp\left[-\frac{(-2.179\ 907 \times 10^{-18} \text{ J})(-0.75)}{(1.380\ 662 \times 10^{-23} \text{ J K}^{-1})(1273 \text{ K})}\right]$$

$$= 4 \times 10^{-41}$$

18.24 Calculate the values of D_0 in Table 18.1 from data in Table A.2 for $H_2(g)$, $O_2(g)$, $Cl_2(g)$, $HCl(g)$, and $CO(g)$.

SOLUTION

$$H_2(g) = 2H(g)$$

$$\Delta H^\circ_0 = 2(216.037) = 432.074 \text{ kJ mol}^{-1} = 4.4781 \text{ eV}$$

$$O_2(g) = 2O(g)$$

$$\Delta H^\circ_0 = 2(246.785) = 493.570 \text{ kJ mol}^{-1} = 5.1155 \text{ eV}$$

$$Cl_2(g) = 2Cl(g)$$

$$\Delta H^\circ_0 = 2(119.608) = 239.216 \text{ kJ mol}^{-1} = 2.479 \text{ eV}$$

$$HCl(g) = H(g) + Cl(g)$$

$$\Delta H^\circ_0 = 216.037 + 119.608 + 92.127 = 427.772 \text{ kJ mol}^{-1}$$

$$= 4.4336 \text{ eV}$$

$$CO(g) = C(g) + O(g)$$

$$\Delta H^\circ_0 = 709.506 + 246.785 + 113.805 = 1070.096 \text{ kJ mol}^{-1}$$

$$= 11.0909 \text{ eV}$$

18.25 Calculate the statistical mechanical values of C°_P, S, H°, and G° for H(g) at 3000 K.

SOLUTION

$$\left(C^\circ_P\right)_t = \frac{5}{2}(8.314 \text{ J K}^{-1} \text{ mol}^{-1}) = 20.786 \text{ J K}^{-1} \text{ mol}^{-1}$$

$$H^\circ_t = \frac{5}{2}(8.314\ 41 \text{ J K}^{-1} \text{ mol}^{-1})(3000 \text{ K}) = 62.358 \text{ kJ mol}^{-1}$$

$$\left(\frac{2\pi mkT}{h^2}\right)^{3/2} \frac{kT}{P^\circ} =$$

$$\left[\frac{2\pi(1.0080\text{x}10^{-3}\ \text{kg mol}^{-1})(1.380662\text{x}10^{-23}\ \text{J K}^{-1})(3000\ \text{K})}{(6.626176\ \text{x}\ 10^{-34}\ \text{J s})^2(6.022045\ \text{x}\ 10^{23}\ \text{mol}^{-1})}\right]^{3/2}$$

$$\text{x}\quad \frac{(1.380662\ \text{x}\ 10^{-23}\ \text{J K}^{-1})(3000\ \text{K})}{10^5\ \text{N m}^{-2}} = 1.294\ 43\ \text{x}\ 10^7$$

$$S^\circ_t = (8.314\ 41\ \text{J K}^{-1}\ \text{mol}^{-1})(2.5 + \ln 1.294\ 43\ \text{x}\ 10^7)$$
$$= 156.944\ \text{J K}^{-1}\ \text{mol}^{-1}$$

$$G^\circ_t = -(8.314\ 41\ \text{J K}^{-1}\ \text{mol}^{-1})(3000\ \text{K})(\ln 1.294\ 43\ \text{x}\ 10^7)$$
$$= -408.474\ \text{kJ mol}^{-1}$$

For the purposes of this calculation $\left(C^\circ_P\right)_e = 0$ and $H^\circ_e = 0.$

$$S^\circ = S^\circ_t + S^\circ_e = (156.944 + 8.314 \ln 2)\ \text{J K}^{-1}\ \text{mol}^{-1}$$
$$= 162.707\ \text{J K}^{-1}\ \text{mol}^{-1}$$

$$G^\circ = G^\circ_t + G^\circ_e = [-408.474 + (8.314\ \text{x}\ 10^{-3})(3000) \ln 2]\text{kJ mol}^{-1}$$
$$= -425.763\ \text{kJ mol}^{-1}$$

18.26 Calculate the statistical mechanical values of C°_P, S°, H°, and G° for $H_2(g)$ at 3000 K.

SOLUTION

$$\left(C^\circ_P\right)_t = \frac{5}{2} R = 20.786\ \text{J K}^{-1}\ \text{mol}^{-1}$$

$$H^\circ_t = \frac{5}{2} RT = 62.358\ \text{kJ mol}^{-1}$$

$$\left(\frac{2\pi mkT}{h^2}\right)^{3/2} \frac{kT}{P^\circ} = 3.661\ 20\ \text{x}\ 10^7$$

$$S^\circ_t = R(2.5 + \ln 3.661\ 20\ \text{x}\ 10^7) = 165.589\ \text{J K}^{-1}\ \text{mol}^{-1}$$

$$G^\circ_t = -(8.314\ 41\ \text{J K}^{-1}\ \text{mol}^{-1})(3000\ \text{K}) \ln 3.661\ 20\ \text{x}\ 10^7$$
$$= -434.409\ \text{kJ mol}^{-1}$$

$$\left(C_P^0\right)_r = R = 8.314 \text{ J K}^{-1} \text{ mol}^{-1}$$

$$H_r^o = RT = 24.943 \text{ kJ mol}^{-1}$$

$$S_r^o = R \ln\left(\frac{eT}{\sigma\theta_r}\right)$$

$$= (8.314\ 41) \ln\left[\frac{(2.7183)(3000 \text{ K})}{(2)(87.547 \text{ K})}\right] = 31.936 \text{ J K}^{-1} \text{ mol}^{-1}$$

$$G_r^o = -RT \ln\left(\frac{T}{\sigma\theta_r}\right) = -70.865 \text{ kJ mol}^{-1}$$

$$\theta_v = \frac{hc\omega}{k} =$$

$$\frac{(6.626176\text{x}10^{-34} \text{ J s})(2.997925\text{x}10^{8} \text{ m s}^{-1})(4405.3 \text{ cm}^{-1})(10^{2} \text{ cm m}^{-1})}{1.380\ 662 \text{ x } 10^{-23} \text{ J K}^{-1}}$$

$$= 6338.3 \text{ K}$$

$$H_v^o = RT \frac{(\theta_v/T)}{e^{\theta_v/R} - 1} = 7.248 \text{ kJ mol}^{-1}$$

$$\left(C_P^0\right)_v = R\left(\frac{\theta_v}{T}\right)^2 \frac{e^{\theta_v/T}}{(e^{\theta_v/T} - 1)^2} = 5.806 \text{ J K}^{-1} \text{ mol}^{-1}$$

$$S_v^o = R\left[\frac{(\theta_v/T)}{e^{\theta_v/T} - 1} - \ln\left(1 - e^{-\theta_v/T}\right)\right] = 3.487 \text{ J K}^{-1} \text{ mol}^{-1}$$

$$G_v^o = RT \ln\left(1 - e^{-\theta_v/T}\right) = -3.214 \text{ kJ mol}^{-1}$$

Since $g_o = 1$, $S_e^o = 0$ $\qquad H_e^o = G_e^o = -432.073 \text{ kJ mol}^{-1}$

$$C_P^0 = \left(C_P^0\right)_t + \left(C_P^0\right)_r + \left(C_P^0\right)_v + \left(C_P^0\right)_e = 34.907 \text{ J K}^{-1} \text{ mol}^{-1}$$

$$H^o = H_t^o + H_r^o + H_v^o + H_e^o = -337.524 \text{ kJ mol}^{-1}$$

$$S^\circ = S_t^\circ + S_r^\circ + S_v^\circ + S_e^\circ = 201.013 \text{ J K}^{-1} \text{ mol}^{-1}$$

$$G^\circ = G_t^\circ + G_r^\circ + G_v^\circ + G_e^\circ = -940.561 \text{ kJ mol}^{-1}$$

18.27 What are the values of ΔC_P°, ΔH°, ΔS°, and ΔG° for $H_2(g) = 2H(g)$ at 3000 K calculated in the preceding two problems? What is the value of K_P? What is the degree of dissociation at 1 bar?

SOLUTION

$$\Delta C_P^\circ = 2C_P^\circ(H) - C_P^\circ(H_2) = [2(20.786) - 34.907] \text{ J K}^{-1} \text{ mol}^{-1}$$
$$= 6.665 \text{ J K}^{-1} \text{ mol}^{-1}$$

$$\Delta H^\circ = 2H^\circ(H) - H^\circ(H_2) = [2(62.358) - (-337.524)] \text{ kJ mol}^{-1}$$
$$= 462.240 \text{ kJ mol}^{-1}$$

$$\Delta S^\circ = 2S^\circ(H) - S^\circ(H_2) = [2(162.707) - 201.013)] \text{ kJ mol}^{-1}$$
$$= 124.401 \text{ J K}^{-1} \text{ mol}^{-1}$$

$$\Delta G^\circ = 2G^\circ(H) - G^\circ(H_2) = [2(-425.763) - (-940.561)] \text{ kJ mol}^{-1}$$
$$= 89.035 \text{ kJ mol}^{-1}$$

$$K_p = e^{-\Delta G^\circ/RT} = e^{-89.035/(8.314)(3000)} = 2.82 \times 10^{-2}$$

$$K_P = \frac{4\alpha^2(P/P^\circ)}{1 - \alpha^2} \qquad \alpha = \left(\frac{K_P}{4 + K_P}\right)^{1/2} = 0.084$$

18.28 Calculate the entropy for chlorine gas at 25 °C and 1 bar pressure.

SOLUTION

$$S^\circ = R\left\{\frac{5}{2} + \ln\left[\left(\frac{2\pi mkT}{h^2}\right)^{3/2} \frac{kT}{P^\circ}\right]\right\} + R\ln\left(\frac{eT}{\sigma\theta_r}\right)$$
$$+ R\left[\frac{x}{e^x - 1} - \ln(1 - e^{-x})\right] \qquad \text{where } x = \theta_v/T$$

Table 18.1 gives $\theta_r = 0.3456$ K, $\theta_v = 807.3$ K, and $\sigma = 2$.

$$S^\circ = 161.889 + 58.752 + 2.218 = 222.859 \text{ J K}^{-1} \text{ mol}^{-1}$$

18.29 Calculate the equilibrium constant for the isotope exchange reaction $D + H_2 = H + DH$ at 25 °C. Assume that the equilibrium distance and force constants of H_2 and DH are the same.

SOLUTION

Since the force constants are the same

$$\nu_{HD} = \nu_{H_2} (\mu_{H_2}/\mu_{HD})^{1/2} = \nu_{H_2} [(m_H + m_D)/2m_D]^{1/2}$$

The change in energy for the reaction is due to the difference in zero point energies for H_2 and HD.

$$\Delta\varepsilon_o = \frac{h}{2}(\nu_{HD} - \nu_{H_2}) = \frac{h}{2}\nu_{H_2}\left[\left(\frac{m_H + m_D}{2m_D}\right)^{1/2} - 1\right]$$

$$= \frac{h}{2}(1.319 \times 10^{14} \text{ s}^{-1})\left[\left(\frac{3}{4}\right)^{1/2} - 1\right] = -5.85 \times 10^{-21} \text{ J}$$

$$K = \frac{\left(f_{HD}/P^\circ\right)\left(f_H/P^\circ\right)}{\left(f_{H_2}/P^\circ\right)\left(f_H/P^\circ\right)} = \frac{q_H q_{HD}}{q_D q_{H_2}} e^{-\Delta\varepsilon_o/kT}$$

$$\frac{q_{tH} q_{tHD}}{q_{tD} q_{tH_2}} = \left(\frac{m_H m_{HD}}{m_D m_{H_2}}\right)^{3/2} = \left(\frac{3}{4}\right)^{3/2} = 0.650$$

$$\frac{q_{rHD}}{q_{rH_2}} = \frac{2\left(\frac{1}{m_H} + \frac{1}{m_H}\right)}{\left(\frac{1}{m_H} + \frac{1}{m_D}\right)} = \frac{2(2/m_H)}{(3/2m_H)} = \frac{4x2}{3} = 2.66$$

$$K = (0.650)(2.66)\, e^{(5.85 \times 10^{-21} \text{ J})/(1.38 \times 10^{-23} \text{ J K}^{-1})(298 \text{ K})}$$

$$= 7.18$$

18.30 Express the equilibrium constant for the reaction $H_2 + I_2 = 2HI$ in terms of molecular properties.

SOLUTION

$$K = \frac{\left(q_{HI}/V\right)^2}{\left(q_{H_2}/V\right)\left(q_{I_2}/V\right)}$$

$$= \left(\frac{m_{HI}^{\ 2}}{m_{H_2} m_{I_2}}\right)^{3/2} \left(\frac{4\theta_{rH_2}\theta_{rI_2}}{\theta_{rHI}^{\ 2}}\right) \frac{\left(1 - e^{-\theta_{vH_2}/kT}\right)\left(1 - e^{-\theta_{vI_2}/kT}\right)}{\left(1 - e^{-\theta_{vHI}/kT}\right)}$$

$$\times \exp\left[\frac{-\left(2D_{oHI} - D_{oH_2} - D_{oI_2}\right)}{RT}\right]$$

18.31 Calculate the equilibrium constant at 25 °C for the reaction $H_2 + D_2 = 2HD$. It may be assumed that the equilibrium distance and force constant k are the same for all three molecular species, so that the additional vibrational frequencies required may be calculated from $2\pi\nu = (k/\mu)^{1/2}$. Because of the zero point vibration, $\Delta\varepsilon_o$ for this reaction is given by

$$\Delta\varepsilon_o = \frac{1}{2} N_A h\left(2\nu_{HD} - \nu_{H_2} - \nu_{D_2}\right)$$

SOLUTION

Since the force constants are the same

$$\nu_{HD} = \nu_{H_2}\left(\mu_{H_2}/\mu_{HD}\right)^{1/2} = \nu_{H_2}\left[(m_H + m_D)/2m_D\right]^{1/2}$$

$$\nu_{D_2} = \nu_{H_2}\,(m_H/m_D)^{1/2}$$

$$\Delta\varepsilon_o = \frac{1}{2} h\nu_{H_2}\left[\sqrt{2}\left(\frac{m_H + m_D}{m_D}\right)^{1/2} - \left(\frac{m_H}{m_D}\right)^{1/2} - 1\right]$$

$$= \frac{h\nu_{H_2}}{2}\left[\sqrt{2}\left(\frac{3}{2}\right)^{1/2} - \left(\frac{1}{2}\right)^{1/2} - 1\right]$$

$$= \frac{(6.626 \times 10^{-34}\ \text{J s})(1.319 \times 10^{14}\ \text{s}^{-1})(0.0249)}{2}$$

$$= 1.09 \times 10^{-21}\ \text{J}$$

$$K = \frac{\left(f_{HD}/P^\circ\right)^2}{\left(f_{H_2}/P^\circ\right)\left(f_{D_2}/P^\circ\right)} = \frac{q^2_{HD}}{q_{H_2}q_{D_2}}\, e^{-\Delta U_o/RT}$$

$$= \frac{q_{t\,HD}^2}{q_{tH_2}q_{tD_2}}\ \frac{q_{r\,HD}^2}{q_{rH_2}q_{rD_2}}\ e^{-\Delta U_o/RT}$$

The vibrational partition functions are essentially equal to unity.

$$\frac{q_{t\,HD}^2}{q_{tH_2}q_{tD_2}} = \frac{m^3_{HD}}{m_{H_2}^{3/2}\ m_{D_2}^{3/2}} = \frac{3^3}{2^{3/2}\ 4^{3/2}} = 1.19$$

The following is obtained by use of equation 18.91 with the symmetry number in the denominator.

$$\frac{q_{r\,HD}^2}{q_{rH_2}\ q_{rD_2}} = \frac{2^2\left(\frac{1}{m_H} + \frac{1}{m_H}\right)\left(\frac{1}{m_D} + \frac{1}{m_D}\right)}{\left(\frac{1}{m_H} + \frac{1}{m_D}\right)^2} = \frac{16\ m_H\ m_D}{\left(m_D + m_H\right)^2}$$

$$= \frac{16(1)(2)}{3^2} = 3.55$$

$$K = (1.19)(3.55)\ e^{-(1.09 \times 10^{-21}\ \text{J})/kT} = 4.22\ e^{-79/T}$$

$$= 4.22\ e^{-79/298} = 3.23$$

18.32 Given the Lennard-Jones parameters for argon in Table 14.5, what values of the second virial coefficient do you expect at 200 K and 500 K? Do these values agree with the experimental values in Fig. 1.4?

SOLUTION

For argon $\varepsilon/k = 120\ K \qquad \sigma = 341 \times 10^{-12}\ m$

At 500 K $T^* = \dfrac{500\ K}{120\ K} = 4.17$

$$B^* = 0$$

$$B = \frac{2\pi}{3} N_A \sigma^3 B^* = 0$$

At 200 K $T^* = \dfrac{200\ K}{120\ K} = 1.67$

$$B^* = -1.0$$

$$B = \frac{2\pi}{3}(6.022 \times 10^{23}\ mol^{-1})(341 \times 10^{-12}\ m)^3(-1.0)$$

$$= -50\ cm^3\ mol^{-1}$$

These values are in agreement with Fig. 1.4.

18.33 Use the law of corresponding states to estimate the Lennard-Jones parameters for methane from $T_c = 191$ K and $V_c = 100\ cm^3\ mol^{-1}$. The values obtained from second virial coefficient data are $\sigma = 0.378$ nm, and $\varepsilon/k = 148.9$ K.

SOLUTION

$$\sigma = \left(\frac{V_c}{2.7N_A}\right) = \left[\frac{(100\ cm^3\ mol^{-1})(10^{-2}\ m\ cm^{-1})^3}{(2.7)(6.022 \times 10^{23}\ mol^{-1})}\right]^{1/3}$$

$$= 0.395\ nm$$

$$\varepsilon/k = T_c/1.3 = \frac{191\ K}{1.3} = 146\ K$$

18.34 Plot the probability density W(r) for random walk in three dimensions after 1000 steps with a step length of unity. Indicate the root-mean-square end-to-end distance on this plot.

SOLUTION

$$W(r) = \left(\frac{3}{2\pi n l^2}\right)^{3/2} \exp\left(-\frac{3r^2}{2nl^2}\right)$$

$$= \left(\frac{3}{2\pi 1000}\right)^{3/2} \exp\left(-\frac{3r^2}{2000}\right)$$

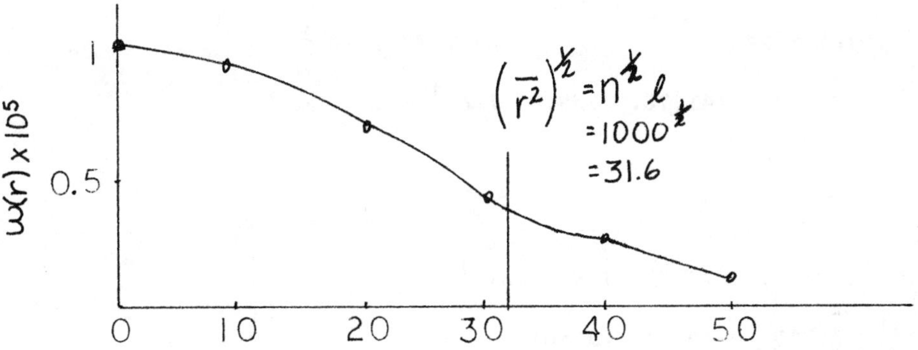

18.35 In polyethene $H(CH_2 - CH_2)_nH$ the bond length 1 is 0.15 nm. What is the root-mean-square end-to-end distance for a molecule with universal joints with a molar mass of 10^5 g mol^{-1}? Taking into account the fact that carbon forms tetrahedral bonds, what is $(\overline{r^2})^{1/2}$?

SOLUTION

$$N = \frac{10^5 \text{ g mol}^{-1}}{14 \text{ g mol}^{-1}}$$

$$\langle L^2\rangle^{1/2} = N^{1/2}\, b = \left(\frac{10^5}{14}\right)^{1/2} 0.15 \text{ nm}$$

$$= 12.7 \text{ nm}$$

$$\langle L^2\rangle^{1/2} = N^{1/2}\, b \left(\frac{1 + \cos\theta}{1 - \cos\theta}\right)^{1/2}$$

$$= 12.7 \text{ nm} \left(\frac{1 + \cos 71^\circ}{1 - \cos 71^\circ}\right)^{1/2} = 17.8 \text{ nm}$$

18.36 (a) 3.85×10^{-173} (b) 2.47×10^{-23}

18.38 1.998×10^{33} compared with 1.414×10^{30}

18.39 (a) 1.23 (b) 2.83×10^{-11}

18.40 108.96, 141.78, 139.85 J K^{-1} mol^{-1}

18.41 126.156 J K^{-1} mol^{-1}

18.42 2.86

18.43 12.472, 47.822 J K^{-1} mol^{-1}

3.718, -10.540 kJ mol^{-1}

18.44 (a) 4 (b) 3 (c) 18 (d) 27 (e) 324

18.45

18.46 2.175×10^{-6}

18.47

18.48 3.369 8.144

18.49 11.526 J K^{-1} mol^{-1} -3.437 kJ mol^{-1}
11.634 J K^{-1} mol^{-1} -34.902 kJ mol^{-1}

18.50 $<10^{-99}$, 1.49×10^{-23}

18.51 251.800 J K^{-1} mol^{-1}

18.52 29.100, 32.926 J K^{-1} mol^{-1}

18.53

18.54 3.20×10^{-3}

18.55 4.41×10^{-4}

18.56 0.211

18.57 111.770 kJ mol^{-1} 1.13×10^{-2}

18.58 -46, 7 cm^3 mol^{-1}

18.60

	ξ	0	25	50	75
1000 steps of 1	$W(\xi) \times 10^4$	126	92	36	7.6
500 steps of 2	$W(\xi) \times 10^4$	89.2	76.3	47.8	21.9

18.61 (a) 1100 nm (b) 13 nm (c) 18.4 nm

18.62 The maximum in the plot is at r = 10 nm.

PART THREE

CHEMICAL DYNAMICS

CHAPTER 19: Kinetic Theory of Gases

19.1 If the diameter of a gas molecule is 0.4 nm and each is imagined to be in a separate cube, what is the length of the side of the cube in molecular diameters at 0 °C and pressures of (a) 1 bar, and (b) 1 Pa.

SOLUTION

(a) $$\left[\frac{(22.4\ \text{L mol}^{-1})(1000\ \text{cm}^3\ \text{L}^{-1})(10^7\ \text{nm cm}^{-1})^3}{6.02 \times 10^{23}\ \text{mol}^{-1}}\right]^{1/3}$$

$$= 3.338\ \text{nm} = \frac{3.338\ \text{nm}}{0.4\ \text{nm}} = 8.3\ \text{molecular diameters}$$

(b) $$\left[\frac{(22.4\ \text{L atm mol}^{-1})(10^{24}\ \text{nm}^3\ \text{L}^{-1})(101{,}325\ \text{Pa atm}^{-1})/(1\ \text{Pa})}{6.02 \times 10^{23}\ \text{mol}^{-1}}\right]^{1/3}$$

$$= 155.6\ \text{nm} = \frac{155.6\ \text{nm}}{0.4\ \text{nm}} = 389\ \text{molecular diameters}$$

19.2 Plot the probability density f(v) of various molecular speeds versus speed for oxygen at 25 °C.

SOLUTION

$$f(v) = 4\pi v^2 \left(\frac{M}{2\pi RT}\right)^{3/2} e^{-Mv^2/2RT}$$

$$= 4\pi v^2 \left[\frac{32 \times 10^{-3}\ \text{kg mol}^{-1}}{2\pi(8.314\ \text{J K}^{-1}\text{mol}^{-1})(298\ \text{K})}\right]^{3/2} e^{-\frac{(32 \times 10^{-3}\ \text{kg mol}^{-1})v^2}{2(8.314\ \text{J K}^{-1}\ \text{mol}^{-1})(298\ \text{K})}}$$

$$= v\ (3.704 \times 10^{-8}\ \text{s m}^{-1})e^{-6.458 \times 10^{-6}\ v^2}$$

Where v is in ms^{-1}

$v/10^2\ ms^{-1}$	$f(v)/10^{-4}\ s\ m^{-1}$
1	3.47
3	18.64
5	18.42
7	7.67
10	0.58

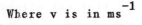

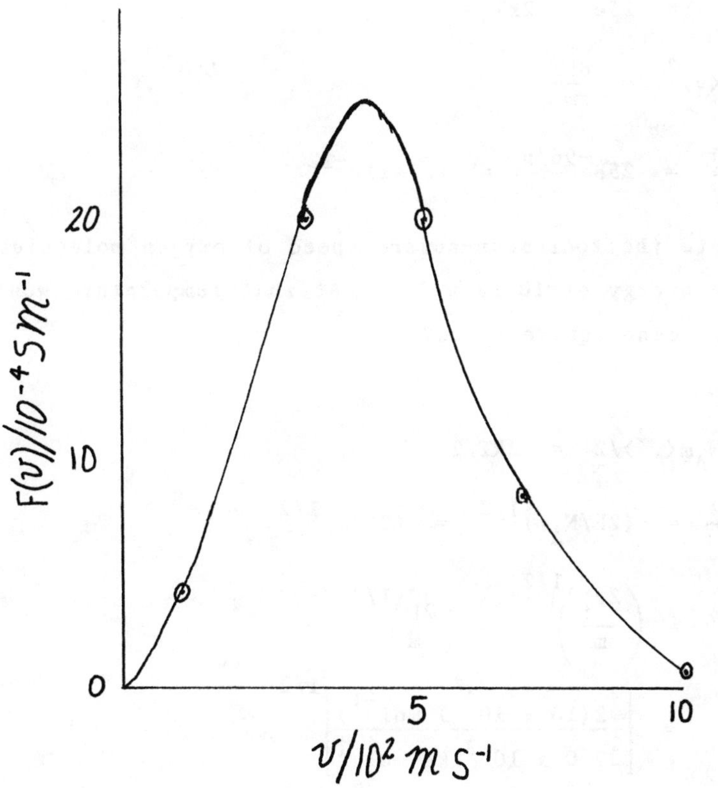

19.3 What is the ratio of the probability that gas molecules have five times the mean speed to the probability that they have the mean speed?

SOLUTION

$$f(\langle v\rangle) \propto \langle v\rangle^2 e^{-\frac{\langle v\rangle^2 m}{2kT}}$$

$$f(5\langle v\rangle) \propto 25\langle v\rangle^2 e^{-\frac{25\langle v\rangle^2 m}{2kT}}$$

$$\frac{f(5\langle v\rangle)}{f(\langle v\rangle)} = 25e^{-\frac{24\langle v\rangle^2 m}{2kT}}$$

Since $\langle v\rangle^2 = \dfrac{8kT}{\pi m}$

$$\frac{f(5\langle v\rangle)}{f(\langle v\rangle)} = 25e^{-96/\pi} = 1.24\text{x}10^{-12}$$

19.4 Calculate the root-mean-square speed of oxygen molecules having a kinetic energy of 10 kJ mol^{-1}. At what temperature would this be the root-mean-square speed?

SOLUTION

$$E = N_A m\langle v^2\rangle/2 = 3RT/2$$

$$\langle v^2\rangle^{1/2} = (2E/N_A m)^{1/2} = (2E/M)^{1/2}$$

$$\langle v^2\rangle^{1/2} = \left(\frac{2\varepsilon_t}{m}\right)^{1/2} = \left(\frac{2E}{M}\right)^{1/2}$$

$$= \left[\frac{2(10 \text{ x } 10^3 \text{ J mol}^{-1})}{32.0 \text{ x } 10^{-3} \text{ kg mol}^{-1}}\right]^{1/2}$$

$$= 7.9 \text{ x } 10^2 \text{ m s}^{-1}$$

$$T = \frac{2E}{3R} = \frac{2(10 \text{ x } 10^3 \text{ J mol}^{-1})}{3(8.314 \text{ J K}^{-1} \text{ mol}^{-1})} = 802 \text{ K}$$

19.5 Derive equations for calculating U and C_V for any monatomic gas from kinetic theory.

SOLUTION

$$U = \frac{N_A}{2} m \int_0^\infty v^2 F(v) dv$$

$$= \frac{N_A}{2} m \langle v^2 \rangle = \frac{N_A}{2} m \left(\frac{3kT}{m} \right) = \frac{3}{2} RT$$

$$C_V = \left(\frac{\partial U}{\partial T} \right)_V = \frac{3}{2} R$$

19.6 Calculate the mean speed and the root-mean-square speed for the following set of molecules: 10 molecules moving 5×10^2 m s^{-1}, 20 molecules moving 10×10^2 m s^{-1}, and 5 molecules moving 15×10^2 m s^{-1}.

SOLUTION

$$\langle v \rangle = \frac{\sum N_i v_i}{\sum N_i} = \frac{10(500) + 20(1000) + 5(1500)}{35}$$

$$= 928 \text{ m s}^{-1}$$

$$\langle v^2 \rangle^{1/2} = \left(\frac{\sum N_i v_i^2}{\sum N_i} \right)^{1/2} = \left[\frac{10(500)^2 + 20(1000)^2 + 5(1500)^2}{35} \right]^{1/2}$$

$$= 982 \text{ m s}^{-1}$$

19.7 Calculate the velocity of sound in nitrogen gas at 25 °C. (See Section 19.4.)

SOLUTION

$$v = \left(\frac{C_P RT}{C_V M}\right)^{1/2}$$

$$= \left[\frac{(29.125\ \text{J K}^{-1}\ \text{mol}^{-1})(8.3144\ \text{J K}^{-1}\ \text{mol}^{-1})(298.15\ \text{K})}{(20.811\ \text{J K}^{-1}\ \text{mol}^{-1})(2)(14.0067 \times 10^{-3}\ \text{kg mol}^{-1})}\right]^{1/2}$$

$$= 352\ \text{m s}^{-1}$$

$$= \frac{(352\ \text{m s}^{-1})(10^2\ \text{cm m}^{-1})(60\ \text{s min}^{-1})(60\ \text{min hr}^{-1})}{(2.54\ \text{cm in}^{-1})(12\ \text{in ft}^{-1})(5280\ \text{ft mile}^{-1})}$$

$$= 787\ \text{miles hr}^{-1}$$

19.8 (a) How many molecules of H_2 strike the wall per unit area per unit time at one bar at 298 K? 1000 K? (b) How many molecules of O_2 strike the wall per unit area per unit time at one bar at 298 K? 1000 K?

SOLUTION

(a) For H_2 at 1 bar and 298 K

$$M_{H_2} = \frac{2.016 \times 10^{-3}\ \text{kg mol}^{-1}}{6.022 \times 10^{23}\ \text{mol}^{-1}} = 3.348 \times 10^{-27}\ \text{kg}$$

$$M_{O_2} = \frac{32 \times 10^{-3}\ \text{kg mol}^{-1}}{6.022 \times 10^{23}\ \text{mol}^{-1}} = 5.314 \times 10^{-26}\ \text{kg}$$

$$J_N = \frac{P}{(2\pi mkT)^{1/2}}$$

$$= \frac{10^5\ \text{N m}^{-2}}{[(2\pi)(3.348 \times 10^{-27}\ \text{kg})(1.381 \times 10^{-23}\ \text{J K}^{-1})(298\ \text{K})]^{1/2}}$$

$$= 1.075 \times 10^{28}\ \text{m}^{-2}\ \text{s}^{-1}$$

At 1000 K, $J_N = (1.075 \times 10^{28}\ m^2\ s^{-1})(298/1000)^{1/2}$

$= 5.866 \times 10^{27}\ m^{-2}\ s^{-1}$

(b) For O_2 at 1 bar and 298 K

$$J_N = \frac{10^5\ N\ m^{-2}}{[(2\pi)(5.314 \times 10^{-26}\ kg)(1.381 \times 10^{-23}\ J\ K^{-1})(298)]^{1/2}}$$

$= 2.698 \times 10^{27}\ m^2\ s^{-1}$

At 1000 K, $J_N = (2.698 \times 10^{27}\ m^2\ s^{-1})(298/1000)^{1/2}$

$= 1.473 \times 10^{27}\ m^2\ s^{-1}$

19.9 The Langmuir is a unit of gas exposure that is defined as 1×10^{-6} Torr s. If a surface with 10^{15} atoms per square centimeter is exposed to a gas at 1×10^{-6} Torr for 1 s, what fraction of a monolayer will be formed? Assume the gas has a molar mass of 28 g mol^{-1} and T = 300 K. (1 Torr = 133.3 Pa.)

SOLUTION

$$m = \frac{0.028\ kg\ mol^{-1}}{6.022 \times 10^{23}\ mol^{-1}} = 4.65 \times 10^{-26}\ kg$$

$$J_N = \frac{P}{(2\pi mkT)^{1/2}}$$

$$= \frac{133.3 \times 10^{-6}\ Pa}{[2\pi(4.65 \times 10^{-26}\ kg)(1.381 \times 10^{-23}\ J\ K^{-1})(300\ K)]^{1/2}}$$

$= 3.83 \times 10^{18}\ m^{-2}\ s^{-1}$

The number of binding sites is

$$(10^{15}\ cm^{-2})(100\ cm\ m^{-1})^2 = 10^{19}\ m^{-2}$$

Thus exposure for 1 s will fill 0.383 of the sites if each gas molecule incident on the surface is trapped by the attractive surface potential (Section 21.8).

19.10 Calculate the number of collisions per square centimeter per second of oxygen molecules with a wall at a pressure of 1 bar and 25 °C.

SOLUTION

$$\rho = \frac{PN_A}{RT} = \frac{(10^5\ \mathrm{N\ m^{-2}})(6.022 \times 10^{23}\ \mathrm{mol^{-1}})}{(8.3144\ \mathrm{J\ K^{-1}\ mol^{-1}})(298.15\ \mathrm{K})}$$

$$= 2.429 \times 10^{25}\ \mathrm{m^{-3}}$$

$$\langle v \rangle = \left(\frac{8RT}{\pi M}\right)^{1/2} = \left[\frac{(8)(8.3144\ \mathrm{J\ K^{-1}\ mol^{-1}})(298.15\ \mathrm{K})}{\pi(32 \times 10^{-3}\ \mathrm{kg\ mol^{-1}})}\right]^{1/2}$$

$$= 444.2\ \mathrm{m\ s^{-1}}$$

$$J_N = \frac{\rho \langle v \rangle}{4} = \frac{1}{4}(2.429 \times 10^{25}\ \mathrm{m^{-3}})(444.2\ \mathrm{m\ s^{-1}})$$

$$= 2.697 \times 10^{27}\ \mathrm{m^{-2}\ s^{-1}}$$

$$= (2.697 \times 10^{27}\ \mathrm{m^{-2}\ s^{-1}})(10^{-2}\ \mathrm{m\ cm^{-1}})^2$$

$$= 2.697 \times 10^{23}\ \mathrm{cm^{-2}\ s^{-1}}$$

19.11 A fresh metal surface with 10^{15} atoms per square centimeter is prepared. This surface is exposed to oxygen at 10^{-2} Pa. If every oxygen molecule that strikes the surface reacts so that there is one oxygen atom per metal atom in the surface, how long will it take for half of the surface to become oxidized at 25 °C?

SOLUTION

$$m = \frac{32 \times 10^{-3}\ \mathrm{kg\ mol^{-1}}}{6.022 \times 10^{23}\ \mathrm{mol^{-1}}} = 5.31 \times 10^{-26}\ \mathrm{kg}$$

$$J_N = \frac{P}{(2\pi mkT)^{1/2}}$$

$$= \frac{10^{-2}\ \mathrm{N\ m^{-2}}}{[2\pi(5.31 \times 10^{-26}\ \mathrm{kg})(1.38 \times 10^{-23}\ \mathrm{J\ K^{-1}})(298\ \mathrm{K})]^{1/2}}$$

$$= 2.70 \times 10^{20}\ m^{-2}\ s^{-1}$$

In the surface there are $(10^{15}\ cm^{-2})(100\ cm\ m^{-1})^2 = 10^{19}$ atoms m^{-2}. Oxidizing half of the surface will require 0.25×10^{19} oxygen molecules

$$\frac{0.25 \times 10^{19}\ m^{-2}}{2.7 \times 10^{20}\ m^{-2}\ s^{-1}} = 9.3 \times 10^{-3}\ s$$

19.12 A Knudsen cell containing crystalline benzoic acid ($M = 122\ g\ mol^{-1}$) is carefully weighed and placed in an evacuated chamber thermostated at 70°C for 1 hr. The circular hole through which effusion occurs is 0.60 mm in diameter. Calculate the sublimation pressure of benzoic acid at 70 °C in Pa from the fact that the weight loss is 56.7 mg.

SOLUTION

$$P = w\left(\frac{2\pi RT}{M}\right)^{1/2}$$

$$= \frac{56.7 \times 10^{-6}\ kg}{(60x60\ s)\pi(0.3x10^{-3}\ m)^2}\left[\frac{2\pi(8.314\ J\ K^{-1}\ mol^{-1})(343\ K)}{122 \times 10^{-3}\ kg\ mol^{-1}}\right]^{1/2}$$

$$= 21.3\ N\ m^{-2}$$

$$= 21.3\ Pa$$

19.13 R. B. Holden, R. Speiser, and H. L. Johnston [J. Am. Chem. Soc., 70, 3897(1948)] found the rate of loss of weight of a Knudsen effusion cell containing finely divided beryllium to be $19.8 \times 10^{-7}\ g\ cm^{-2}\ s^{-1}$ at 1320 K and $1210 \times 10^{-7}\ g\ cm^{-2}\ s^{-1}$ at 1537 K. Calculate ΔH_{sub} for this temperature range.

SOLUTION

According to equations 19.29 and 19.30 the two vapor pressures are proportional to $\Delta gT^{1/2}$.

$$\Delta H^\circ = \frac{RT_1T_2}{(T_2 - T_1)} \ln \frac{P_2}{P_1}$$

$$= \frac{(8.314)(1320)(1537)}{(217)} \ln \frac{1210(1537)^{1/2}}{19.8(1320)^{1/2}}$$

$$= 326 \text{ kJ mol}^{-1}$$

19.14 A 5 mL container with a hole 10 μm in diameter is filled with hydrogen. This container is placed in an evacuated chamber at 0 °C. How long will it take for 90% of the hydrogen to effuse out?

SOLUTION

$J_N = \frac{\rho \langle v \rangle}{4}$ may be written $-\frac{1}{A}\frac{dN}{dt} = \frac{N}{V}\frac{\langle v \rangle}{4}$

$\langle v \rangle = 1.69 \times 10^3 \text{ m s}^{-1}$ from Example 19.3

$$-\frac{dN}{dt} = \frac{\pi(5 \times 10^{-6} \text{ m})(1.69 \times 10^3 \text{ m s}^{-1})N}{(5 \times 10^{-6} \text{ m}^3)(4)}$$

$$= (6.64 \times 10^{-3} \text{ s}^{-1})N$$

$$\int_{N_o}^{N_t} \frac{dN}{N} = -(6.64 \times 10^{-3} \text{ s}^{-1}) \int_o^t dt$$

$$\ln \frac{N_t}{N_o} = -(6.64 \times 10^{-3} \text{ s}^{-1})t$$

$$t = \frac{\ln 0.1}{-6.64 \times 10^{-3} \text{ s}^{-1}} = 347 \text{ s}$$

19.15 (a) Calculate the number of collisions per second undergone by a single nitrogen molecule in nitrogen at 1 bar pressure and 25 °C. (b) What is the number of collisions per cubic centimeter per second? What is the effect on the number of collisions (c) of doubling the absolute temperature at constant pressure, and (d) of doubling the pressure at constant temperature?

SOLUTION

(a) $$Z_{1(1)} = 2^{1/2} \rho \pi d^2 \langle v \rangle$$

$$\rho = \frac{PN_A}{RT} = \frac{(10^5 \text{ N m}^{-2})(6.022 \times 10^{23} \text{ mol}^{-1})}{(8.314 \text{ J K}^{-1} \text{ mol}^{-1})(298.15 \text{ K})}$$

$$= 2.429 \times 10^{25} \text{ m}^{-3}$$

$$\langle v \rangle = \left(\frac{8RT}{\pi M}\right)^{1/2} = \left[\frac{8(8.314)(298.15)}{\pi(28 \times 10^{-3})}\right]^{1/2} = 475 \text{ m s}^{-1}$$

$$Z_{1(1)} = 2^{1/2} (2.429 \times 10^{25})\pi(0.375 \times 10^{-9})^2(475)$$

$$= 7.21 \times 10^9 \text{ s}^{-1}$$

(b) $$Z_{11} = \frac{1}{2^{1/2}} \rho^2 \pi d^2 \langle v \rangle$$

$$= 2^{-1/2} (2.429 \times 10^{25})^2 \pi(0.375 \times 10^{-9})^2(475)$$

$$= 8.75 \times 10^{34} \text{ m}^{-3} \text{ s}^{-1}$$

$$= (8.75 \times 10^{34} \text{ m}^{-3} \text{ s}^{-1})(10^{-2} \text{ m cm}^{-1})^3$$

$$= 8.75 \times 10^{28} \text{ cm}^{2} \text{ s}^{-1}$$

(c) $$\rho \propto T^{-1} \qquad \langle v \rangle \propto T^{1/2}$$

$$Z_{11} \propto \rho^2 \langle v \rangle \propto T^{-2}\, T^{1/2} \propto T^{-1.5}$$

$$\frac{Z_{11}(2T)}{Z_{11}(T)} = \frac{T^{1.5}}{(2T)^{1.5}} = \frac{1}{2^{1.5}} = 0.354$$

(d) $$Z_{11} \propto P^2$$

$$\frac{Z_{11}(2P)}{Z_{11}(P)} = 4$$

19.16 Methyl radicals appear to combine without activation energy $2CH_3 \longrightarrow C_2H_6$. The second order rate constant at 300 K is $10^{10.5}$ L mol^{-1} s^{-1}. Assuming the collision diameter is 0.40 nm, what rate constant is expected from the rate of collisions?

SOLUTION

$$-\frac{d(CH_3)}{dt} = 2k(CH_3)^2$$

The rate constant is equal to 1/2 the rate of disappearance of CH_3 when $(CH_3) = 1$ mol L^{-1} .

$$\rho = 6.02 \times 10^{23} \text{ m}^{-3}$$

$$\langle v \rangle = \left(\frac{8RT}{\pi M}\right)^{1/2} = \left[\frac{8(8.314)(300)}{\pi(15 \times 10^{-3})}\right]^{1/2} = 651 \text{ m s}^{-1}$$

$$Z_{11} = 2^{-1/2} \rho^2 \pi d^2 \langle v \rangle$$

$$= 2^{-1/2} (6.02 \times 10^{23})^2 \pi (0.4 \times 10^{-9})^2 (651)$$

$$= 8.39 \times 10^{37} \text{ m}^{-3} \text{ s}^{-1}$$

$$= \frac{(8.39 \times 10^{37} \text{ m}^{-3} \text{ s}^{-1})(10^{-3} \text{ m}^{-3} \text{ L}^{-1})}{(6.02 \times 10^{23} \text{ mol}^{-1})}$$

$$= 1.39 \times 10^{11} \text{ mol L}^{-1} \text{ s}^{-1}$$

$$-\frac{d(CH_d)}{dt} = 2.78 \times 10^{11} \text{ mol L}^{-1} \text{ s}^{-1} = 2K\ (1 \text{ mol L}^{-1})^2$$

$$k = 1.39 \times 10^{11} \text{ L mol}^{-1} \text{ s}^{-1} = 10^{11.1} \text{ L mol}^{-1} \text{ s}^{-1}$$

19.17 What is the average time between collisions of an oxygen molecule in oxygen at 298 K and (a) 1 bar, (b) 10^{-6} bar, and (c) 10^{-12} bar? ($d = 0.36 \times 10^{-9}$ m)

SOLUTION

At 1 bar an oxygen molecule is involved in 6.24×10^9 collisions per second, according to Example 19.5. Since the number of collisions is proportional to the pressure, a single oxygen molecule is involved in 6.24×10^3 collisions s^{-1} at 10^{-6} bar, and 6.24×10^{-3} collisions s^{-1} at 10^{-12} bar. The average times between collisions are therefore

P/bar	t/s
1	1.60×10^{-10}
10^{-6}	1.60×10^{-4}
10^{-12}	160

19.18 What is the mean free path of nitrogen at 1 bar and 25 °C? What is the average time between collisions?

SOLUTION

$$\rho = \frac{PN_A}{RT} = \frac{(10^5 \text{ N m}^{-2})(6.02 \times 10^{23} \text{ mol}^{-1})}{(8.314 \text{ J K}^{-1} \text{ mol}^{-1})(298 \text{ K})}$$

$$= 2.43 \times 10^{25} \text{ m}^{-3}$$

$$l = [2^{1/2} \rho\pi d^2]^{-1}$$

$$= [2^{1/2} (2.43 \times 10^{25} \text{ m}^{-3})\pi(0.375 \times 10^{-9} \text{ m})^2]^{-1}$$

$$= 65.9 \text{ nm}$$

$\langle v \rangle = 475 \text{ m s}^{-1}$ from Problem 19.15

$$t = \frac{65.9 \times 10^{-9} \text{ m}}{475 \text{ m s}^{-1}} = 1.4 \times 10^{-10} \text{ s}$$

19.19 (a) Calculate the mean free path for hydrogen gas ($\sigma = 0.247$ nm) at 1 bar and 0.1 Pa at 25 °C. (b) Repeat the calculation for chlorine gas ($\sigma = 0.496$ nm).

SOLUTION

At 1 bar

$$\rho = \frac{N_A P}{RT} = \frac{(6.022 \times 10^{23}\ \mathrm{mol^{-1}})(10^5\ \mathrm{N\ m^{-2}})}{(8.314\ \mathrm{J\ K^{-1}\ mol^{-1}})(298\ \mathrm{K})} = 2.44 \times 10^{25}\ \mathrm{m^{-3}}$$

At 0.1 Pa

$$\rho = \frac{(6.022 \times 10^{23}\ \mathrm{mol^{-1}})(0.1\ \mathrm{N\ m^{-2}})}{(8.314\ \mathrm{J\ K^{-1}\ mol^{-1}})(298\ \mathrm{K})} = 2.43 \times 10^{19}\ \mathrm{m^{-3}}$$

(a) $$l = \frac{1}{2^{1/2}\pi\sigma^2\rho}$$

$$= \frac{1}{2^{1/2}\pi(0.247 \times 10^{-9}\ \mathrm{m})^2(2.44 \times 10^{25}\ \mathrm{m^{-3}})}$$

$$= 1.52 \times 10^{-7}\mathrm{m\ at\ 1\ bar}$$

$$l = \frac{1}{2^{1/2}\pi(0.247 \times 10^{-9}\ \mathrm{m})^2(2.43 \times 10^{19}\ \mathrm{m^{-3}})}$$

$$= 0.152\ \mathrm{m\ at\ 0.1\ Pa}$$

(b) $$l = \frac{1}{2^{1/2}\pi(0.496 \times 10^{-9}\ \mathrm{m})^2(2.44 \times 10^{25}\ \mathrm{m^{-3}})}$$

$$= 3.77 \times 10^{-8}\mathrm{m\ at\ 1\ bar}$$

$$l = \frac{1}{2^{1/2}\pi(0.496 \times 10^{-9}\ \mathrm{m})^2(2.43 \times 10^{19}\ \mathrm{m^{-3}})}$$

$$= 0.037\ \mathrm{m\ at\ 0.1\ Pa}$$

19.20 Large vacuum chambers have been built for testing space vehicles at 10^{-6} Pa. Calculate (a) the mean-free path of nitrogen at this pressure, and (b) the number of molecular impacts per square meter of wall per second at 25 °C. $\sigma_{N_2} = 0.375$ nm.

SOLUTION

$$n = \frac{PN_A}{RT} = \frac{(10^{-6}\ \text{Pa})(6.022 \times 10^{23}\ \text{mol}^{-1})}{(8.314\ \text{J K}^{-1}\ \text{mol}^{-1})(298\ \text{K})} = 2.43 \times 10^{14}\ \text{m}^{-3}$$

(a) $$l = \frac{1}{\sqrt{2}\ \pi\sigma^2 n}$$

$$= \frac{1}{\sqrt{2}\ \pi(3.75 \times 10^{-10}\ \text{m})^2(2.43 \times 10^{14}\ \text{m}^{-3})} = 6590\ \text{m}$$

(b) $$Z = \rho\left(\frac{RT}{2\pi M}\right)^{1/2}$$

$$= (2.43 \times 10^{14}\ \text{m}^{-3})\left[\frac{(8.314\ \text{J K}^{-1}\ \text{mol}^{-1})(298\ \text{K})}{2\pi\ 28 \times 10^{-3}\ \text{kg mol}^{-1}}\right]^{1/2}$$

$$= 2.88 \times 10^{16}\ \text{m}^{-2}\ \text{s}^{-1}$$

19.21 The pressure in interplanetary space is estimated to be of the order of 10^{-14} Pa. Calculate (a) the average number of molecules per cubic centimeter, (b) the number of collisions per second per molecule, and (c) the mean free path in miles . Assume that only hydrogen atoms are present and that the temperature is 1000 K. Assume σ = 0.2 nm.

SOLUTION

(a) $$\rho = \frac{N_A P}{RT} = \frac{(6.022 \times 10^{23}\ \text{mol}^{-1})(10^{-14}\ \text{N m}^{-2})}{(8.314\ \text{J K}^{-1}\ \text{mol}^{-1})(10^3\ \text{K})}$$

$$= 0.724 \times 10^6\ \text{m}^{-3} = 0.724\ \text{cm}^{-3}$$

(b) $$\langle v \rangle = \left(\frac{8RT}{\pi M}\right)^{1/2} = \left[\frac{8(8.314\ \text{J K}^{-1}\ \text{mol}^{-1})(10^3\ \text{K})}{\pi(1 \times 10^{-3}\ \text{kg mol}^{-1})}\right]^{1/2}$$

$$= 4600\ \text{m s}^{-1}$$

$$Z_1 = 2^{1/2}\ \rho\pi\sigma^2\ \langle v \rangle$$

$$= 2^{1/2}\ (0.724 \times 10^6\ m^{-3})(0.2 \times 10^{-9}\ m)^2(4600\ ms^{-1})$$

$$= 5.92 \times 10^{-10} s^{-1}$$

(c) $$1 = \frac{1}{2^{1/2}\pi\sigma^2\rho} = \frac{1}{2^{1/2}\pi(0.2 \times 10^{-9}\ m)^2(0.724 \times 10^6)}$$

$$= 7.77 \times 10^{12}\ m$$

$$= \frac{(7.77 \times 10^{12}\ m)(10^2\ cm\ m^{-1})}{(2.54\ cm\ m^{-1})(12\ in\ ft^{-1})(5280\ ft\ mile^{-1})}$$

$$= 4.83 \times 10^9\ miles$$

19.22 Consider an atomic beam of potassium passing through a scattering gas of Ar contained in a cell of 1 cm length at 0 °C. Assuming a collision cross section of $6 \times 10^{-18}\ m^2$ for potassium-argon collisions, calculate the pressure of argon required to produce an attenuation of the beam of 25%.

SOLUTION

$$I_A = I_A^o\ e^{-Q_{AB}n_BL}$$

$$n_B = \frac{\ln(I_A^o/I_A)}{Q_{AB}L} = \frac{2.303 \log (100/75)}{(600 \times 10^{-20}\ m^2)(10^{-2}\ m)}$$

$$= 4.80 \times 10^{18}\ m^{-3} = 4.80 \times 10^{15}\ L^{-1}$$

$$P = \frac{n_BRT}{N_A} = \frac{(4.80 \times 10^{15}\ L^{-1})(8.314\ J\ K^{-1}\ mol^{-1})(273\ K)}{(6.022 \times 10^{23}\ mol^{-1})}$$

$$= 1.81 \times 10^{-5}\ Pa$$

19.23 The viscosity of helium is 1.88×10^{-5} Pa s at °C. Calculate (a) the collision diameter, and (b) the diffusion coefficient at 1 bar.

SOLUTION

(a) $$d = \left[\frac{5}{16}\ \frac{(\pi mkT)^{1/2}}{\pi\eta}\right]^{1/2}$$

$$m = \frac{4.0026 \times 10^{-3}\ \text{kg mol}^{-1}}{6.022 \times 10^{23}\ \text{mol}^{-1}} = 6.647 \times 10^{-27}\ \text{kg}$$

$$d = \left[\frac{5}{16}\ \frac{(\pi 6.647 \times 10^{-27} \times 1.38 \times 10^{-23} \times 273)^{1/2}}{\pi(1.88 \times 10^{-5})}\right]^{1/2}$$

$$= 0.217\ \text{nm}$$

(b) $$\rho = \frac{PN_A}{RT} = \frac{(10^5\ \text{Pa m}^{-2})(6.022 \times 10^{23}\ \text{mol}^{-1})}{(8.314\ \text{J K}^{-1}\ \text{mol}^{-1})(273\ \text{K})}$$

$$= 2.65 \times 10^{25}\ \text{m}^{-3}$$

$$D = \frac{3}{8}\ \frac{(\pi mkT)^{1/2}}{\pi d^2 \rho m}$$

$$= \frac{3}{8}\ \frac{[\pi(6.647 \times 10^{-27}\ \text{kg})(1.38 \times 10^{-23}\ \text{J K}^{-1})(273\ \text{K})]^{1/2}}{\pi(0.217\text{x}10^{-9}\ \text{m})^2(2.65\text{x}10^{25}\ \text{m}^{-3})(6.647\text{x}10^{-27}\ \text{kg})}$$

$$= 1.28 \times 10^{-4}\ \text{m}^2\ \text{s}^{-1}$$

19.24 What is the self-diffusion coefficient of radioactive CO_2 in ordinary CO_2 at 1 bar and 25 °C? The collision diameter is 0.40 nm.

SOLUTION

$$m = \frac{44 \times 10^{-3}\ \text{kg mol}^{-1}}{6.022 \times 10^{23}\ \text{mol}^{-1}} = 7.307 \times 10^{-26}\ \text{kg}$$

$$\rho = \frac{N}{V} = \frac{PN_A}{RT} = \frac{(1\ \text{bar})(6.022 \times 10^{23}\ \text{mol}^{-1})(10^3\ \text{L mol}^{-1})}{(0.083\ 14\ \text{L bar K}^{-1}\ \text{mol}^{-1})(298\ \text{K})}$$

$$= 2.43 \times 10^{25}\ \text{m}^{-3}$$

$$D = \frac{3}{8} \frac{(\pi m k T)^{1/2}}{\pi d^2 \rho m}$$

$$= \frac{3[\pi(7.307 \times 10^{-26}\ \text{kg})(1.38 \times 10^{-23}\ \text{J K}^{-1})(298\ \text{K})]^{1/2}}{\pi(0.4 \times 10^{-9}\ \text{m})^2(2.43 \times 10^{25}\ \text{m}^{-3})(7.307 \times 10^{-26}\ \text{kg})}$$

$$= 1.26 \times 10^{-5}\ \text{m}^2\ \text{s}^{-1}$$

19.25 3.89×10^{-2} eV

19.26 0.1991

19.27 418, 400, and 440 m s^{-1}

19.28 281 m s^{-1}

19.29 4.18×10^2, 4.00×10^2, 4.40×10^2 m s^{-1}

19.30 481, 445, and 394 m s^{-1}

19.31 (a) 1020 m s^{-1} (b) 352 m s^{-1}

19.32 3.7×10^3 s

19.33 (a) 1.136×10^{22} molecules cm^{-2} s^{-1}

(b) 0.0629 g cm^{-2} min^{-1}

19.34 0.0713 g

19.35 8.1×10^{-12}, 2×10^{-10} kg

19.36 78 g mol^{-1}

19.37 (a) 4.75×10^{29} m^{-3} s^{-1}

(b) 2.86×10^{-5} m

19.38 5.8×10^{28} mL^{-1} s^{-1}

19.39 1.58×10^{-10} s 7489 vibrations

19.40 (a) $2.65 \times 10^6\ \mathrm{cm}^{-3}$ (b) $6.52 \times 10^5\ \mathrm{m}$

19.41 $$l_1 = \frac{\langle v_1 \rangle}{2^{1/2}\ \rho_1 \pi d_1^{\ 2} \langle v \rangle + \rho_2 \pi d_{12}^{\ 2} \langle v_{12} \rangle}$$

$$l_2 = \frac{\langle v_2 \rangle}{2^{1/2}\ \rho_2 \pi d_2^{\ 2} \langle v \rangle + \rho_1 \pi d_{12}^{\ 2} \langle v_{12} \rangle}$$

19.42 (a) $28\ \mathrm{m\,m}^2$ (b) $14\ \mathrm{nm}^2$

(c) The cis compound has a large dipole moment while the trans compound has no permanent dipole moment. Since CsCl is a dipolar molecule, it interacts more strongly with the polar isomer of dichloroethylene.

19.43 646 Pa

19.44 1.66×10^{-5} Pa s

19.45 1.302×10^3 Pa s

CHAPTER 20: Experimental Gas Kinetics

20.1 The half-life of a first-order chemical reaction A ⟶ B is 10 min. What percent of A remains after 1 hr?

SOLUTION

$$[A] = [A_o]e^{-kt} = [A]_o\, e^{-0.693\, t/t_{1/2}}$$

$$\frac{[A]}{[A]_o} = \exp\left[\frac{-0.693\ (60\ \text{m})}{10\ \text{m}}\right] = 0.0156$$

Thus 1.56% remains after one hour.

20.2 The following data were obtained on the rate of hydrolysis of 17% sucrose in 0.099 mol L^{-1} HCl aqueous solutions at 35 °C.

t/min	9.82	59.60	93.18	142.9	294.8	589.4
Sucrose remaining, %	96.5	80.3	71.0	59.1	32.8	11.1

What is the order of the reaction with respect to sucrose and the value of the rate constant k?

SOLUTION

$$\log\frac{[A]}{[A]_o} = \frac{-kt}{2.303}$$

$$\text{slope} = \frac{-1}{612\ \text{m}} = \frac{-k}{2.303}$$

$$k = \frac{2.303}{612\ \text{m}} = 3.76 \times 10^{-3}\ \text{m}^{-1}$$

$$= (3.76 \times 10^{-3}\ \text{m}^{-1})(\tfrac{1}{60}\ \text{m s}^{-1})$$

$$= 6.27 \times 10^{-5}\ \text{s}^{-1}$$

(Solution concluded on page 309)

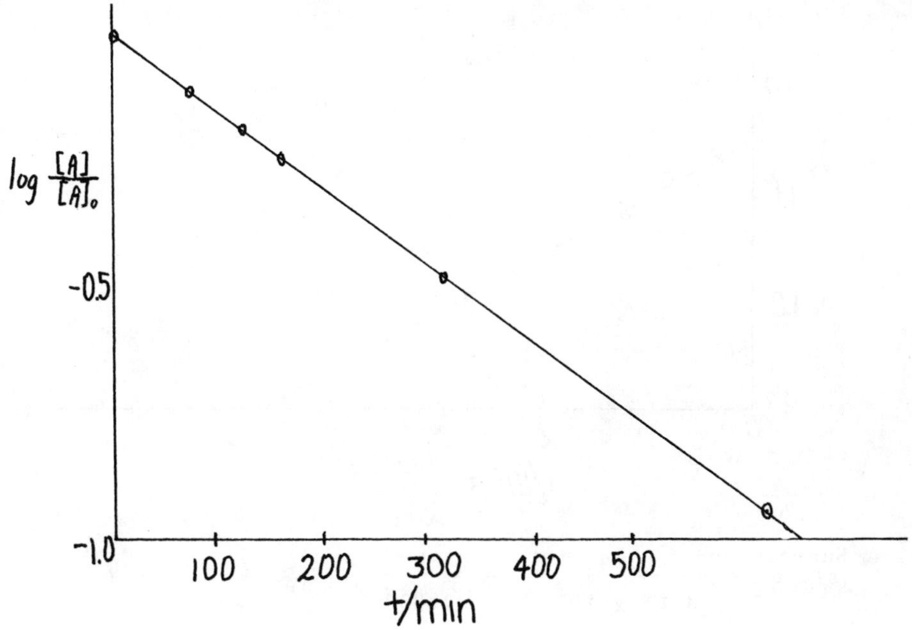

20.3 Methyl acetate is hydrolyzed in approximately 1 mol L^{-1} HCl at 25 °C. Aliquots of equal volume are removed at intervals and titrated with a solution of NaOH. Calculate the first-order rate constant from the following experimental data.

t/s	339	1231	2745	4546	∞
v/cm^3	26.34	27.80	29.70	31.81	39.81

SOLUTION

Any quantity proportional to the concentration of reactant A that remains may be used in equation

$$\log[A] = \frac{-kt}{2.303} + \log[A]_o$$

In this case the concentration of A remaining is proportional to $V-39.81\ cm^3$. Therefore $\log(V-39.81\ cm^3)$ is plotted versus t.

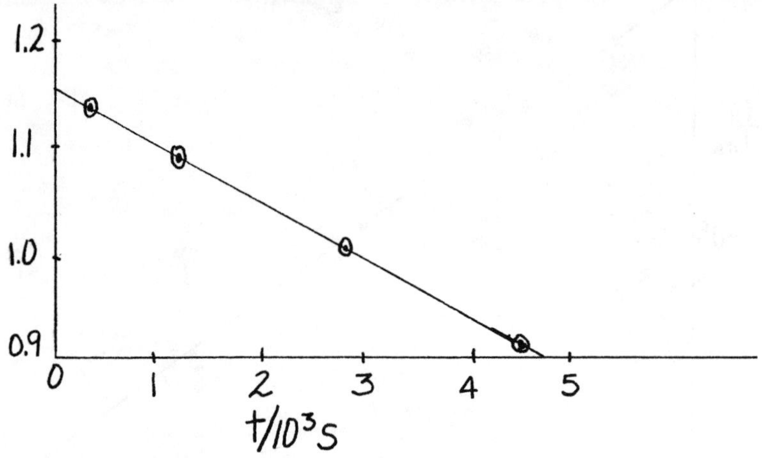

$$\text{slope} = \frac{-0.247}{4.65 \times 10^3\ s} = \frac{-k}{2.303}$$

$$k = \frac{(2.303)(0.247)}{4.65 \times 10^3\ s} = 1.22 \times 10^{-4}\ s^{-1}$$

20.4 Prove that in a first-order reaction, where $dn/dt = -kn$, the average life, that is the average life expectancy of the molecules, is equal to $1/k$.

SOLUTION

$$\text{Average life} = \frac{\int_0^\infty n\,dt}{n_o}$$

$$\text{Average life} = \frac{-\frac{1}{k}\int_0^\infty \frac{dn}{dt}\,dt}{n_o}$$

$$= \frac{-\frac{1}{k}(0 - n_o)}{n_o}$$

$$= \frac{1}{k}$$

20.5 Since radioactive decay is a first-order process, the decay rate for a particular nuclide is commonly given as the half life. Given that potassium contains 0.0118% ^{40}K which has a half life of 1.27×10^9 years, how many disintegrations per second are there in a gram of KCl?

SOLUTION

$$t_{1/2} = (1.27 \times 10^9\ y)(365\ d\ y^{-1})(24\ h\ d^{-1})(60\ min\ h^{-1})(60\ s\ min^{-1})$$

$$= 4.01 \times 10^{16}\ s$$

$$\frac{dN}{dt} = \frac{0.693}{t_{1/2}} N$$

$$= \frac{(0.6930)[(39.1/76.6)g](6.02 \times 10^{23}\ mol^{-1})(1.18 \times 10^{-4})}{(4.01 \times 10^{16}\ s)(40.0\ g\ mol^{-1})}$$

$$= 15.7\ s^{-1}$$

20.6 The decomposition of HI to $H_2 + I_2$ at 508 °C has a half-life of 135 min when the initial pressure of HI is 0.1 atm and 13.5 min when the pressure is 1 atm. (a) Show that this proves that the reaction is second order. (b) What is the value of the rate constant in L mol^{-1} s^{-1}? (c) What is the value of the rate constant in bar^{-1} s^{-1}? (d) What is the value of the rate constant in cm^3 s^{-1}?

SOLUTION

(a) For a second order reaction $t_{1/2} = 1/k[A]_o$ where $[A]_o$ is the initial concentration or pressure of the reactant. This is in agreement with the fact that the half-life is reduced by a factor of 10 when the pressure is increased by a factor of 10.

(b) $$c = \frac{P}{RT} = \frac{(101{,}325\ \text{Pa})}{(8.314\ \text{J K}^{-1}\ \text{mol}^{-1})(781.15\ \text{K})}$$

$$= 1.56 \times 10^{-2}\ \text{mol L}^{-1}$$

$$k = \frac{1}{[A]_o t_{1/2}} = \frac{1}{(1.56 \times 10^{-2}\ \text{mol L}^{-1})(13.5 \times 60\ \text{s})}$$

$$= 7.91 \times 10^{-2}\ \text{L mol}^{-1}\ \text{s}^{-1}$$

(c) $$k = \frac{1}{t_{1/2}P_o} = \frac{1}{(13.5 \times 60\ \text{s})(1.013\ \text{bar})}$$

$$= 1.22 \times 10^{-3}\ \text{bar}^{-1}\ \text{s}^{-1}$$

(d) $$k = \frac{(7.91 \times 10^{-2}\ \text{L mol}^{-1}\ \text{s}^{-1})(10^3\ \text{cm}^3\ \text{L}^{-1})}{(6.022 \times 10^{23}\ \text{mol}^{-1})}$$

$$= 1.31 \times 10^{-22}\ \text{cm}^3\ \text{s}^{-1}$$

20.7 The reaction between propionaldehyde and hydrocyanic acid has been studied at 25 °C by W. J. Svirbely and J. F. Roth [J. Am. Chem. Soc., 75, 3106 (1953)]. In a certain aqueous solution at 25 °C the concentrations at various times were as follows.

t/min	2.78	5.33	8.17	15.23	19.80	∞
[HCN]/mol L^{-1}	0.0990	0.0906	0.0830	0.0706	0.0653	0.0424
[C_3H_7CHO]/mol L^{-1}	0.0566	0.0482	0.0406	0.0282	0.0229	0.0000

What is the order of the reaction and the value of the rate constant K?

SOLUTION

The data does not give a linear log versus t plot, and so the data are tested in the integrated equation for a second order reaction. Equation 20.27 is

$$\frac{1}{([A]_o - [B]_o)} \ln \frac{[A][B]_o}{[A]_o[B]} = kt$$

In order to get $[A]_o$ and $[B]_o$ for t = 0, the clock is started at the first experimental point. This yields the following data to be plotted.

t/min	0	2.55	5.39	12.45	17.02
$\frac{1}{([A]_o - [B]_o)} \ln \frac{[A][B]_o}{[A]_o[B]}$	0	1.695	3.677	8.458	11.523

The slope of this plot, 0.675 L mol^{-1} min^{-1}, is the second order rate constant.

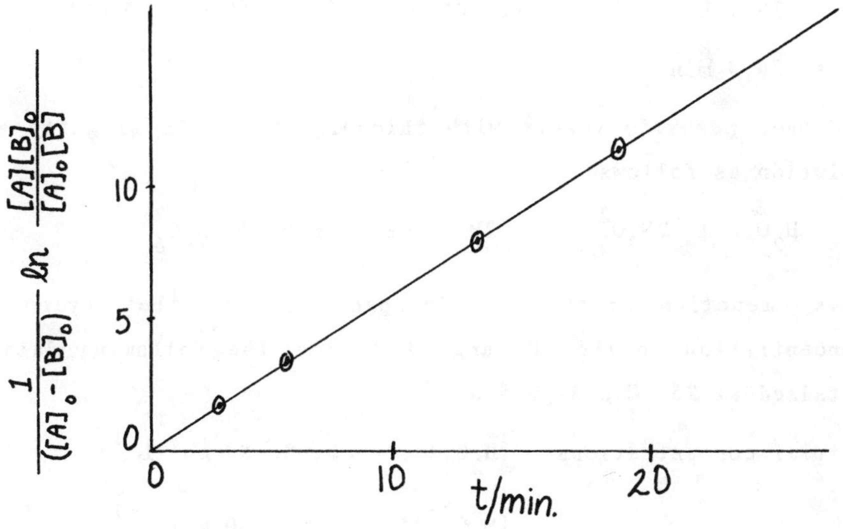

20.8 The second order rate constant for the reaction of ClO and NO is $6.2 \times 10^{-12}\ cm^3\ s^{-1}$. What is its value in $L\ mol^{-1}\ s^{-1}$?

SOLUTION

$$k = (6.2 \times 10^{-12} cm^3\ s^{-1})(10^{-3}\ L\ cm^{-3})(6.022 \times 10^{23}\ mol^{-1})$$

$$= 3.73 \times 10^{9}\ L\ mol^{-1}\ sec^{-1}$$

20.9 The reaction $CH_3CH_2NO_2 + OH^- \longrightarrow H_2O + CH_3CHNO_2^-$ is of second order, and k at 0 °C is $39.1\ L\ mol^{-1}\ min^{-1}$. An aqueous solution is 0.004 molar in nitroethane and 0.005 molar in NaOH. How long will it take for 90% of the nitroethane to react?

SOLUTION

$$t = \frac{1}{k(a - b)} \ln \frac{b(a - x)}{a(b - x)}$$

$$= \frac{1}{(39.1\ L\ mol^{-1} min^{-1})(0.001\ mol\ L^{-1})} \ln \frac{(0.004)(0.005 - 0.0036)}{(0.005)(0.004 - 0.0036)}$$

$$= 26.3\ min$$

20.10 Hydrogen peroxide reacts with thiosulfate ion in slightly acidic solution as follows

$$H_2O_2 + 2S_2O_3^{2-} + 2H^+ \longrightarrow 2H_2O + S_4O_6^{2-}$$

This reaction rate is independent of the hydrogen-ion concentration in the pH range 4 to 6. The following data were obtained at 25 °C and pH 5.0.

Initial concentrations: $[H_2O_2] = 0.036\ 80\ mol\ L^{-1}$;

$[S_2O_3^{2-}] = 0.020\ 40\ mol\ L^{-1}$

t/min	16	36	43	52
$[S_2O_3^{2-}]/10^{-3}$ mol L^{-1}	10.30	5.18	4.16	3.13

(a) What is the order of reaction? (b) What is the rate constant?

SOLUTION

(a) In the first 16 minutes, $[S_2O_3^{2-}]$ is approximately halved. In the next 20 minutes, $[S_2O_3^{2-}]$ is approximately halved. In the next 16 minutes, $[S_2O_3^{2-}]$ is considerably less than halved. Therefore, the order is higher than one. The next section shows that the reaction is first order in H_2O_2, first order in $S_2O_3^{2-}$, and second order overall.

(b) Let $A = H_2O_2$ and $B = S_2O_3^{2-}$. Equation 20.26 becomes

$$kt = \frac{1}{(2[A]_o - [B]_o)} \ln \frac{[A][B]_o}{[A]_o[B]}$$

$$\ln \frac{[A]}{[B]} = \ln \frac{[A]_o}{[B]_o} + (2[A]_o - [B]_o)\, kt$$

t/min	0	16	36	43	52
$[B]/10^{-3}$ mol L^{-1}	20.40	10.30	5.18	4.16	3.13
$[A]/10^{-3}$ mol L^{-1}	36.80	31.75	29.19	28.68	28.17
$\ln \frac{[A]}{[B]}$	0.590	1.126	1.729	1.931	2.197

(Solution continued on page 316)

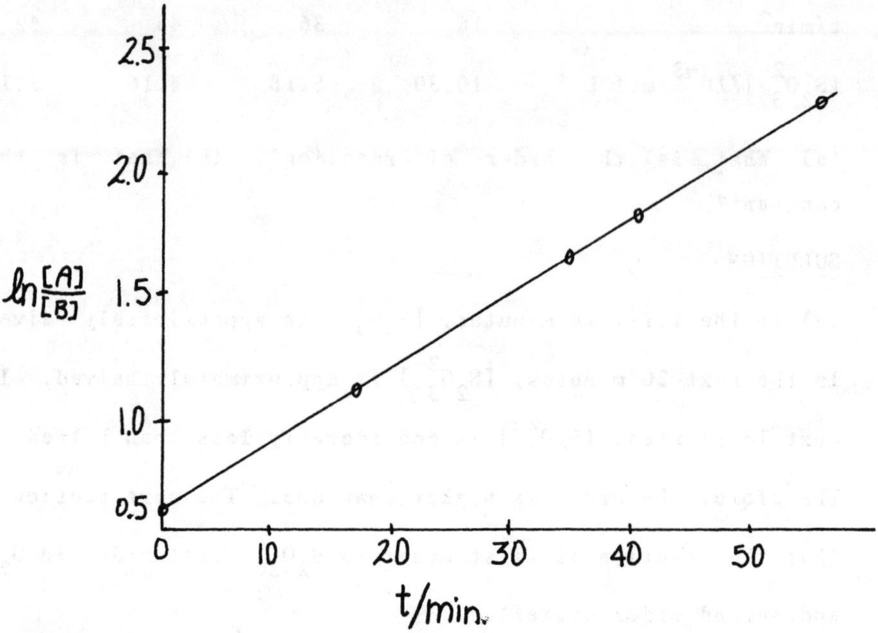

$$\text{slope} = \frac{2.46 - 0.59}{60\ \text{min}} = 0.0312\ \text{min}^{-1}$$

$$= (2[A]_o - [B]_o)\,k$$

$$k = \frac{0.0312\ \text{min}^{-1}}{2(36.80 \times 10^{-3}\ \text{mol L}^{-1}) - 20.40 \times 10^{-3}\ \text{mol L}^{-1}}$$

$$= 0.59\ \text{L mol}^{-1}\ \text{min}^{-1}$$

20.11 The pre-exponential factor for the trimolecular reaction $2NO + O_2 \longrightarrow 2NO_2$ is 10^9 cm^6 mol^{-2} s^{-1}. What is the value in L^2 mol^{-2} s^{-1} and cm^6 s^{-1}?

SOLUTION

$$(10^9\ \text{cm}^6\ \text{mol}^{-2}\ \text{s}^{-1})(10^{-1}\ \text{dm cm}^{-1})^6 = 10^3\ \text{dm}^6\ \text{mol}^{-2}\ \text{s}^{-1}$$

$$= 10^3\ L^2\ mol^{-2}\ s^{-1}$$

$$= \frac{(10^3\ L^2\ mol^{-2}\ s^{-1})(10^3\ cm^3\ L^{-1})^2}{(6.02 \times 10^{23}\ mol^{-1})^2}$$

$$= 2.76 \times 10^{-39}\ cm^6\ s^{-1}$$

20.12 A solution of A is mixed with an equal volume of a solution of B containing the same number of moles, and the reaction A + B = C occurs. At the end of 1 hr A is 75% reacted. How much of A will be left unreacted at the end of 2 hr if the reaction is (a) first order in A and zero order in B; (b) first order in both A and B; and (c) zero order in both A and B?

SOLUTION

(a) If the reaction is first order in A and zero order in B, the half time is 1/2 hour, and the following table may be constructed doing calculations in your head.

t/hours	0	1/2	1	1-1/2	2
% Unreacted	100	50	25	12.5	6.25
% Reacted	0	50	75	8.75	93.75

(b) If the reaction is first order in both A and B, and the initial concentrations are equal, and the stoichiometry is 1:1, the concentration of A will follow

$$k = \frac{1}{t}\left(\frac{1}{[A]} - \frac{1}{[A]_o}\right) = \frac{1}{t[A]_o}\left(\frac{[A]_o}{[A]} - 1\right)$$

$$k[A]_o = \frac{1}{t}\left(\frac{[A]_o}{[A]} - 1\right) = \frac{1}{1\ hr}\left[\frac{100}{25} - 1\right] = 3\ hr^{-1}$$

After 2 hr

$$3\ hr^{-1} = \frac{1}{2\ hr}\left[\frac{100}{[A]} - 0\right]$$

$$(A) = \frac{100}{7} = 14.3\%$$

(c) If the reaction is zero order in both A and B, both will be completely gone in 1-1/3 hr.

20.13 Derive the integrated rate equation for a reaction of 1/2 order. Derive the expression for the half-life of such a reaction.

SOLUTION

$$\frac{d[A]}{dt} = -k[A]^{1/2}$$

$$\int_{[A]_o}^{[A]} \frac{d[A]}{[A]^{1/2}} = -k\int_o^t dt = -kt = \left[2[A]^{1/2}\right]_{[A]_o}^{[A]} = \left(2\ [A]^{1/2} - [A]_o^{\ 1/2}\right)$$

$$[A]_o^{\ 1/2} - [A]^{1/2} = \frac{k}{2}\ t$$

At $t_{1/2}$, $[A] = [A]_o/2$

$$t_{1/2} = \left[[A]_o^{\ 1/2} - \left(\frac{[A]_o}{2}\right)^{1/2}\right]\frac{2}{k} = \frac{2}{k}\left(1 - \frac{1}{\sqrt{2}}\right)[A]_o^{1/2}$$

$$t_{1/2} = \frac{\sqrt{2}}{k}\ (\sqrt{2} - 1)\ [A]_o^{\ 1/2}$$

20.14 Show that for a reaction following the rate law $-d[A]/dt = k[A]^{\alpha}$, the half life is given by

$$t_{1/2} = \frac{2^{\alpha - 1} - 1}{k[A]_o^{\ \alpha - 1}\ (\alpha - 1)}$$

where α is an integer not equal to unity.

SOLUTION

$$-\frac{d[A]}{[A]^{\alpha}} = kdt = -[A]^{-\alpha}\ d[A] \qquad \left.\frac{-[A]^{-\alpha + 1}}{1 - \alpha}\right]_{[A]_o}^{[A]} = kt$$

$$\frac{1}{[A]^{\alpha - 1}} - \frac{1}{[A_o]^{\alpha - 1}} = (\alpha - 1)kt$$

If $[A] = \frac{1}{2}[A]_o$ $$\frac{1}{[A]_o^{\alpha - 1}}(2^{\alpha - 1} - 1) = (\alpha - 1)kt_{1/2}$$

20.15 Show that for a first order reaction R ⟶ P the concentration of product can be represented as a function of time by $[P] = a + bt + ct^2 + \ldots$ and express a, b, and c in terms of $[R]_o$ and k.

SOLUTION

$$[P] = [R]_o - [R] = [R]_o(1 - e^{-kt})$$

Expanding e^{-kt} as a power series gives

$$[P] = [R]_o\left(kt - \frac{(kt)^2}{2!} + \frac{(kt)^3}{3!} - \ldots\right)$$

Thus a = 0, $b = [R]_o k$, and $c = [R]_o k^2/2$.

20.16 For a reaction A ⟶ X, the following concentrations of A were found in a single kinetics experiment

$[A]/\text{mol L}^{-1}$	1.000	0.952	0.909	0.870	0.833	0.800
t/hours	0	0.05	0.10	0.15	0.20	0.25

What is the rate v of this reaction at $[A] = 1.000\ \text{mol L}^{-1}$.

SOLUTION

$$v = -\frac{d[A]}{dt}$$

As a first approximation $v = \frac{0.048}{0.05} = 0.960\ \text{mol L}^{-1}\ \text{hr}^{-1}$

A better value may be obtained by using

$[P]/t = b + ct$

for the first 10% of the reaction.

$[X]/mol\ L^{-1}$	.048	.091	.130	.167	.200
t/hr	.05	.10	.15	.10	.25

$$\frac{0.048}{.05} = b + c\ (0.05) = 0.960$$

$$\frac{0.091}{0.10} = b + c\ (0.10) = 0.910$$

Taking the difference between these equations

$$0.050 = -0.05\ c$$

$$c = -1.00, \qquad b = 1.010 = v$$

More experimental points can be included by least squaring. Least squaring through 13% reaction yields

$$v = 1.005\ mol\ L^{-1}\ hr^{-1}$$

Least squaring through 16.7% reaction yields

$$v = 0.998\ mol\ L^{-1}\ hr^{-1}$$

Least squaring through 20% reaction yields

$$v = 0.993\ mol\ L^{-1}\ hr^{-1}$$

As the extent of reaction increases, this truncated series expansion becomes less adequate, and so the initial velocity is $1.000 \pm 0.005\ mol\ L^{-1}\ hr^{-1}$, rather than $0.96\ mol\ L^{-1}\ hr^{-1}$, obtained as a first approximation.

20.17 The following table gives kinetic data [Y. T. Chia and R. E. Connick, J. Phys. Chem., 63, 1518 (1959)] for the following reaction at 25 °C.

$$OCl^- + I^- = OI^- + Cl^-$$

$[OCl^-]$	$[I^-]$	$[OH^-]$	$\frac{d[IO^-]}{dt}/10^{-4}$
	$mol\ L^{-1}$		$mol\ L^{-1}\ s^{-1}$
0.0017	0.0017	1.00	1.75
0.0034	0.0017	1.00	3.50
0.0017	0.0034	1.00	3.50
0.0017	0.0017	0.5	3.50

What is the rate law for the reaction and what is the value of the rate constant?

SOLUTION

When other concentrations are held constant, doubling $[OCl^-]$ doubles the rate, doubling $[I^-]$ doubles the rate, and halving $[OH^-]$ doubles the rate. Therefore the rate law is

$$\frac{d[OI^-]}{dt} = \frac{k[OCl^-][I^-]}{[OH^-]}$$

Substituting the values for the first experiment

$$1.75 \times 10^{-4}\ \text{mol L}^{-1}\ \text{s}^{-1} = \frac{k(0.0017\ \text{mol L}^{-1})(0.0017\ \text{mol L}^{-1})}{(1\ \text{mol L}^{-1})}$$

$$k = 61\ \text{s}^{-1}$$

20.18 When an optically active substance is isomerized, the optical rotation decreases from that of the original isomer to zero in a first-order manner. In a given case the half-time for this process is found to be 10 min. Calculate the rate constant for the conversion of one isomer to another.

SOLUTION

$$t_{1/2} = \frac{0.693}{k_1 + k_{-1}} = 600\ \text{s}$$

Since $K = 1$, $k_1 = k_{-1}$

$$t_{1/2} = \frac{0.693}{2k_1} = 600\ \text{s}$$

$$k_1 = \frac{0.693}{2(600\ \text{s})} = 5.78 \times 10^{-4}\ \text{s}^{-1}$$

20.19 The first three steps in the decay of ^{238}U are

$$^{238}U \xrightarrow[4.5 \times 10^9\ y]{\alpha} {}^{234}Th \xrightarrow[24.1\ d]{\beta} {}^{234}Pa \xrightarrow[1.14\ m]{\beta} {}^{234}U$$

If we start with pure ^{238}U, what fraction will be ^{234}Th after 10, 20, 40, and 80 days?

SOLUTION

$$k_1 = \frac{0.693}{(4.5 \times 10^9\ y)(365\ d\ y^{-1})} = 4.2 \times 10^{-13}\ d^{-1}$$

$$k_2 = \frac{0.693}{24.1\ d} = 0.0288\ d^{-1}$$

$$F = \frac{[^{234}Th]}{[^{238}U]_o} = \frac{k_1}{k_2 - k_1}\left(e^{-k_1 t} - e^{-k_2 t}\right)$$

t/d	10	20	40	80
F	3.65×10^{-12}	6.39×10^{-12}	9.98×10^{-12}	13.14×10^{-12}

20.20 For the reaction 2A = B + C the rate law for the forward reaction is

$$-\frac{d[A]}{dt} = k[A]$$

Give two possible rate laws for the reverse reaction.

SOLUTION

$$K = \frac{[B][C]}{[A]^2}$$

At equilibrium $-\frac{d[A]}{dt} = 0 = k_f[A]^2 - k_r[B][C]$

$$0 = k_f[A] - k_r[B][C]/[A]$$

Possible rate law for the reverse reaction $k_r[B][C]/[A]$

$$K = \frac{[B]^{1/2}[C]^{1/2}}{[A]}$$

At equilibrium $-\frac{d[A]}{dt} = 0 = k_f[A] - k_r[B]^{1/2}[C]^{1/2}$

Possible rate law for the reverse reaction $k_r[B]^{1/2}[C]^{1/2}$

$$K = \frac{[B]^2[C]^2}{[A]^4}$$

At equilibrium $\frac{-d[A]}{dt} = 0 = k_f[A]^4 - k_r[B]^2[C]^2$

$$0 = k_f[A] - k_r\frac{[B]^2[C]^2}{[A]^3}$$

Possible rate law for the reverse reaction $k_r[B]^2[C]^2/[A]^3$

20.21 Suppose the transformation of A to B occurs by both a reversible first-order reaction and a reversible second-order reaction involving hydrogen ion.

$$A \underset{k_2}{\overset{k_1}{\rightleftharpoons}} B \qquad\qquad A + H^+ \underset{k_4}{\overset{k_3}{\rightleftharpoons}} B + H^+$$

What is the relationship between these four rate constants?

SOLUTION

$$\frac{[B]_{eq}}{[A]_{eq}} = \frac{k_1}{k_2}$$

$$\frac{[B]_{eq}[H^+]_{eq}}{[A]_{eq}[H^+]_{eq}} = \frac{k_3}{k_4} \quad \text{or} \quad \frac{[B]_{eq}}{[A]_{eq}} = \frac{k_3}{k_4}$$

Therefore $\frac{k_1}{k_2} = \frac{k_3}{k_4}$ or $k_1k_4 = k_2k_3$

20.22 The hydrolysis of $(CH_2)_6C\langle^{Cl}_{CH_3}$ in 80% ethanol follows the first-order rate equation. The values of the specific reaction-rate constants, as determined by H. C. Brown and M. Borkowski [J. Am. Chem. Soc., 74, 1896 (1952)], are as follows.

t/°C	0	25	35	45
k/s^{-1}	1.06×10^{-5}	3.19×10^{-4}	9.86×10^{-4}	2.92×10^{-3}

(a) Plot log k against 1/T; (b) calculate the activation energy; (c) calculate the pre-exponential factor.

SOLUTION

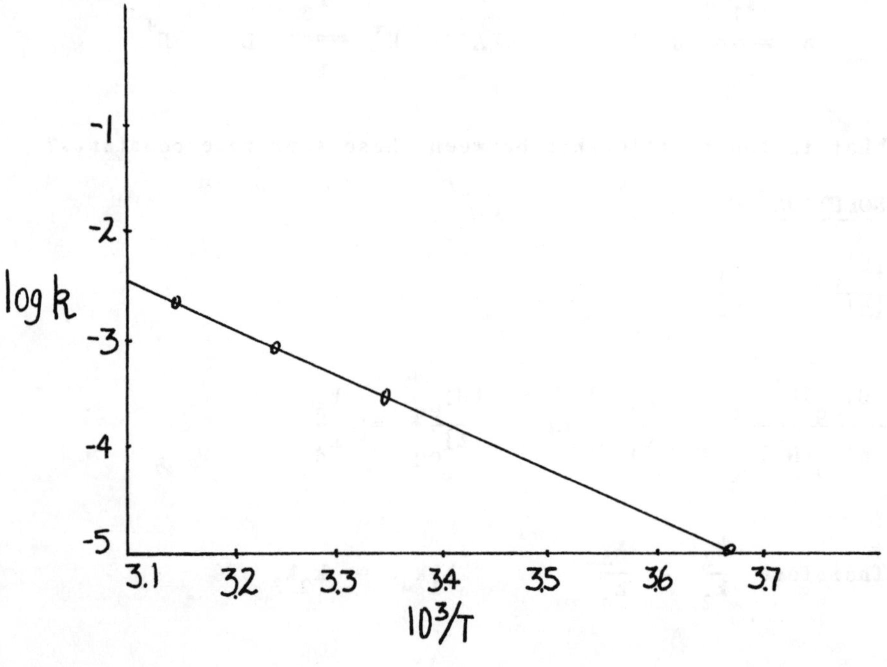

(b) slope $= \frac{-E_a}{2.303\ R} = -4600\ K$

$E_2 = (4600\ K)(2.303)(8.314\ K^{-1}\ mol^{-1})$

$= 88.1\ kJ\ mol^{-1}$

(c) Taking the 35 °C point and $E_a = 88.1\ kJ\ mol^{-1}$

$$\log k = \frac{-E_a}{2.303\ RT} + \log A$$

$$-3.006 = \frac{-88{,}100\ J\ mol^{-1}}{2.303(8.314\ K^{-1}\ mol^{-1})(308.15\ K)} + \log A$$

$$A = 8.4 \times 10^{11}\ s^{-1}$$

20.23 (a) The viscosity of water changes about 2% per degree at room temperature. What is the activation energy for this process? (b) The activation energy for a reaction is 62.8 kJ mol^{-1}. Calculate k_{35}/k_{25}.

SOLUTION

(a) $\frac{1}{\eta} = Ae^{-E_a/RT}$

$$-\ln \eta = \ln A - \frac{E_a}{RT}$$

$$-\frac{d\eta}{\eta} = \frac{E_a}{RT^2}\,dT$$

$$E_a = -RT^2\,\frac{\Delta\eta/\eta}{\Delta T}$$

$$= -(8.314\ J\ K^{-1}\ mol^{-1})(300\ K)^2(-0.02)/(1\ K)$$

$$= 15.1\ kJ\ mol^{-1}$$

(b) $\ln \frac{k_{35}}{k_{25}} = \frac{\Delta H(T_2 - T_1)}{RT_1T_2}$

$$= \frac{(62{,}800 \text{ J mol}^{-1})(10 \text{ K})}{(8.314 \text{ J K}^{-1} \text{ mol}^{-1})(298.15 \text{ K})(308.15 \text{ K})}$$

$$\frac{k_{35}}{k_{25}} = 2.27$$

20.24 If a first-order reaction has an activation energy of 104,600 J mol^{-1} and, in the equation $k = Ae^{-E_a/RT}$, A has a value of 5×10^{13} s^{-1}, at what temperature will the reaction have a half-life of (a) 1 min, and (b) 30 days?

SOLUTION

$$k = \frac{0.693}{t_{1/2}} = Ae^{-E_a/RT}$$

$$T = \frac{-E_a}{R \ln\left(\frac{0.693}{At_{1/2}}\right)}$$

$$\text{(a) } T = \frac{-104{,}600 \text{ J mol}^{-1}}{(8.314 \text{ J K}^{-1} \text{ mol}^{-1}) \ln\left[\frac{0.693}{(5 \times 10^{13} \text{ s}^{-1})(60 \text{ s})}\right]}$$

$$= 349 \text{ K} \quad \text{or} \quad 76\ ^\circ\text{C}$$

$$\text{(b) } T = \frac{-104{,}600 \text{ J mol}^{-1}}{(8.314 \text{ J K}^{-1} \text{ mol}^{-1}) \ln\left[\frac{0.693}{(5 \times 10^{13})(60)^2(24)(30)}\right]}$$

$$= 270 \text{ K} \quad \text{or} \quad -3\ ^\circ\text{C}$$

20.25 Isopropenyl allyl ether in the vapor state isomerizes to allyl acetone according to a first-order rate equation. The following equation gives the influence of temperature on the rate constant (in s^{-1}).

$$k = 5.4 \times 10^{11}\ e^{-123{,}000/RT}$$

where the activation energy is expressed in J mol^{-1}. At 150 °C, how long will it take to build up a partial pressure of 0.395 bar of allyl acetone, starting with 1 bar of isopropenyl allyl ether [L. Stein and G. W. Murphy, J. Am. Chem. Soc., 74, 1041 (1952)]?

SOLUTION

$$k = (5.4 \times 10^{11}\ \text{s}^{-1})e^{-\frac{123{,}000}{(8.314)(423.15)}}$$

$$= 3.53 \times 10^{-4}\ \text{s}^{-1}$$

$$t = \frac{1}{k} \ln \frac{P^{\circ}}{P} = \frac{1}{3.53 \times 10^{-4}\ \text{s}^{-1}} \ln \frac{1\ \text{bar}}{0.605\ \text{bar}}$$

$$= 1420\ \text{s}$$

20.26 For the mechanism

$$A + B \underset{k_2}{\overset{k_1}{\rightleftharpoons}} C \qquad C \xrightarrow{k_3} D$$

(a) Derive the rate law using the steady-state approximation to eliminate the concentration of C. (b) Assuming that $k_3 \ll k_2$, express the pre-exponential factor A and E_a for the apparent second-order rate constant in terms of A_1, A_2, and A_3 and E_{a1}, E_{a2}, and E_{a3} for the three steps.

SOLUTION

(a) $$\frac{d[C]}{dt} = k_1[A][B] - (k_2 + k_3)[C] = 0$$

$$\frac{d[D]}{dt} = k_3[C] = \frac{k_1 k_3 [A][B]}{k_2 + k_3}$$

(b) For $k_2 \gg k_3$,

$$k_{app} = \frac{k_1 k_3}{k_2} = \frac{A_1 e^{-E_{a1}/RT} A_3 e^{-E_{a3}/RT}}{A_2 e^{-E_{a2}/RT}}$$

$$= \frac{A_1 A_3}{A_2} e^{-(E_{a1} + E_{a3} - E_{a2})/RT}$$

$$A_{app} = \frac{A_1 A_3}{A_2} \qquad E_{app} = E_{a1} + E_{a3} - E_{a2}$$

20.27 For the two parallel reactions $A \xrightarrow{k_1} B$ and $A \xrightarrow{k_2} C$, show that the activation energy E' for the disappearance of A is given in terms of the activation energies E_1 and E_2 for the two paths by

$$E' = \frac{k_1E_1 + k_2E_2}{k_1 + k_2}$$

SOLUTION

The rate equation for A is

$$\frac{-d[A]}{dt} = k_1[A] + k_2[A] = (k_1 + k_2)[A] = k'[A]$$

where $k' = k_1 + k_2 = Ae^{-E'/RT}$

$$\frac{d \ln k'}{dT} = \frac{E'}{RT^2} = \frac{d \ln(k_1 + k_2)}{dT} = \frac{d(k_1 + k_2)}{(k_1 + k_2)dT} = \frac{1}{(k_1 + k_2)}\left(\frac{dk_1}{dT} + \frac{dk_2}{dT}\right)$$

$$= \frac{1}{(k_1 + k_2)} k_1\left[\frac{d \ln k_1}{dT} + k_2 \frac{d \ln k_2}{dT}\right]$$

$$= \frac{1}{(k_1 + k_2)}\left[\frac{k_1E_1}{RT^2} + \frac{k_2E_2}{RT^2}\right]$$

$$E' = \frac{k_1E_1 + k_2E_2}{k_1 + k_2}$$

20.28 The reaction $2NO + O_2 \longrightarrow 2NO_2$ is third order. Assuming that a small amount of NO_3 exists in rapid reversible equilibrium with NO and O_2 and that the rate-determining step is the slow bimolecular reaction $NO_3 + NO \longrightarrow 2NO_2$, derive the rate equation for this mechanism.

SOLUTION

$$NO + O_2 \rightleftharpoons NO_3 \qquad K = \frac{[NO_3]}{[NO][O_2]}$$

$$NO_3 + NO \xrightarrow{k_2} 2NO_2 \quad \text{(rate determining)}$$

$$\frac{d[NO_2]}{dt} = 2k_2[NO_3][NO] = 2k_2\, K[NO]^2[O_2]$$

$$= k'[NO]^2[O_2]$$

20.29 Set up the rate expressions for the following mechanism

$$A \underset{k_2}{\overset{k_1}{\rightleftharpoons}} B \qquad B + C \xrightarrow{k_3} D$$

If the concentration of B is small compared with the concentrations of A, C, and D, the steady-state approximation may be used to derive the rate law. Show that this reaction may follow the first-order equation at high pressures and the second-order equation at low pressures.

SOLUTION

Since the concentration of B is small, it is assumed to be in a steady state.

$$\frac{d[B]}{dt} = k_1[A] - (k_2 + k_3[C])[B] = 0$$

$$[B] = \frac{k_1[A]}{k_2 + k_3[C]}$$

$$\frac{d[D]}{dt} = k_3[B][C] = \frac{k_1 k_3 [A][C]}{k_2 + k_3[C]}$$

At high pressures, $k_3[C] >> k_2$, $\frac{d[D]}{dt} = k_1[A]$

At low pressures, $k_2 >> k_3[C]$, $\frac{d[D]}{dt} = \frac{k_1 k_3}{k_2}[A][C]$

20.30 The reaction $NO_2Cl = NO_2 + \frac{1}{2} Cl_2$ is first order and appears to follow the mechanism

$$NO_2Cl \xrightarrow{k_1} NO_2 + Cl \qquad NO_2Cl + Cl \xrightarrow{k_2} NO_2 + Cl_2$$

(a) Assuming a steady state for the chlorine atom concentration, show that the empirical first-order rate constant can be identified with $2k_1$. (b) The following data were obtained by H. F. Cordes and H. S. Johnston [J. Am. Chem. Soc., 76, 4264 (1954)] at 180 °C. In a single experiment the reaction is first order, and the empirical rate constant is represented by k. Show that the reaction is second order at these low gas pressures and calculate the second-order rate constant.

$c/10^{-8}$ mol cm^{-3}	5	10	15	20
$k/10^{-4}$ s^{-1}	1.7	3.4	5.2	6.9

SOLUTION

(a) $\frac{d[Cl]}{dt} = k_1[NO_2Cl] - k_2[NO_2Cl][Cl] = 0$

Therefore $[Cl] = \frac{k_1}{k_2}$

$$\frac{-d[NO_2Cl]}{dt} = k_1[NO_2Cl] + k_2[NO_2Cl][Cl]$$

$$= k_1[NO_2Cl] + k_2[NO_2Cl]k_1/k_2$$

$$= 2k_1[NO_2Cl]$$

(b) At low pressures the rate determining step is the first step, and the reaction becomes second order.

$$NO_2Cl + NO_2Cl \xrightarrow{k_1'} NO_2Cl + NO_2 + Cl$$

$$\frac{-d[NO_2Cl]}{dt} = k_1'[NO_2Cl]^2$$ assuming $[NO_2Cl]$ is approximately constant

$$k_{obs} = k_1'[NO_2Cl]$$

$$1.7 \times 10^{-4} = k_1'(5 \times 10^{-8})$$

$$k_1' = 3.4 \times 10^3 \text{ cm}^3 \text{ mol}^{-1} \text{ s}^{-1}$$

20.31 The apparent activation energy for the recombination of iodine atoms in argon is -5.9 kJ mol^{-1}. This negative temperature coefficient may result from the following mechanism.

$$I + M = IM \qquad K = \frac{[IM]}{[I][M]} \qquad IM + I \underset{k_{-1}}{\overset{k_1}{\rightleftharpoons}} I_2 + M$$

Assuming that the first step remains at equilibrium, derive the rate equation that includes both the forward and reverse reactions. Show that the reverse reaction is bimolecular and the equilibrium constant expression for the dissociation of iodine is independent of the concentration of the third body.

SOLUTION

$$\frac{d[I_2]}{dt} = k_1[I][IM] - k_{-1}[I_2][M]$$

Since $[IM] = K[I][M]$

$$\frac{d[I_2]}{dt} = k_1K[I]^2[M] - k_1[I_2][M]$$

Thus the rate law for the reverse reaction is $k_{-1}[I_2][M]$.

At equilibrium $d[I_2]/dt = 0$ and $\frac{[I_2]}{[I]^2} = \frac{k_1}{k_{-1}}K$

20.32 What is the rate constant for the following reaction at 500 k?

$$H + HCl \longrightarrow Cl + H_2$$

The data required are to be found in Tables 20.3 and A.2.

SOLUTION

The rate constant for the backward reaction is given by

$$k_b = (10^{10.9}\ L\ mol^{-1}\ s^{-1})e^{-23{,}000/8.314 \times 500}$$

$$= 3.1 \times 10^8\ L\ mol^{-1}\ s^{-1}$$

$$\Delta G^o = \Delta_f G^o(Cl) - \Delta_f G^o(H) - \Delta_f G^o(HCl)$$

$$= 94.191 - 192.955 + 97.169$$

$$= -1.595\ kJ\ mol^{-1}$$

$$K_p = e^{1595/8.314 \times 500} = 0.384$$

$$K_p = K_c = \frac{k_f}{k_b}$$

$$k_f = (0.384)(3.1 \times 10^8\ L\ mol^{-1} s^{-1}) = 1.2 \times 10^8\ L\ mol^{-1}\ s^{-1}$$

20.33 What are the Arrhenius parameters for the following elementary reaction?

$$H + HCl \longrightarrow Cl + H_2$$

The data required are to be found in Tables 20.3 and A.2.

<u>SOLUTION</u>

At 500 K

$$\Delta H^\circ = \Delta_f H^\circ(1) - \Delta_f H^\circ(H) - \Delta_f H^\circ(HCl)$$

$$= 122.261 - 219.254 + 92.914 = -4.079 \text{ kJ mol}^{-1}$$

$$\Delta S^\circ = S^\circ(Cl) + S^\circ(H_2) - S^\circ(H) - S^\circ(Cl)$$

$$= 176.749 + 145.737 - 125.462 - 201.995$$

$$= -4.971 \text{ J K}^{-1} \text{ mol}^{-1}$$

$$K_P = K_c = e^{-\Delta H^\circ/RT} e^{\Delta S^\circ/R}$$

$$= e^{4079/8.314\ T} e^{-4.971/8.314} = 0.550\ e^{490.6/T}$$

$$k_b = (10^{10.9} \text{ L mol}^{-1} \text{ s}^{-1}) e^{-23,000/8.314\ T}$$

$$K_c = k_f/k_b$$

$$k_f = k_b K_c = (10^{10.9} \text{ L mol}^{-1} \text{ s}^{-1}) e^{-23,000/RT} (0.550) e^{4079/RT}$$

$$= (4.4 \times 10^{10} \text{ L mol}^{-1} \text{ s}^{-1}) e^{-18,900/RT}$$

20.34 For the gas reaction $O + O_2 + M \underset{k'}{\overset{k}{\rightleftharpoons}} O_3 + M$

When M = O_2, Benson and Axworthy [<u>J</u>. <u>Chem</u>. <u>Phys</u>., <u>26</u>, 1718 (1957)] obtained $k = (6.0 \times 10^7 \text{ L}^2 \text{ mol}^{-2} \text{ s}^{-1}) e^{2.5/RT}$ where the activation energy is in kJ mol^{-1}. Calculate the values of the parameters in the Arrhenius equation for the reverse reaction assuming ΔH^o and ΔS are independent of temperature.

SOLUTION

$$\Delta H^o = 142.7 - 249.170 = -106.5 \text{ kJ mol}^{-1}$$

$$\Delta S^o = 238.82 - 160.946 - 205.029 = -127.16 \text{ J K}^{-1} \text{ mol}^{-1}$$

$$K_P = e^{-\Delta H^o/RT} \; e^{-\Delta S^o/R}$$

$$= e^{106,500/RT} \; e^{-127.16/R}$$

$$K_c = K_p \left(\frac{P^o}{c^o RT}\right)^{\Sigma -\nu_i} = K_P \left(\frac{1}{24.46}\right)^{1} \text{ L mol}^{-1} = \frac{k}{k'}$$

$$k' = \frac{k}{24.46 \, K_P}$$

$$k' = \frac{6 \times 10^7 \; e^{2500/RT}}{24.46 \; e^{106,500/RT} \; e^{-127.16/R}}$$

$$= \left[\frac{6 \times 10^7}{24.46} \; e^{127.16R}\right] e^{2500 - 106,500)/RT}$$

$$= (1.1 \times 10^{13} \text{ L mol}^{-1} \text{ s}^{-1}) \; e^{-104,000/RT}$$

20.35 Derive the steady-state rate equation for the following mechanism for a trimolecular reaction.

$$A + A \underset{k_{-1}}{\overset{k_1}{\rightleftharpoons}} A_2^* \qquad\qquad A_2^* + M \xrightarrow{k_2} A_2 + M$$

SOLUTION

$$\frac{d[A_2^*]}{dt} = k_1[A]^2 - (k_{-1} + k_2[M])[A_2^*] = 0$$

$$\frac{d[A_2]}{dt} = k_2[A_2^*][M] = \frac{k_1 k_2[M][A]^2}{k_{-1} + k_2[M]}$$

20.36 For the mechanism

$$H_2 + X_2 \underset{k_{-1}}{\overset{k_1}{\rightleftharpoons}} 2HX \qquad X + H_2 \underset{k_{-3}}{\overset{k_3}{\rightleftharpoons}} HX + H$$

$$X_2 \underset{k_{-2}}{\overset{k_2}{\rightleftharpoons}} 2X \qquad H + X_2 \underset{k_{-4}}{\overset{k_4}{\rightleftharpoons}} HX + X$$

show that the steady-state rate law is

$$\frac{d[HX]}{dt} = 2k_1[H_2][X_2]\left(1 - \frac{[HX]^2}{K[H_2][X_2]}\right)\left(1 + \frac{\frac{k_3}{k_1}\sqrt{\frac{2k_2}{k_{-2}[X_2]}}}{1 + \frac{k_{-3}[X_2]}{k_4[X_2]}}\right)$$

SOLUTION

There are five species, but there are two conservation equations.

$$[H] + [HX] + 2[H_2] = \text{const.}$$

$$[X] + [HX] + 2[X_2] = \text{const.}$$

Therefore there are only three independent rate equations.

(1) $$\frac{d[HX]}{dt} = 2k_1[H_2][X_2] + k_3[X][H_2] + k_4[H][X_2] - k_{-1}[HX]^2 - k_{-3}[HX][H] - k_{-4}[HX][X]$$

(2) $$\frac{d[X]}{dt} = 2k_2[X_2] + k_{-3}[HX][H] + k_4[H][X_2] - k_{-2}[X]^2 - k_3[X][H_2] - k_{-4}[HX][X] = 0$$

(3) $$\frac{d[H]}{dt} = k_3[X][H_2] + k_{-4}[HX][X] - k_{-3}[HX][H] - k_4[H][X_2] = 0$$

Adding equations 2 and 3 $$[X] = \sqrt{\frac{2k_2}{k_{-2}}[X_2]}$$

Substituting in equation 3 yields

$$[H] = \frac{k_3[H_2] + k_{-4}[HX]}{k_{-3}[HX] + k_4[X_2]} \sqrt{\frac{2k_2[X_2]}{k_{-2}}}$$

Substituting in equation 1

$$\frac{d[HX]}{dt} = 2k_1[H_2][X_2] - k_{-1}[HX]^2(k_4[X_2] - k_3[HX])$$

$$\times \frac{(k_3[H_2] + k_{-4}[HX]) \sqrt{\frac{2k_2[X_2]}{k_{-2}}}}{k_{-3}[HX] + k_4[X_2]}$$

$$+ (k_3[H_2] - k_{-4}[HX]) \sqrt{\frac{2k_2[X_2]}{k_{-2}}} \quad \frac{(k_{-3}[HX] + k_4[X_2]}{(k_3[HX] + k_4[X_2]}$$

$$\frac{d[HX]}{dt} = 2k_1[H_2][X_2] - 2k_{-1}[HX]^2 + (k_3k_4[H_2][X_2]$$

$$+ k_4k_{-4}[X_2][HX] - k_3k_{-3}[HX][H_2] - k_{-3}k_{-4}[HX]^2$$

$$+ k_3k_{-3}[H_2][HX] + k_3k_4[H_2][X_2] - k_{-3}k_{-4}[HX]^2$$

$$- k_4k_{-4}[HX][X_2]) \times \frac{\sqrt{\frac{2k_2[X_2]}{k_{-2}}}}{k_{-3}[HX] + k_4[X_2]}$$

$$= 2k_1[H_2][X_2] - 2k_{-1}[HX]^2 + (2k_3k_4[H_2][X_2] - 2k_{-3}k_{-4}[HX]^2)$$

$$\times \frac{\sqrt{\frac{2k_2[X_2]}{k_{-2}}}}{k_{-3}[HX] + k_4[X_2]}$$

$$= 2k_1[H_2][X_2]\left(1 - \frac{k_{-1}[HX]^2}{k_1[H_2][X_2]}\right) + 2k_3k_4[H_2][X_2]\left(1 - \frac{k_{-3}k_{-4}[HX]^2}{k_3k_4[H_2][X_2]}\right)$$

$$\times \frac{\sqrt{\dfrac{2k_2[X_2]}{k_{-2}}}}{k_3[HX] + k_4[X_2]}$$

$$\frac{d[HX]}{dt} = 2k_1[H_2][X_2]\left(1 - \frac{[HX]^2}{K[H_2][X_2]}\right)\left(1 + \frac{k_3k_4}{k_1}\ \frac{\sqrt{\dfrac{2k_2[X_2]}{k_{-2}}}}{k_{-3}[HX] + k_4[X_2]}\right)$$

$$= 2k_1[H_2][X_2]\left(1 - \frac{[HX]^2}{K[H_2][X_2]}\right)\left(1 + \frac{\dfrac{k_3}{k_1}\sqrt{\dfrac{2k_2}{k_{-2}[X_2]}}}{1 + \dfrac{k_{-3}[HX]}{k_4[X_2]}}\right)$$

20.37 The mechanism of the pyrolysis of acetaldehyde at 520 °C and 0.2 bar is

$$CH_3CHO \xrightarrow{k_1} CH_3 + CHO$$

$$CH_3 + CH_3CHO \xrightarrow{k_2} CH_4 + CH_3CO$$

$$CH_3CO \xrightarrow{k_3} CO + CH_3$$

$$CH_3 + CH_3 \xrightarrow{k_4} C_2H_6$$

What is the rate law for the reaction of acetaldehyde, using the usual assumptions? (As a simplification further reactions of the radical CHO have been omitted and its rate equation may be ignored.)

SOLUTION

(1) $$\frac{d[CH_3CHO]}{dt} = -(k_1 + k_2[CH_3])[CH_3CHO] = 0$$

(2) $$\frac{d[CH_3]}{dt} = k_1[CH_3CHO] - k_2[CH_3][CH_3CHO] + k_3[CH_3CO] - 2k_4[CH_3]^2 = 0$$

(3) $$\frac{d[CH_3CO]}{dt} = k_2[CH_3][CH_3CHO] - k_3[CH_3CO] = 0$$

Equation 3 yields $[CH_3CO] = k_2[CH_3][CH_3CHO]/k_3$

Substituting this in equation 2 yields

$$[CH_3] = \left(\frac{k_1[CH_3CHO]}{2k_4}\right)^{1/2}$$

Substituting this in equation 1 yields

$$\frac{d[CH_3CHO]}{dt} = -\left[k_1 + k_2\left(\frac{k_1}{2k_4}\right)^{1/2} [CH_3CHO]^{1/2}\right][CH_3CHO]$$

If k_1 is small,

$$\frac{d[CH_3CHO]}{dt} = -k_2(k_1/2k_4)^{1/2}[CH_3CHO]^{3/2}$$

20.38 (a) at 27 min $k = 0.0348\ min^{-1}$
at 60 min $k = 0.0347\ min^{-1}$

(b) 20.0 min

20.39 $0.25\ min^{-1}$ 2.77 min 4.00 min

20.40 (a) first (b) $3.59 \times 10^{-5}\ s^{-1}$ (c) 0.354

20.41 $4.03 \times 10^{-9}\ min^{-1}$

20.42 14.6%

20.43 (a) 4.97×10^{-7} % (b) 6950 g

20.44 (a) 0.107 L mol^{-1} s^{-1} (b) 2850 s

20.45 (a) 221 s (b) 82.5 s

20.46 40.4 bar^{-1} s^{-1}

20.47 (a) 80% (b) 67.2% (c) 61.7%

20.48 (a) 3.5% (b) 9%

20.49 (a) 395 s (b) 32,900 ft^3

20.50 $t_{1/2} = \frac{2(\sqrt{2} - 1)}{k[A_o]^{1/2}}$

20.51 3.3×10^{-3} s

20.52 $-\frac{d[H_2SeO_3]}{dt} = k[H_2SeO_3][H^+]^2[I^-]^3$

20.53 0.00750 mol L^{-1}

20.55 $[B] = k[A]_o te^{-kt}$

$[C] = [A]_o[1 - e^{-kt}(1 + kt)]$

20.56 $\frac{d[BrO_3^-]}{dt} = k'[SO_3^{2-}]^2[SO_4^{2-}]^3[H^+][Br^-]$

20.57 (a) 97.1 kJ mol^{-1} (b) 9.3×10^{13} s^{-1}

(c) 1.7 s

20.58 (a) 447 °C (b) 388 °C

20.59 (a) 5.3×10^6 s (b) 2740 s

20.60 550 s

20.61 (a) 1091 K (b) 2180 K (c) 727 K

(d) The rate of the reaction with the higher activation energy increases more rapidly with increasing temperature than the rate of the reaction with the lower activation energy.

20.62 $A = 3 \times 10^9\ L^2\ mol^{-2}\ s^{-1}$ $E_a = 5730\ J\ mol^{-1}$

20.63 (a) $1.07\ bar^{-1}\ s^{-1}$ $2.06 \times 10^{-3}\ bar^{-1}\ s^{-1}$

(b) 0.013 s 0.62 s

20.64 $$K = [AB]/[A][B] = \exp\left(\frac{\Delta S^\circ}{R}\right) \exp\left(-\frac{\Delta H^\circ}{RT}\right)$$

$$\frac{d[D]}{dt} = k[C][AB] = k[C][A][B] \exp\left(\frac{\Delta S^\circ}{R}\right) \exp\left(-\frac{\Delta H^\circ}{RT}\right)$$

$$= s \exp\left(-\frac{E_a}{RT}\right) [C][A][B] \exp\left(\frac{\Delta S^\circ}{R}\right) \exp\left(-\frac{\Delta H^\circ}{RT}\right)$$

$$k = A \exp\left[\frac{-E_a + \Delta H^\circ}{RT}\right] \qquad \text{where } A = S \exp\left(\frac{\Delta S^\circ}{R}\right)$$

20.66 $$\frac{d[O_2]}{dt} = 2k_1[NO][O_3]$$

20.67 $3.0 \times 10^{16}\ L\ mol^{-1}\ s^{-1}$

20.68 $$-\frac{d[A]}{dt} = \left(\frac{3}{4} k_1 + \frac{1}{4}\sqrt{k_1^{\,2} + 8\, k_1 k_2 k_3/k_4}\right)[A]$$

CHAPTER 21: Theoretical Gas Kinetics

21.1 Use equation 20.68 to derive the expression for the activation energy for a reaction following simple collision theory.

SOLUTION

$$E_a = RT^2 \frac{d \ln k}{dT}$$

$$k = N_A \pi d_{AB}^{\;2} (8k_B T/\pi\mu)^{1/2} \, e^{-\varepsilon_0/k_B T}$$

$$E_a = RT^2 \frac{d}{dT} \left\{ \ln \left[N_A \pi d_{AB}^{\;2} (8k_B/\pi\mu)^{1/2} \right] + \frac{1}{2} \ln T - \varepsilon_0/k_B T \right\}$$

$$= RT^2 \left(\frac{1}{2T} + \frac{\varepsilon_0}{k_B T^2} \right)$$

$$= N_A \varepsilon_0 + \frac{1}{2} RT$$

21.2 Use the pre-exponential factor $A = 3 \times 10^{13}$ $cm^3\ mol^{-1}\ s^{-1}$ for the reaction $Br + H_2 \longrightarrow HBr + H$ to calculate the cross section and collision diameter for this reaction at 400 K.

SOLUTION

$$\mu N_A = \frac{1}{\dfrac{1}{79.904 \times 10^{-3}\ kg} + \dfrac{1}{2.0158 \times 10^{-3}\ kg}}$$

$$= 1.966 \times 10^{-3}\ kg\ mol^{-1}$$

$$\left(\frac{8RT}{\pi\mu N_A}\right)^{1/2} = \left[\frac{8(8.314\ J\ K^{-1}\ mol^{-1})(400\ K)}{\pi(1.966 \times 10^{-3}\ kg\ mol^{-1})}\right]^{1/2}$$

$$= 2075\ m\ s^{-1}$$

$$A = \pi d_{12}^{\;2} \left(\frac{8RT}{\pi\mu N_A}\right)^{1/2}$$

$$d_{12} = \left[\frac{A}{\pi\left(\frac{8RT}{\pi\mu N_A}\right)^{1/2}} \right]^{1/2}$$

$$= \left[\frac{(3 \times 10^{13}\ \text{cm}^3\ \text{mol}^{-1}\ \text{s}^{-1})(10^{-2}\ \text{m cm}^{-1})^3}{\pi(6.02 \times 10^{23}\ \text{mol}^{-1})(2075\ \text{m s}^{-1})} \right]^{1/2} = 87.4\ \text{pm}$$

The cross section is $\pi d_{12}^2 = 2.40 \times 10^{-20}\ \text{m}^2$.

21.3 The thermal decomposition of gaseous acetaldehyde is a second-order reaction. The value of E_a is 190,400 J mol^{-1}, and the molecular diameter of the acetaldehyde molecule is 5×10^{-8} cm. (a) Calculate the number of molecules colliding per cm^3 per second at 800 K and 1 bar pressure. (b) Calculate k in L $\text{mol}^{-1}\ \text{s}^{-1}$.

SOLUTION

$$\text{(a)}\quad \rho = \frac{PN_A}{RT} = \frac{(10^5\ \text{N m}^{-2})(6.02 \times 10^{23}\ \text{mol}^{-1})}{(8.314\ \text{J K}^{-1}\ \text{mol}^{-1})(800\ \text{K})}$$

$$= 9.05 \times 10^{24}\ \text{m}^{-3}$$

$$\langle v \rangle = \left(\frac{\pi RT}{M}\right)^{1/2} = 689\ \text{m s}^{-1}$$

$$Z_{11} = 2^{-1/2}\rho^2 \pi d^2 \langle v \rangle$$

$$= \frac{(9.05 \times 10^{24}\ \text{m}^{-3})^2\ \pi(5 \times 10^{-10}\ \text{m})^2(689)}{2^{1/2}}$$

$$= 3.13 \times 10^{34}\ \text{m}^{-3}\ \text{s}^{-1}$$

$$= (3.13 \times 10^{34}\ \text{m}^{-3}\ \text{s}^{-1})(10^{-2}\ \text{m cm}^{-1})^3$$

$$= 3.13 \times 10^{28}\ \text{cm}^{-3}\ \text{s}^{-1}$$

(b) $k = \frac{(10^3\ L\ m^{-3})N_A\ Z_{11}}{\rho^2} e^{-E_o/RT}$

$= \frac{(10^3\ L\ m^{-3})(6.022 \times 10^{23}\ mol^{-1})(3.13 \times 10^{34}\ m^{-3}\ s^{-1})}{(9.17 \times 10^{24}\ m^{-3})} \times e^{-\frac{190,400}{(8.314)(800)}}$

$= 0.083\ L\ mol^{-1}\ s^{-1}$

21.4 The pre-exponential factor for the reaction $H_2 + I_2 = 2\ HI$ is $10^{11}\ L\ mol^{-1}\ s^{-1}$ and the activation energy is 165 kJ mol^{-1} in the range 300 - 500 °C. If the collision diameter is 320 pm, what value of the pre-exponential factor is expected from collision theory at 600 K and what is the value of the steric factor p? (At higher temperatures this reaction goes by the unbranched chain mechanism described in Section 20.15.)

SOLUTION

$\mu N_A = \frac{1}{\frac{1}{2(1.0079 \times 10^{-3}\ kg)} + \frac{1}{2(126.9045 \times 10^{-3}\ kg)}}$

$= 1.999 \times 10^{-3}$ kg

$A = \pi d_{12}^{\ 2} \left(\frac{8RT}{\pi\mu N_A}\right)^{1/2}$

$A = \pi(320 x 10^{-12} m)^2 \left[\frac{8(8.314\ J\ K^{-1} mol^{-1})(600\ K)}{\pi(1.999 \times 10^{-3}\ kg)}\right]^{1/2} (6.02 x 10^{23}\ mol^{-1}) \times (10^3\ L\ m^{-3})$

$= 4.87 \times 10^{11}\ L\ mol^{-1}\ s^{-1}$

$A_{exp} = 10^{11}\ L\ mol^{-1}\ s^{-1} = (4.87 \times 10^{11}\ L\ mol^{-1}\ s^{-1})p$

$p = 0.21$

21.5 Show that the activated complex theory yields the simple collision theory result when it is applied to the reaction of two rigid spherical molecules.

SOLUTION

Assuming that the molecules react on their first collision (that is, the activation energy is zero), activated complex theory yields

$$k = \frac{RT}{h} \frac{q''_{AB}{}^{\ddagger}}{q_A' \, q_B'} \qquad (1)$$

where q_A' and q_B' are molecular partition functions without the volume factor, and $q''_{AB}{}^{\ddagger}$ is the molecular partition function for the activated complex without the volume factor and without the vibrational factor.

$$q_A' = \left(\frac{2\pi m_A kT}{h^2}\right)^{3/2} \qquad (2) \qquad\qquad q_B' = \left(\frac{2\pi m_B kT}{h^2}\right)^{3/2} \qquad (3)$$

$$q_{AB}'' = \left[\frac{2\pi(m_A + m_B)kT}{h^2}\right]^{3/2} \frac{8\pi^2 \mu (R_A + R_B)^2 kT}{h^2} \qquad (4)$$

$$\mu = \frac{m_A \, m_B}{m_A + m_B} \qquad (5)$$

Substituting equations 2, 3, 4 and 5 in equation 1 yields

$$k = N_A \left(\frac{8\pi kT}{\mu}\right)^{1/2} (R_A + R_B)^2$$

which may be compared with equation 21.2

$$k = \pi d_{12}{}^2 \left(\frac{8RT}{\pi\mu N_A}\right)^{1/2} = \left(\frac{8\pi kT}{\mu}\right)^{1/2} (R_A + R_B)^2$$

The difference between these expressions of a factor of N_A is simply a matter of units.

21.6 What is the expression for the pre-exponential factor of a second order gas reaction according to the thermodynamic formulation of transition-state theory? What is the value of the pre-exponential factor at 500 K if the entropy of activation is zero? When this value is compared with values in Table 20.3, what do you conclude about the sign of $\Delta S^\ddagger$?

SOLUTION

$$A = \frac{e^2 RT}{c^\circ N_A h} = \frac{(2.718)^2(8.314\ \text{J K}^{-1}\ \text{mol}^{-1})(500\ \text{K})}{(1\ \text{mol L}^{-1})(6.022 \times 10^{23}\ \text{mol}^{-1})(6.63 \times 10^{-34}\ \text{J s})}$$

$$= 7.70 \times 10^{13}\ \text{L mol}^{-1}\ \text{s}^{-1}$$

$$\log(A/\text{L mol}^{-1}\ \text{s}^{-1}) = 13.9$$

$\Delta S^\ddagger$ is negative for all of these reactions, and is more negative for metathesis reactions not involving atoms. This means that the transition state has a lower entropy than the reactants.

21.7 F. W. Schuler and G. W. Murphy [J. Am. Chem. Soc., 72, 3155 (1950)] studied the thermal rearrangement of vinyl allyl ether to allyl acetaldehyde in the range 150 to 200 °C and found that

$$k = 5 \times 10^{11}\ e^{-128,000/RT}$$

where k is in s^{-1} and the activation energy is in J mol^{-1}. Calculate (a) the enthalpy of activation, and (b) the entropy of activation, and (c) give an interpretation of the latter.

SOLUTION

The values of $\Delta H^\ddagger$ and $\Delta S^\ddagger$ are calculated for the mean temperature of the range 273 + 175 = 448 k.

(a) $\Delta H^\ddagger = E_a - RT$

$$= 128.0\ \text{kJ mol}^{-1} - (8.314 \times 10^{-3}\ \text{kJ K}^{-1}\ \text{mol}^{-1})(448\ \text{K})$$

$$= 124.3 \text{ kJ mol}^{-1}$$

(b) $$\Delta S^{\ddagger} = R\left[\ln \frac{N_A hA}{RT} - 1\right]$$

$$= (8.314 \text{ J K}^{-1}\text{mol}^{-1})\left[\ln \frac{(6.022\times10^{23} \text{ mol}^{-1})(6.626\times10^{-34} \text{ J s})A}{(8.314 \text{ J K}^{-1} \text{ mol}^{-1})(448 \text{ K})} - 1\right.$$

$$= -32.6 \text{ J K}^{-1} \text{ mol}^{-1}$$

$$A = 5 \times 10^{11} \text{ s}^{-1}$$

(c) The activated complex may be an improbable ring structure which is formed with a decrease in entropy.

21.8 The vapor-phase decomposition of di-t-butyl peroxide is first order in the range 110 to 280 °C and follows the equation

$$k = 3.2 \times 10^{16}\, e^{-163{,}600/RT}$$

where k is in s^{-1} and E_a is in J mol^{-1}. Calculate (a) $\Delta H^{\ddagger}$, and (b) $\Delta S^{\ddagger}$.

SOLUTION

(a) $$\Delta H^{\ddagger} = E_a - RT$$

$$= 163{,}600 \text{ J mol}^{-1} - (8.314 \text{ J K}^{-1} \text{ mol}^{-1})(448 \text{ K})$$

$$= 159{,}900 \text{ J mol}^{-1}$$

(b) $$\Delta S^{\ddagger} = R \ln\left(\frac{N_A hA}{eRT}\right)$$

$$= (8.314 \text{ J K}^{-1}\text{mol}^{-1})\left[\ln \frac{(6.02\times10^{23} \text{ mol}^{-1})(6.626\times10^{-34} \text{ J s})A}{(2.718)(8.314 \text{ J K}^{-1} \text{ mol}^{-1})(448 \text{ K})}\right]$$

$$= 59.4 \text{ J K}^{-1} \text{ mol}^{-1}$$

$$A = 3.2 \times 10^{16} \text{ s}^{-1}$$

21.9 (a) 2.0×10^{11} L mol^{-1} s^{-1}

(b) 3.2×10^{-10}

(c) 2.4×10^{-4} s

21.10 $10^{14.45}$ cm^3 mol^{-1} s^{-1}

21.12 -598, -544 and -523 J K^{-1} mol^{-1} when the standard concentration is taken as 1 mol L^{-1}. The activated complex is more organized (lower entropy) than the reactants. This is especially true for the first reaction.

21.13 (a) 6.25×10^{14} s^{-1} (b) 9×10^{-7} s^{-1}

21.14 (a) 224 kJ mol^{-1} (b) 18.0 J K^{-1} mol^{-1}

CHAPTER 22: Kinetics: Liquid Phase

22.1 Show that if A and B can be represented by spheres of the same radius that react when they touch, the second-order rate constant is given by

$$k_a = \frac{8 \times 10^3 \; RT}{3\eta} \; L \; mol^{-1} \; s^{-1}$$

where R is in J K^{-1} mol^{-1}. To obtain this result the diffusion coefficient is expressed in terms of the radius of a spherical particle by use of equation 22.45. For water at 25 °C, $\eta = 8.95 \times 10^{-4}$ kg m^{-1} s^{-1}. Calculate k at 25 °C.

SOLUTION

The diffusion coefficient for a spherical particle of radius r is given by

$$D = \frac{RT}{N_A 6\pi\eta r}$$

Substituting in equation 22.4

$$k_a = 4\pi(10^3 \; L \; m^{-3})N_A(D_1 + D_2)R_{12}$$

$$= 4\pi(10^3 \; L \; m^{-3})\left(\frac{2RT}{6\pi\eta r}\right)(2r)$$

$$= \frac{8 \times 10^3 \; RT}{3\eta} \; L \; mol^{-1} \; s^{-1}$$

where RT is expressed in J mol^{-1} and η is expressed in kg m^{-1} s^{-1}.

For water at 25 °C

$$k_a = \frac{8(10^3 \; L \; m^{-3})(8.314 \; J \; K^{-1} \; mol^{-1})(298 \; K)}{3(8.95 \times 10^{-4} \; kg \; m^{-1} \; s^{-1})} = 7.4 \times 10^9 \; L \; mol^{-1} \; s^{-1}$$

22.2 The diffusion coefficient D of an ion is related to its ionic mobility u by

$$D = \frac{uRT}{zF}$$

The ionic mobilities of H^+ and OH^- are

$3.63 \times 10^{-7}\ m^2\ V^{-1}\ s^{-1}$ and

$2.05 \times 10^{-7}\ m^2\ V^{-1}\ s^{-1}$ at 25°C. What is the rate constant for the reaction $H^+ + OH^- \longrightarrow H_2O$? The reaction radius is 0.75 nm, because once the proton is this close the reaction can proceed very rapidly by quantum mechanical tunneling. The electrostatic factor f is 1.70.

SOLUTION

$$D_1 = \frac{(3.63 \times 10^{-7}\ m^2\ V^{-1}\ s^{-1})(8.314\ J\ K^{-1}\ mol^{-1})(298\ K)}{96{,}500\ C\ mol^{-1}}$$

$$= 9.31 \times 10^{-9}\ m^2\ s^{-1}$$

$$D_2 = \frac{(2.05 \times 10^{-7}\ m^2\ V^{-1}\ s^{-1})(8.314\ J\ K^{-1}\ mol^{-1})(298\ K)}{96{,}500\ C\ mol^{-1}}$$

$$= 5.26 \times 10^{-9}\ m^2\ s^{-1}$$

$$k_a = 4\pi(10^3\ L\ m^{-3})N_A(D_1 + D_2)R_{12}f$$

$$= 4\pi(10^3\ L\ m^{-3})(6.022 \times 10^{23}\ mol^{-1})(14.57 \times 10^{-9}\ m^2\ s^{-1})$$

$$\times\ (0.75 \times 10^{-9}\ m)1.7$$

$$= 1.4 \times 10^{11}\ L\ mol^{-1}\ s^{-1}$$

22.3 What is the reaction radius for the reaction

$$H^+ + OH^- \xrightarrow{1.4 \times 10^{11}\ L\ mol^{-1}\ s^{-1}} H_2O$$

at 25 °C given that the diffusion coefficients of H^+ and OH^- at this temperature are $9.1 \times 10^{-9}\ m^2\ s^{-1}$ and $5.2 \times 10^{-9}\ m^2\ s^{-1}$.

SOLUTION

$$k_a = 4\pi N_A(D_1 + D_2)R_{12}f$$

$$R_{12}f = \frac{k_a}{4\pi N_A(D_1 + D_2)}$$

$$= \frac{(1.4 \times 10^{11}\ L\ mol^{-1}\ s^{-1})(10^{-3}\ m^3\ L^{-1})}{4\pi(6.02 \times 10^{23}\ mol^{-1})(14.3 \times 10^{-9}\ m^2\ s^{-1})}$$

$$= 1.29 \times 10^{-9}\ m$$

To calculate the electrostatic factor f, we have to know R_{12}. Successive approximations have to be made. Let us begin by estimating R_{12} = 0.9 nm.

$$\frac{z_1 z_2 e^2}{4\pi\varepsilon_o \kappa kTR_{12}} = \frac{-(1.602 \times 10^{-19}\ C)^2(8.988 \times 10^9\ N\ C^{-2}\ m^2)}{(78.3)(1.38 \times 10^{-23}\ J\ K^{-1})(298\ K)(0.9 \times 10^{-9}\ m)}$$

$$= -0.796 = x$$

$$f = x(e^x - 1)^{-1} = -0.796(e^{-0.796} - 1)^{-1}$$

$$= 1.45$$

$$\text{Thus } R_{12} = \frac{1.29 \times 10^{-9}\ m}{1.45} = 0.89\ nm$$

22.4 Derive the relation between the relaxation time τ and the rate constants for the reaction

$$A \underset{k_2}{\overset{k_1}{\rightleftharpoons}} B,$$ which is subjected to a small displacement from equilibrium.

SOLUTION

$$\frac{d[B]}{dt} = k_1[A] - k_2[B]$$

At equilibrium

$$\frac{d[B]}{dt} = 0 = k_1[A]_{eq} - k_2[B]_{eq}$$

Let $[A] = [A]_{eq} - \Delta[B]$

$[B] = [B]_{eq} + \Delta[B]$

$$\frac{d[B]}{dt} = \frac{d\Delta[B]}{dt} = k_1([A]_{eq} - \Delta[B]) - k_2([B]_{eq} + \Delta[B])$$

$$= k_1[A]_{eq} - k_2[B]_{eq} - (k_1 + k_2)\Delta[B]$$

$$= -(k_1 + k_2)\Delta[B] = -\frac{\Delta[B]}{\tau}$$

$$\tau = (k_1 + k_2)^{-1}$$

22.5 For acetic acid in dilute aqueous solution at 25 °C $K = 1.73 \times 10^{-5}$, and the relaxation time is 8.5×10^{-9} seconds for a 0.1 M solution. Calculate k_a and k_d in

$$CH_3CO_2H \underset{k_a}{\overset{k_d}{\rightleftharpoons}} CH_3CO_2^- + H^+$$

SOLUTION

$$K = \frac{[H^+][CH_3CO_2^-]}{[CH_3CO_2H]} = \frac{x^2}{0.10} = 1.73 \times 10^{-5} = \frac{k_d}{k_a}$$

$$x = (1.73 \times 10^{-6})^{1/2} = 1.32 \times 10^{-3} = [H^+] = [CH_3CO_2^-]$$

$$\tau = \frac{1}{k_d + k_a([H^+] + [CH_3CO_2^-])}$$

$$8.5 \times 10^{-9}\ s = \frac{1}{1.73 \times 10^{-5}\ k_a + k_a(2.64 \times 10^{-3})}$$

$$k_a = 4.42 \times 10^{10}\ L\ mol^{-1}\ s^{-1}$$

$$k_d = (1.73 \times 10^{-5})k_a = 7.65 \times 10^5\ s^{-1}$$

22.6 Calculate the first-order rate constants for the dissociation of the following weak acids: acetic acid, acid form of imidazole $C_3N_2H_5^+$, NH_4^+. The corresponding acid dissociation constants are 1.75×10^{-5}, 1.2×10^{-7}, and 5.71×10^{-10}, respectively. The second-order rate constants for the formation of the acid forms from a proton plus the base are 4.5×10^{10}, and 4.3×10^{10} L mol^{-1} s^{-1}, respectively.

SOLUTION

For $HA \underset{k_2}{\overset{k_1}{\rightleftharpoons}} H^+ + A^-$

$$K = \frac{[H^+][A^-]}{[HA]} = \frac{k_1}{k_2}$$

For acetic acid $k_1 = k_2K$

$$= (4.5 \times 10^{10}\ L\ mol^{1-}\ s^{1-})(1.75 \times 10^{-5}\ mol\ L^{-1})$$

$$= 7.9 \times 10^5\ s^{-1}$$

For imidazole $k_1 = k_2K$

$$= (1.5 \times 10^{10}\ L\ mol^{-1}\ s^{-1})(1.2 \times 10^{-7}\ mol\ L^{-1})$$

$$= 1.8 \times 10^3\ s^{-1}$$

For NH_4^+ $k_1 = k_2K$

$$= (4.3 \times 10^{10}\ L\ mol^{-1}\ s^{-1})(5.71 \times 10^{-10}\ mol\ L^{-1})$$

$$= 25\ s^{-1}$$

22.7 The mutarotation of glucose is first order in glucose concentration and is catalyzed by acids (A) and bases (B). The first-order rate constant may be expressed by an equation of the type that is encountered in reactions with parallel paths.

$$k = k_o + k_{H^+}[H^+] + k_A[A] + k_B[B]$$

where k_o is the first-order rate constant in the absence of acids and bases other than water. The following data were obtained by J. N. Brönsted and E. A. Guggenheim [J. Am. Chem. Soc., 49, 2554 (1927)] at 18 °C in a medium containing 0.02 mol L^{-1} sodium acetate and various concentrations of acetic acid.

$[CH_3CO_2H]/mol\ L^{-1}$	0.020	0.105	0.199
$k/10^{-4}\ min^{-1}$	1.36	1.40	1.46

Calculate k_o and k_A. The term involving k_{H^+} is negligible under these conditions.

SOLUTION

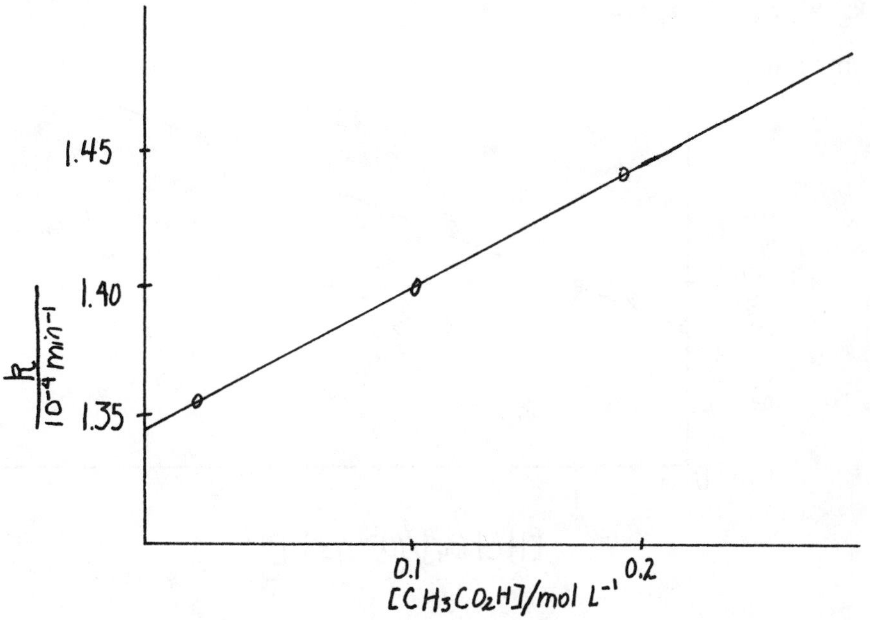

$$\text{Intercept} = k_o = 1.35 \times 10^{-4}\ \text{min}^{-1}$$

$$\text{Slope} = \frac{(1.46 - 1.35) \times 10^{-4}}{0.2\ \text{mol L}^{-1}} = 5.5 \times 10^{-5}\ \text{L mol}^{-1}\ \text{min}^{-1} = k_A$$

22.8 The mutarotation of glucose is catalyzed by acids and bases and is first order in the concentration of glucose. When perchloric acid is used as a catalyst, the concentration of hydrogen ions may be taken to be equal to the concentration of perchloric acid, and the catalysis by perchlorate ion may be ignored since it is such a weak base. The following first-order constants were obtained by J. N. Brönsted and E. A. Guggenheim [J. Am. Chem. Soc., 49, 2554 (1927)] at 18 °C.

$[HClO_4]/\text{mol L}^{-1}$	0.0010	0.0048	0.0099	0.0192	0.0300	0.0400
$k/10^{-4}\text{min}^{-1}$	1.25	1.38	1.53	1.90	2.15	2.59

Calculate the values of the constants in the equation

$$k = k_o + k_{H^+}[H^+]$$

SOLUTION

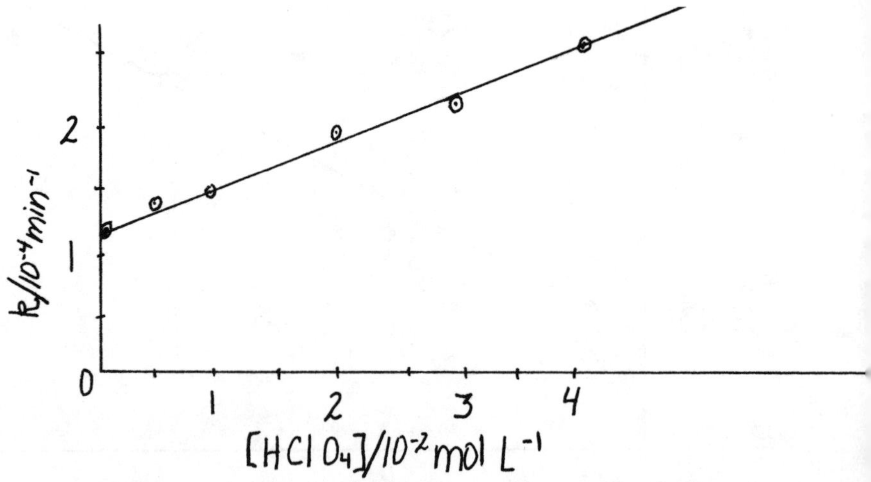

Intercept $= k_o = 1.21 \times 10^{-4}\ \text{min}^{-1}$

Slope $= k_{H^+} = 3.4 \times 10^{-3}\ \text{L mol}^{-1}\ \text{min}^{-1}$

22.9 Adipic acid and 1,10-decamethylene glycol were mixed in equimolar amounts and polymerized at 161 °C and 191 °C [S. D. Hamann, D. H. Solomon, and J. D. Swift, J. Macromol. Sci. Chem., A2, 153 (1968)]. The extents of reaction, as determined by acid titration, are as follows:

t/min	0	200	400	600	800
p(161 °C)	0.820	0.900	0.927	0.940	0.947
p(191 °C)	0.820	0.937	0.955	0.963	0.968

Time was measured from the point at which there was 85% esterification. Taking $[CO_2H]_0 = 1.25$ mol kg^{-1} at both temperatures, what are the two values of k and the activation energy?

SOLUTION

$$\frac{1}{(1 - p)^2} - \frac{1}{(1 - 0.82)^2} = 2[CO_2H]_0^2\, kt$$

At 161 °C

$$k = \frac{(1 - p)^{-2} - (1 - 0.82)^{-2}}{2[CO_2H]_0^2\, t}$$

$= 0.111,\ 0.125,\ 0.132,\ 0.130$ for the 4 successive times

$= 0.13\ \text{kg}^2\ \text{mol}^{-2}\ \text{min}^{-1}$

At 191 °C

$k = 0.36\ \text{kg}^2\ \text{mol}^{-2}\ \text{min}^{-1}$

$$E_a = \frac{RT_1T_2}{T_2 - T_1} \ln\frac{k_2}{k_1} = \frac{(8.314)(459)(489)}{30} \ln\frac{0.36}{0.13}$$

$= 63\ \text{kJ mol}^{-1}$

22.10 Adipic acid and diethylene glycol were polymerized at 109 °C using p-toluene sulfuric acid as a catalyst [P. J. Flory, J. Am. Chem. Soc., 61, 3334 (1939)]. The extents of reaction at various times are as follows:

t/min	0	40	80	120
p	0.800	0.909	0.944	0.960

Note that the time is taken as zero at p = 0.80. At what time will the number average molar mass reach 10,000 g mol^{-1} for this concentration of reactants and catalyst?

SOLUTION

$$\frac{1}{1-p} = 1 + k_2[CO_2H]_o t$$

$$\frac{1}{1-0.800} = 1 + k_2[CO_2H]_o 0$$

$$\frac{\frac{1}{1-p} - \frac{1}{1-0.800}}{t} = k_2[CO_2H]_o$$

$$k_2[CO_2H]_o = 0.16 \text{ min}^{-1}$$

$$p = 1 - \frac{M_o}{M_n} = 1 - \frac{216 \text{ g mol}^{-1}}{10^4 \text{ g mol}^{-1}} = 0.9784$$

$$\frac{1}{1-0.9784} - \frac{1}{1-0.800} = (0.16 \text{ min}^{-1})t$$

$$t = 258 \text{ min}$$

22.11 For a condensation polymerization of a hydroxyacid in which 99% of the acid groups are used up, calculate (a) the average number of monomer units in the polymer molecules, (b) the probability that a given molecule will have the number of residues given by this value, and (c) the weight fraction having this particular number of monomer units.

SOLUTION

(a) The number-average degree of polymerization is given by

$$\overline{X}_n = \frac{1}{1 - p} = \frac{1}{1 - 0.99} = 100$$

(b) The probability that a given molecule will have 100 residues is given by

$$\pi_i = p^{i - 1}(1 - p)$$

$$\pi_{100} = 0.99^{100 - 1}(1 - 0.99) = 3.70 \times 10^{-3}$$

(c) The weight fraction having this number of residues is given by

$$W_i = ip^{i - 1}(1 - p)^2$$

$$W_{100} = (100)(0.99)^{100 - 1}(1 - 0.99)^2$$

$$= 3.70 \times 10^{-3}$$

22.12 In the condensation polymerization of a hydroxyacid with a residue mass of 200 it is found that 99% of the acid groups are used up. Calculate (a) the number-average molar mass, and (b) the mass-average molar mass.

SOLUTION

(a) $$\overline{X}_n = \frac{1}{1 - p} = \frac{1}{1 - 0.99} = 100$$

$$M_n = (100)(200 \text{ g mol}^{-1}) = 20{,}000 \text{ g mol}^{-1}$$

(b) $$\frac{M_m}{M_n} = 1 + p = 1.99$$

$$M_m = 39{,}800 \text{ g mol}^{-1}$$

22.13 A hydroxyacid $HO-(CH_2)_5-CO_2H$ is polymerized and it is found that the product has a number average molar mass of 20,000 g mol^{-1}. (a) What is the extent of reaction p? (b) What is the degree of polymerization $\bar{X}_n$? (c) What is the mass-average molar mass?

SOLUTION

The molar mass of a monomer unit is 114 g mol^{-1} = M_o

(a) $M_n = M_o/(1 - p)$

$$p = 1 - M_o/M_n = 1 - 114/20{,}000 = 0.9943$$

(b) $$\bar{X}_n = \frac{1}{1 - p} = \frac{1}{1 - 0.9943} = 175$$

(c) $$M_m = M_o \frac{1 + p}{1 - p} = 114 \frac{1.9943}{1 - 0.9943}$$

$$= 39{,}900 \text{ g mol}^{-1}$$

22.14 Suppose an enzyme has a turnover number of 10^4 min^{-1} and a molar mass of 60,000 g mol^{-1}. How many moles of substrate can be turned over per hour per gram of enzyme if the substrate concentration is twice the Michaelis constant? It is assumed that the substrate concentration is maintained constant by a preceding enzymatic reaction and that products do not accumulate and inhibit the reaction.

SOLUTION

$$v = \frac{V}{1 + K/[S]} = \frac{(10^4 \text{ min}^{-1})[E]_o}{1 + 0.5}$$

$$= \frac{(10^4 \text{ min}^{-1})(1 \text{ g}/60{,}000 \text{ g mol}^{-1})(60 \text{ min hr}^{-1})}{1.5}$$

$$= 6.7 \text{ mol hr}^{-1}$$

22.15 The kinetics of the fumarase reaction fumarate + H_2O = L-malate is studied at 25 $^{\circ}C$ using an 0.01 ionic strength buffer of pH 7. The rate of reaction is obtained using a recording ultraviolet spectrometer to measure the fumarate concentration. The following rates of the forward reaction are obtained using a fumarase concentration of 5×10^{-10} mol L^{-1}.

$[F]/10^{-6}$ mol L^{-1}	$v_F/10^{-7}$ mol L^{-1} s^{-1}
2	2.2
40	5.9

The following rates of the reverse reaction are obtained using a fumarase concentration of 5×10^{-10} mol L^{-1}.

$[M]/10^{-6}$ mol L^{-1}	$v_M/10^{-7}$ mol L^{-1} s^{-1}
5	1.3
100	3.6

(a) Calculate the Michaelis constants and turnover numbers for the two substrates. In practice many more concentrations would be studied. (b) Calculate the four rate constants in the mechanism

$$E + F \underset{k_{-1}}{\overset{k_1}{\rightleftharpoons}} X \underset{k_{-2}}{\overset{k_2}{\rightleftharpoons}} E + M$$

where E represents the catalytic site. There are four catalytic sites per fumarase molecule. (c) Calculate K_{eq} for the reaction catalyzed. The concentration of H_2O is omitted in the expression for the equilibrium constant because its concentration cannot be varied in dilute aqueous solutions.

<u>**SOLUTION**</u>

<u>(a)</u> $$v_F = \frac{V_F}{1 + K_F/[F]}$$

$$v_F + \frac{V_F}{[F]} K_F = V_F$$

$$2.2 \times 10^{-7} + 0.110\ K_F = V_F$$

$$5.9 \times 10^{-7} + 0.0148\ K_F = V_F$$

$$-3.7 \times 10^{-7} + 0.0952\ K_F = 0$$

$$K_F = \frac{3.7 \times 10^{-7}}{0.0952} = 3.9 \times 10^{-6}\ \text{mol L}^{-1}$$

$$v_F = 2.2 \times 10^{-7} + (0.110)(3.9 \times 10^{-6}) = 6.5 \times 10^{-7}\ \text{mol L}^{-1}\ \text{s}^{-1}$$

$$v_M = \frac{V_M}{1 + K_M/[M]}$$

$$v_M = \frac{v_M}{[M]} K_M = V_M$$

$$1.3 \times 10^{-7} + 0.0260\ K_M = V_M$$

$$3.6 \times 10^{-7} + 0.0036\ K_M = V_M$$

$$-2.3 \times 10^{-7} + 0.0224\ K_M = 0$$

$$K_M = \frac{2.3 \times 10^{-7}}{0.0224} = 1.03 \times 10^{-5}\ \text{mol L}^{-1}$$

$$v_M = 1.3 \times 10^{-7} + 0.026(1.03 \times 10^{-5}) = 4.0 \times 10^{-7}\ \text{mol L}^{-1}\ \text{s}^{-1}$$

The rate constants should be expressed in terms of the concentration of enzymatic sites and so

$$k_2 = \frac{V_F}{[E]_o} = \frac{6.5 \times 10^{-7}\ \text{mol L}^{-1}\ \text{s}^{-1}}{4(5 \times 10^{-10}\ \text{mol L}^{-1})} = 3.3 \times 10^2\ \text{s}^{-1}$$

$$k_{-1} = \frac{V_M}{[E]_o} = \frac{4.0 \times 10^{-7}\ \text{mol L}^{-1}\ \text{s}^{-1}}{4(5 \times 10^{-10}\ \text{mol L}^{-1})} = 2.0 \times 10^2\ \text{s}^{-1}$$

(b) $$K_F = \frac{k_2 + k_1}{k_1}$$

$$k_1 = \frac{k_2 + k_{-1}}{K_F} = \frac{(3.3 + 2.0) \times 10^2\ \text{s}^{-1}}{3.9 \times 10^{-6}\ \text{mol L}^{-1}}$$

$$= 1.4 \times 10^8\ \text{L mol}^{-1}\ \text{s}^{-1}$$

$$K_M = \frac{k_2 + k_1}{k_{-2}}$$

$$k_{-2} = \frac{k_2 + k_{-1}}{K_M} = \frac{(3.3 + 2.0) \times 10^2\ \text{s}^{-1}}{1.03 \times 10^{-5}\ \text{mol L}^{-1}}$$

$$= 5.1 \times 10^7\ \text{L mol}^{-1}\ \text{s}^{-1}$$

(c) $$K = \frac{[M]_{eq}}{[F]_{eq}} = \frac{V_F\ K_M}{V_M\ K_F}$$

$$= \frac{(6.5 \times 10^{-7})(1.03 \times 10^{-5})}{(4.0 \times 10^{-7})(3.9 \times 10^{-6})} = 4.3$$

$$K = \frac{k_1\ k_2}{k_{-1}\ k_{-2}} = \frac{(1.4 \times 10^8)(3.3 \times 10^2)}{(2.0 \times 10^2)(5.1 \times 10^7)} = 4.5$$

22.16 At pH 7 the measured Michaelis constant and maximum velocity for the enzymatic conversion of fumarate to L-malate

$$\text{fumarate} + H_2O = \text{L-malate}$$

are 4.0×10^{-6} mol L^{-1} and $(1.3 \times 10^3\ s^{-1})[E]_o$, where $[E]_o$ is the total molar concentration of the enzyme. The Michaelis constant and maximum velocity for the reverse reaction are 10×10^{-6} mol L^{-1} and $(0.80 \times 10^3\ s^{-1})[E]_o$. What is the equilibrium constant for this hydration reaction? (The activity of water is set equal to unity since it is in excess.)

SOLUTION

$$K = \frac{[M]_{eq}}{[F]_{eq}} = \frac{V_F\ K_M}{V_M\ K_F}$$

$$= \frac{(1.3 \times 10^3\ s^{-1})[E]_o(10 \times 10^{-6}\ \text{mol L}^{-1})}{(0.8 \times 10^3\ s^{-1})[E]_o(4.0 \times 10^{-6}\ \text{mol L}^{-1})} = 4.1$$

22.17 Derive the steady-state rate equation for the mechanism

$$E + S \underset{k_2}{\overset{k_1}{\rightleftharpoons}} X \xrightarrow{k_3} E + P \qquad E + I \underset{k_5}{\overset{k_4}{\rightleftharpoons}} EI$$

for the case that $[S] \gg [E]_o$ and $[I] \gg [E]_o$.

SOLUTION

$$[E]_o = [E] + [EI] + [X]$$

$$= [E](1 + [I]/K_I) + [X] \qquad (1)$$

since $K_I = \dfrac{[E][I]}{[EI]} = \dfrac{k_5}{k_4}$

Assuming that X is in a steady state

$$\frac{d[X]}{dt} = k_1[E][S] - (k_2 + k_3)[X] = 0 \qquad (2)$$

Solving equation 1 for [E] and substituting this expression in equation 2 yields

$$\frac{k_1[S][E]_o}{1 + [I]/K_I} = [X]\left[\frac{k_1[S]}{1 + [I]/K_I} + k_2 + k_3\right]$$

$$\frac{d[P]}{dt} = k_3[X] = \frac{k_3[E]_o}{1 + \frac{k_2 + k_3}{k_1[S]}\left(1 + \frac{[I]}{K_I}\right)}$$

$$= \frac{V_s}{1 + \frac{K_s}{[S]}\left(1 + \frac{[I]}{K_I}\right)}$$

22.18 The following initial velocities were determined spectrophotometrically for solutions of sodium succinate to which a constant amount of succinoxidase was added. The velocities are given as the change in absorbancy at 250 nm in 10 x. Calculate V, K_s, and K_I for malonate.

[Succinate] 10^{-3} mol L^{-1}	$\frac{A \times 10^3}{10\ s}$ No Inhibitor	15×10^{-6} mol L^{-1} malonate
10	16.7	14.9
2	14.2	10.0
1	11.3	7.7
0.5	8.8	4.9
0.33	7.1	--

SOLUTION

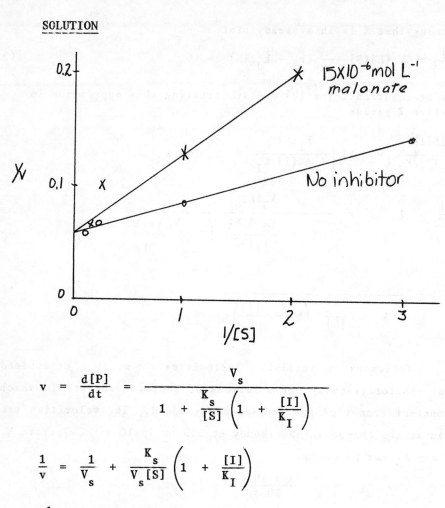

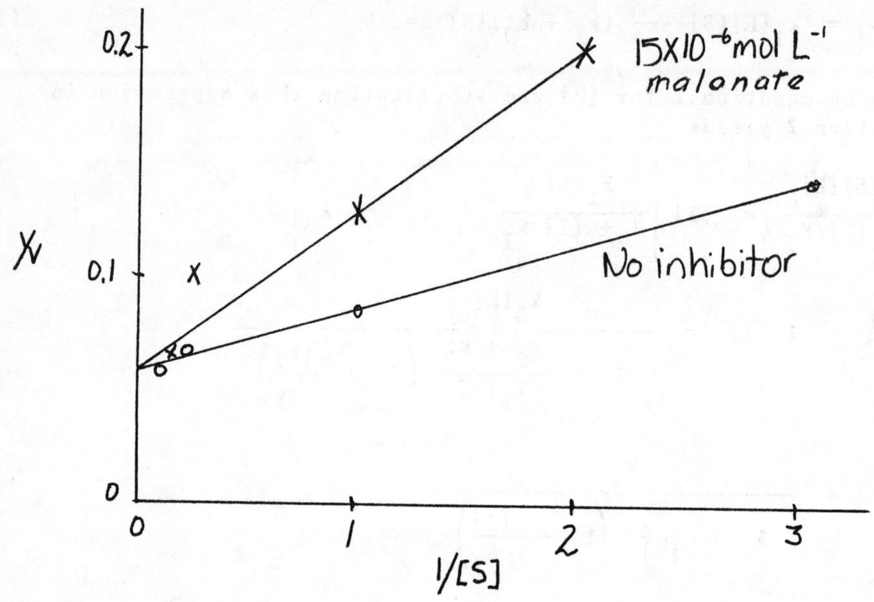

$$v = \frac{d[P]}{dt} = \frac{V_s}{1 + \frac{K_s}{[S]}\left(1 + \frac{[I]}{K_I}\right)} \qquad (22.85)$$

$$\frac{1}{v} = \frac{1}{V_s} + \frac{K_s}{V_s[S]}\left(1 + \frac{[I]}{K_I}\right)$$

$$\frac{1}{V_s} = 0.0662 \qquad V_s = 15.1$$

Slope without inhibitor

$$= \frac{0.141 - 0.066}{3 \times 10^3 \text{ L mol}^{-1}} = 25 \times 10^{-6} = \frac{K_s}{V_s}$$

$$K_s = (25 \times 10^{-6})(15.1) = 0.38 \times 10^{-3} \text{ mol L}^{-1}$$

Slope with inhibitor

$$= \frac{0.204 - 0.066}{2 \times 10^3 \text{ L mol}^{-1}} = 0.069 = \frac{K_s}{V_s}\left[1 + \frac{[I]}{K_I}\right]$$

$$0.069 = \frac{0.38}{15.1}\left[1 + \frac{15 \times 10^{-6} \text{ mol L}^{-1}}{K_I}\right]$$

$$K_I = \frac{15 \times 10^{-6} \text{ mol L}^{-1}}{\frac{(15.1)(0.069)}{0.38} - 1} = 8.6 \times 10^{-6} \text{ mol L}^{-1}$$

22.19 The maximum initial velocities for an enzymatic reaction are determined at a series of pH values.

pH	6.0	7.0	7.5	8.0	8.5	9.0
V	11	74	129	147	108	53

Calculate the values of the parameters V', K_a, K_b in

$$V = \frac{V'}{1 + [H^+]/K_a + K_b/[H^+]}$$

Hint. A plot of V versus pH may be constructed and the hydrogen ion concentration at the midpoint on the acid side referred to as $[H^+]_a$ and the hydrogen ion concentration at the midpoint on the basic side is referred to as $[H^+]_b$. Then

$$K_a = [H^+]_a + [H^+]_b - 4\sqrt{[H^+]_a[H^+]_b}$$

$$K_b = \frac{[H^+]_a[H^+]_b}{K_a}$$

SOLUTION

(Please see top of page 366)

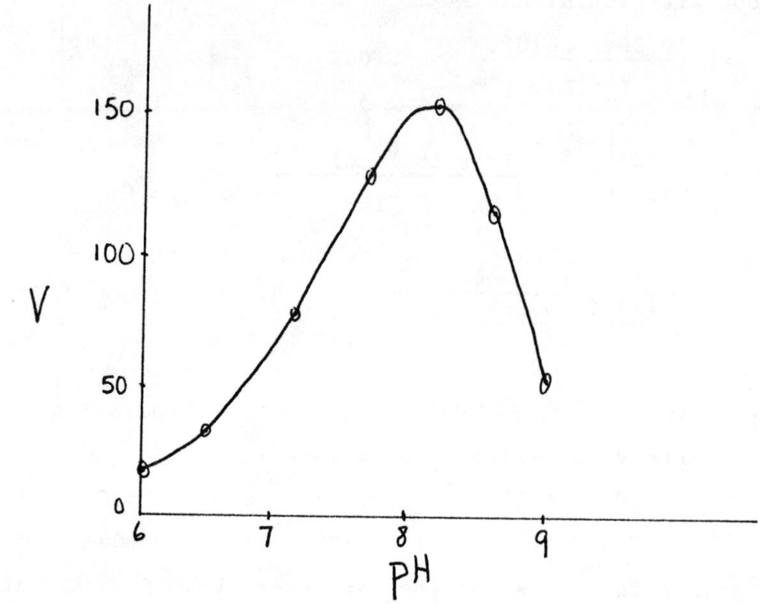

$$[H^+]_a = 10^{-7.03} = 9.3 \times 10^{-8}$$

$$[H^+]_b = 10^{-8.76} = 1.74 \times 10^{-9}$$

$$K_a = [H^+]_a + [H^+]_b - 4\sqrt{[H^+]_a [H^+]_b}$$

$$= 9.3 \times 10^{-9} + 1.74 \times 10^{-9} - 4\sqrt{(9.3 \times 10^{-8})(1.74 \times 10^{-9})}$$

$$= 4.41 \times 10^{-8} \text{ mol L}^{-1}$$

$$pK_a = 7.36$$

$$K_b = \frac{[H^+]_a [H^+]_b}{K_a} = \frac{(9.3 \times 10^{-8})(1.74 \times 10^{-9})}{4.41 \times 10^{-8}}$$

$$= 3.67 \times 10^{-9} \text{ mol L}^{-1}$$

$$pK_b = 8.45$$

$$V = \frac{V'}{1 + [H^+]/K_a + K_b/[H^+]}$$

$$147 = \frac{V'}{1 + (10^{-8})/(4.41 \times 10^{-8}) + (0.367 \times 10^{-8})/(10^{-8})}$$

$$V' = 234$$

22.20 Alcohol dehydrogenase catalyzes the reaction

$$\text{alcohol} + NAD^+ = \text{aldehyde} + NADH + H^+$$

where NAD^+ is nicotinamide adenine dinucleotide in the oxidized form and NADH in the reduced form. For enzyme isolated from yeast the maximum velocity for ethyl alcohol is given by

$$V(\text{mol L}^{-1}\ \text{min}^{-1}) = 2.7 \times 10^4 [E]_o$$

where $[E]_o$ is the molar concentration of enzyme of molar mass 150,000 g mol^{-1}. The reaction is conveniently followed spectrophotometrically at 340 nm because NADH has a molar absorbancy coefficient of 620 $(\text{mol L}^{-1})^{-1}\ \text{cm}^{-1}$.

$$\text{Absorbancy} = \log \frac{I_o}{I} = 6200[\text{NADH}]d$$

where d is the thickness of the spectrophotometer cuvette. How many grams of enzyme must be placed in a 3-mL reaction mixture to produce a rate of change of absorbancy of 0.001 per minute when a 1.0-cm cuvette is used?

SOLUTION

$$\frac{d[\text{NADH}]}{dt} = V = (2.7 \times 10^4\ \text{min}^{-1})[E]_o$$

$$\frac{dA}{dt} = [6200(\text{mol L}^{-1})^{-1}\ \text{cm}^{-1}](1\ \text{cm})\ \frac{d[\text{NADH}]}{dt}$$

$$10^{-3}\ \text{min}^{-1} = [6200(\text{mol L}^{-1})^{-1}](2.7 \times 10^{4}\ \text{min}^{-1})[E]_o$$

$$= [6200(\text{mol L}^{-1})](2.7 \times 10^{4}\ \text{min}^{-1})\frac{m}{(0.003\ \text{L})(150{,}000\ \text{g mol}^{-1})}$$

$$m = 2.7 \times 10^{-9}\ \text{g of enzyme in 3 mL}$$

22.21 What current density i will be obtained for the evolution of hydrogen gas on (a) Pt, and (b) Fe from 1 mol L^{-1} HCl at 25 °C at an overpotential of 0.1 volt?

SOLUTION

(a) $i = i_o\, e^{E_\eta/2.303a}$

$= (10^{-2.6}\ \text{A cm}^{-2})\ e^{0.1/(2.303)(0.028)}$

$= 0.012\ \text{A cm}^{-2}$

(b) $i = (10^{-6}\ \text{A cm}^{-2})\ e^{0.1/(2.303)(0.130)}$

$= 1.4 \times 10^{-6}\ \text{A cm}^{-2}$

22.22 0.6 nm

22.23 2.7×10^{-4} s

22.24 8.5 ns

22.25 $k_1 = 1.8 \times 10^{10}\ \text{L mol}^{-1}\ \text{s}^{-1}$

$k_{-1} = 1.1 \times 10^{3}\ \text{s}^{-1}$

22.26 $1/\tau = k_1([A]_{eq} + [B]_{eq}) + k_2([C]_{eq} + [D]_{eq})$

22.27 $1/\tau = k_1 + k_1' K_A + k_{-1} + k_{-1}' K_B$

22.29 H^+ production: $k = 1.75 \times 10^5,\ 1.2 \times 10^2,\ 5.71\ \text{s}^{-1}$

OH^- production: $k = 5.71,\ 0.83 \times 10^3,\ 1.8 \times 10^5\ \text{s}^{-1}$
Imidazole can play both roles about equally well.

22.30 $K_1 = [HOCl][OH^-]/[OCl^-]$

$$\frac{d[Cl^-]}{dt} = k[I^-][HOCl] = k[I^-]K_1[OCl^-]/[OH^-]$$

22.31 $0.57 \text{ kg mol}^{-1} \text{ min}^{-1}$

22.32 (a) 20 (b) 0.0189 (c) 0.0189

22.33 $20\ M_o$, $39.0\ M_o$

22.34 179 hr

22.35 $V = 1.2 \times 10^{-6} \text{ mol L}^{-1} \text{ s}^{-1}$

$K_s = 0.48 \times 10^{-3} \text{ mol L}^{-1}$

22.36 (a) $k_1 = 1.3 \times 10^8 \text{ L mol}^{-1} \text{ s}^{-1}$

$k_{-1} = 0.2 \times 10^3 \text{ s}^{-1}$

$k_2 = 0.33 \times 10^3 \text{ s}^{-1}$

$k_{-2} = 5.3 \times 10^7 \text{ L mol}^{-1} \text{ s}^{-1}$

(b) $-3.43 \text{ kJ mol}^{-1}$

22.37 $k_1 = 4.9 \times 10^8 \text{ L mol}^{-1} \text{ s}^{-1}$

$k_{-1} = 2.6 \times 10^3 \text{ s}^{-1}$

$k_2 = 0.8 \times 10^3 \text{ s}^{-1}$

$k_{-2} = 3.4 \times 10^7 \text{ L mol}^{-1} \text{ s}^{-1}$

22.38
$$\frac{d[P_2]}{dt} = \frac{(k_1 k_2[S] - k_{-1} k_{-2}[P_1][P_2])[E]_o}{k_1[S] + k_{-1}[P_1] + k_2 + k_{-2}[P_2]}$$

$$v_f = \frac{k_2 [E]_o}{1 + \dfrac{k_2}{k_1[S]}}$$

$$v_r = \frac{k_{-1}[P_1][E]_o}{1 + \frac{k_2}{k_{-2}[P_2]} + \frac{k_{-1}[P_1]}{k_{-2}[P_2]}}$$

Thus all four rate constants can be determined.

22.39 $V = k_2k_3[E]_o/(k_1 + k_3)$ $K_s = k_2/(k_1 + k_3)$

Since no enzyme-substrate complex is formed, this mechanism does not account for the selectivity of enzyme-catalyzed reactions.

22.40 The product completely inhibits the enzyme, putting it out of action before all the S is used up. Since the binding of product is reversible, the reaction can be restarted by adding more substrate and displacing the product.

22.41 33%

22.42 $V_s = k_2[E]_o/[1 + [H^+]/K_{EHS}]$

$$K_s = \frac{k_{-1} + k_{-2}}{k_1} \frac{[1 + [H^+]/K_{EH}]}{[1 + [H^+]/K_{EHS}]}$$

22.43 $$v = \frac{k_2[alc][E]_o}{1 + \frac{k[alc]}{k_1[NAD^+]}}$$

22.44 (a) 0.045 (b) 1.31 V

22.45 (a) 78.2 kJ mol^{-1} (b) 73.6 kJ mol^{-1}

(c) 14.1 s^{-1} (d) 0.929 s^{-1}

CHAPTER 23: Photochemistry

23.1 A certain photochemical reaction requires an excitation energy of 126 kJ mol^{-1}. To what values does this correspond in the following units: (a) frequency of light, (b) wave number, (c) wavelength in nanometers, and (d) electron volts?

SOLUTION

(a) $$\nu = \frac{E}{h} = \frac{1.26 \times 10^{5}\ J\ mol^{-1}}{(6.63 \times 10^{-34}\ J\ s)(6.02 \times 10^{23}\ mol^{-1})}$$

$$= 3.16 \times 10^{14}\ s^{-1}$$

(b) $$\tilde{\nu} = \frac{\nu}{c} = \frac{(3.16 \times 10^{14}\ s^{-1})(10^{-2}\ m\ cm^{-1})}{2.998 \times 10^{8}\ ms^{-1}}$$

$$= 10{,}500\ cm^{-1}$$

(c) $$\lambda = \frac{1}{\tilde{\nu}} \quad \frac{1}{10{,}500\ cm^{-1}} = 9.52 \times 10^{-5}\ cm = 925\ nm$$

(d) $$\frac{126\ kJ\ mol^{-1}}{96{,}485\ C\ mol^{-1}} = 1.31\ eV$$

23.2 How many moles of photons does a 560 nm laser with an intensity of 0.1 watt at this wavelength produce in one hour?

SOLUTION

$$\frac{E\lambda}{N_A hc} = \frac{(0.1\ J\ s^{-1})(60 \times 60\ s)(560 \times 10^{-9}\ m)}{(6.022 \times 10^{23}\ mol^{-1})(6.626 \times 10^{-34}\ J\ s)(2.998 \times 10^{8}\ m\ s^{-1})}$$

$$= 1.7 \times 10^{-3}\ mol$$

23.3 A sample of gaseous acetone is irradiated with monochromatic light having a wavelength of 313 nm. Light of this wavelength decomposed the acetone according to the equation

$$(CH_3)_2CO \longrightarrow C_2H_6 + CO$$

The reaction cell used has a volume of 59 cm^3. The acetone vapor absorbs 91.5% of the incident energy. During the experiment the following data are obtained.

Temperature of reaction	=	56.7 °C
Initial pressure	=	102.16 kPa
Final pressure	=	104.42 kPa
Time of radiation	=	7 hr
Incident energy	=	48.1×10^{-4} J s^{-1}

What is the quantum yield?

SOLUTION

Since $PV = nRT$, $V\Delta P = RT\Delta n$

$$\Delta n = \frac{V\Delta P}{RT} = \frac{(59 \times 10^{-6}\ m^{-3})[(104.42 - 102.16) \times 10^3\ Pa]}{(8.314\ J\ K^{-1}\ mol^{-1})(329.85\ K)}$$

$$= 4.86 \times 10^{-5}\ mol$$

$$= (4.86 \times 10^{-5})(6.02 \times 10^{23}\ mol^{-1})$$

$$= 2.93 \times 10^{19}\ \text{molecules reacting}$$

$$E = \frac{hc}{\lambda} = \frac{(6.626 \times 10^{-34}\ J\ s)(2.998 \times 10^8\ m\ s^{-1})}{313 \times 10^{-9}\ m}$$

$$= 6.36 \times 10^{-19}\ J$$

Number of quanta absorbed

$$= \frac{(48.1 \times 10^{-4}\ J\ s^{-1})(7 \times 60 \times 60\ s)(0.915)}{6.36 \times 10^{-19}\ J} = 1.75 \times 10^{20}$$

$$\phi = \frac{2.93 \times 10^{19}}{1.75 \times 10^{20}} = 0.167$$

23.4 A 100-cm^3 vessel containing hydrogen and chlorine was irradiated with light of 400 nm. Measurements with a thermopile showed that 11×10^{-7} J of light energy was absorbed by the chlorine per second. During an irradiation of 1 min the partial pressure of chlorine, as determined by the absorption of light and the application of Beer's law, decreased from 27.3 to 20.8 kPa (corrected to 0 °C). What is the quantum yield?

SOLUTION

$$E = \frac{hc}{\lambda} = \frac{(6.626 \times 10^{-34}\ \text{J s})(2.998 \times 10^{8}\ \text{m s}^{-1})}{400 \times 10^{-9}\ \text{m}}$$

$$= 4.97 \times 10^{-19}\ \text{J}$$

$$\text{Number of quanta absorbed} = \frac{(11 \times 10^{-7}\ \text{J s}^{-1})(60\ \text{s})}{4.97 \times 10^{-19}\ \text{J}} = 1.33 \times 10^{14}$$

Since $PV = nRT$, $V\Delta P = RT\Delta n$

$$\Delta n = \frac{V\Delta P}{RT} = \frac{(0.1 \times 10^{-3}\ \text{m}^3)[(27.3 - 20.8) \times 10^{3}\ \text{Pa}](6.02 \times 10^{23}\ \text{mol}^{-1})}{(8.314\ \text{J K}^{-1}\ \text{mol}^{-1})(273\ \text{K})}$$

$$= 1.72 \times 10^{20}$$

Since $H_2 + Cl_2 = 2\ HCl$

$$\phi = \frac{1.72 \times 10^{20}}{1.33 \times 10^{14}} \times 2 = 2.6 \times 10^{6}$$

23.5 Discuss the economic possibilities of using photochemical reactions to produce valuable products with electricity at 5 cents per kilowatt-hour. Assume that 5% of the electric energy consumed by a quartz-mercury-vapor lamp goes into light, and 30% of this is photochemically effective. (a) How much will it cost to produce 1 lb (453.6 g) of an organic compound having a molar mass of 100 g

mol^{-1}, if the average effective wavelength is assumed to be 400 nm and the reaction has a quantum yield of 0.8 molecule per photon? (b) How much will it cost if the reaction involves a chain reaction with a quantum yield of 100?

SOLUTION

(a) Number of quanta required

$$= \frac{(453.6\ g)(6.022 \times 10^{23}\ mol^{-1})}{(100\ g\ mol^{-1})(0.8)} = 3.41 \times 10^{24}$$

$$E = \frac{hc}{\lambda} = \frac{(6.26 \times 10^{-34}\ J\ s)(2.998 \times 10^{8}\ m\ s^{-1})}{4000 \times 10^{-10}\ m}$$

$$= 4.97 \times 10^{-19}\ J$$

$$\text{Energy required} = \frac{(3.41 \times 10^{24})(4.97 \times 10^{-19}\ J)}{(0.05)(0.3)} = 1.13 \times 10^{8}\ J$$

$$\text{Cost} = \frac{(1.13 \times 10^{8}\ J)(0.5\$/KW\ hr)}{36 \times 10^{5}\ J/KW\ hr} = \$1.55$$

$$1\ KW\ hr = (10^{3}\ J\ s^{-1})(60 \times 60\ s) = 36 \times 10^{5}\ J/KW\ hr$$

(b) $\$1.55\ \frac{0.8}{100} = 1.2¢$

23.6 The quantum yield is 2 for the photolysis of gaseous HI to $H_2 + I_2$ by light of 253.7 nm wavelength. Calculate the number of moles of HI that will be decomposed if 300 J of light of this wavelength is absorbed.

SOLUTION

$$E = \frac{hc}{\lambda} = \frac{(6.626 \times 10^{-34}\ J\ s)(3 \times 10^{8}\ m\ s^{-1})}{(253.7 \times 10^{-9}\ m)} = 7.84 \times 10^{-19}\ J$$

$$n = \frac{(2)(300\ J)}{(7.84 \times 10^{-19}\ J)(6.02 \times 10^{23}\ mol^{-1})} = 1.27 \times 10^{-3}\ \text{mol of HI}$$

23.7 Show that if a solute follows the Beer–Lambert law, the intensity of absorbed radiation I_a in moles of photons per unit volume per second is given by

$$I_a = \frac{I_a}{1 N_A h\nu}(1 - e^{-\kappa c l})$$

where 1 is the length of the cell in the direction of the incident monochromatic radiation and I_o is in energy per unit area per unit time.

SOLUTION

Consider that monochromatic radiation of intensity I_o is perpendicular to the face of a reaction cell of area A and optical path length 1. The rate that energy enters the cell is I_oA, and the rate that energy leaves the cell is IA, where $I = I_o \exp(-\kappa c l)$. The rate I_a of absorption of electromagnetic radiation in moles of photons per unit volume per second is given by

$$I_a = \frac{I_o A - IA}{V N_A h\nu} = \frac{I_o}{1 N_A h\nu}(1 - e^{-\kappa c l})$$

If the cell is thick enough that the radiation is essentially all absorbed, $I_a = I_o/1N_a h\nu$.

23.8 In the stratosphere molecular oxygen absorbs solar radiation in the 185 - 220 nm wavelength region.

$$O_2 + h\nu = O + O$$

If the absorption cross section is 1.1×10^{-23} cm^2, what thickness of a layer of O_2 at 298 K and 1 bar is required to absorb half of the radiation.

SOLUTION

$$P = \frac{N}{N_A} RT \quad \text{where N is the number of molecules per m}^3$$

$$N = \frac{PN_A}{RT} = \frac{(10^5 \text{ Pa})(6.02 \times 10^{23} \text{ mol}^{-1})}{(8.314 \text{ J K}^{-1} \text{ mol}^{-1})(298 \text{ K})}$$

$$= 2.43 \times 10^{25} \text{ m}^{-3}$$

$$\frac{I}{I_o} = \exp(-\sigma N x)$$

$$0.50 = \exp[-(1.1 \times 10^{-23} \text{ cm}^2)(2.43 \times 10^{25} \text{ m}^{-3})(0.01 \text{ m}^{-1} \text{ cm})^2 \text{ x}]$$

$$x = \frac{0.693}{(1.1 \times 10^{-27} \text{ m}^2)(2.34 \times 10^{25} \text{ m}^{-3})} = 26.9 \text{ m}$$

23.9 The fluorescence quantum yield for benzene at 25 °C is 0.070. The lifetime of the excited state is 26 ns. What is the radiative lifetime τ_o?

SOLUTION

$$\tau = \frac{\tau_s}{\phi_F} = \frac{26 \times 10^{-9} \text{ s}}{0.070} = 370 \text{ ns}$$

23.10 The phosphorescence of butyrophenone in acetonitrile is quenched by 1,3-pentadiene (P). The following quantum yields were measured at 25 °C (see N. J. Turro, Modern Molecular Photochemistry, The Benjamin/Cummings Publ. Co., Menlo Park, CA (1978), p. 248).

$[P]/10^{-3}$ mol L^{-1}	0	1.0	2.0
ϕ/ϕ_o	1	0.61	0.43

Assuming that the quenching reaction is diffusion controlled and the rate constant has a value of 10^{10} L mol^{-1} s^{-1}, what is the lifetime of the triplet state?

SOLUTION

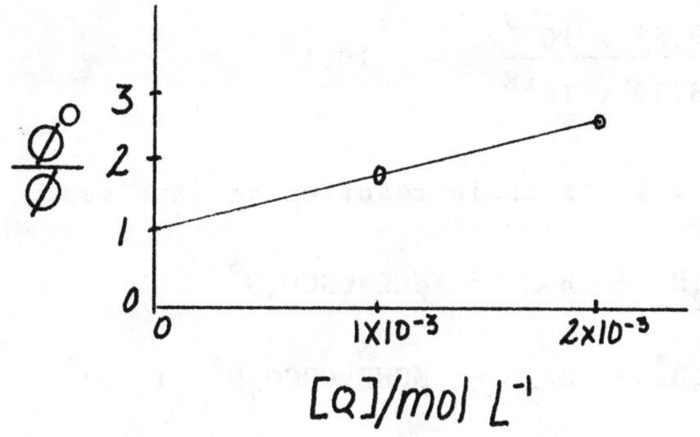

$$\text{slope} \;=\; k_Q\tau_{T_1} \;=\; \frac{1.33}{2 \times 10^{-3}\ \text{mol L}^{-1}} \;=\; 640\ \text{L mol}^{-1}$$

$$\tau_{T_1} \;=\; \frac{640\ \text{L mol}^{-1}}{10^{10}\ \text{L mol}^{-1}\ \text{s}^{-1}} \;=\; 6.4 \times 10^{-8}\ \text{s}$$

23.11 For 900 s, light of 436 nm was passed into a carbon tetrachloride solution containing bromine and cinnamic acid. The average power absorbed was $19.2 \times 10^{-4}\ \text{J s}^{-1}$. Some of the bromine reacted to give cinnamic acid dibromide, and in this experiment the total bromine content decreased by 3.83×10^{19} molecules. (a) What was the quantum yield? (b) State whether or not a chain reaction was involved. (c) If a chain mechanism was involved, suggest suitable reactions that might explain the observed quantum yield.

SOLUTION

$$\text{(a)}\ E \;=\; \frac{hc}{\lambda} \;=\; \frac{(6.63 \times 10^{-34}\ \text{J s})(3 \times 10^{8}\ \text{m s}^{-1})}{(436 \times 10^{-9}\ \text{m})}$$

$$= 4.56 \times 10^{-19}\ \text{J}$$

$$\frac{19.2 \times 10^{4}\ \text{J s}^{-1}}{4.56 \times 10^{-19}\ \text{J}} = 4.21 \times 10^{15}\ \text{s}^{-1},\quad \text{rate quanta are absorbed}$$

$$(4.21 \times 10^{15}\ s^{-1})(900\ s) = 3.79 \times 10^{18} \text{ quanta}$$

$$\phi = \frac{3.83 \times 10^{19}}{3.79 \times 10^{18}} = 10.1$$

(b) Since $\phi > 1$, a chain reaction is involved.

(c) $\phi CH{=}CHCO_2H + h\nu = \phi CH{=}CHCO_2H^*$

$\phi CH{=}CHCO_2H^* + Br_2 = \phi CHBrCHCO_2H^* + Br^*$

$\phi CHBrCHCO_2H^* + Br_2 = \phi CHBrCHBrCO_2H + Br^*$

$\phi CH{=}CHCO_2H + Br^* = CHBrCHCO_2H^*$

23.12 Ketone A dissolved in t-butyl alcohol is excited to a triplet state A^* by light of 320 - 380 nm. The triplet may then return to the ground state A or rearrange to give the isomer B, thus

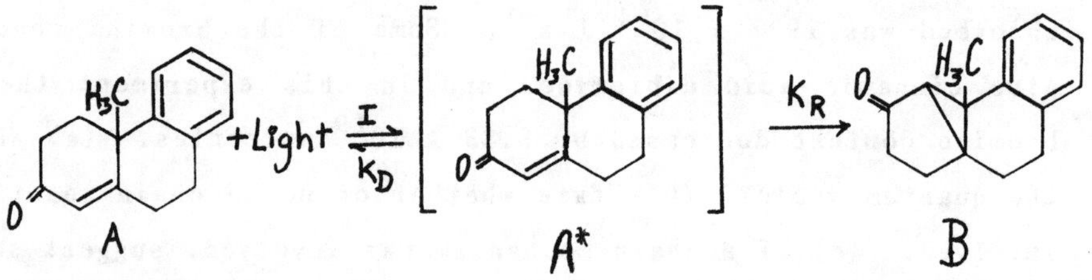

The unimolecular rate constant k determines the rate of deactivation of excited molecules and the unimolecular rate constant k determines the rate of rearrangement. The symbol I represents the number of moles of photons per second and it is assumed that each photon absorbed produces one excited molecule of A^*. The quantum yield ϕ for the formation of B is given by

$$\phi = \frac{d[B]/dt}{I} = \frac{k_R[A^*]}{I}$$

Zimmerman, McCullough, Staley, and Padwa measured the quenching effect of dissolved napthalene and report the following data.

Moles naphthalene L^{-1}	0.0099	0.0330	0.0620	0.0680	0.0775	0.0960
ϕ	0.0049	0.0023	0.0020	0.0017	0.0014	0.0012

The quenching rate by napthalene is controlled by the bimolecular quenching constant of A^*, which is equal to the diffusion-controlled constant $k_Q = 1.2 \times 10^9$ L mol^{-1} s^{-1}. Assuming a steady state,

$$\frac{d[A^*]}{dt} = I - k_R[A^*] - k_D[A^*] - k_Q[A^*][N] = 0$$

where [N] is the concentration of naphthalene. Calculate k_R and k_D by plotting $1/\phi$ versus [N] and determining the slope and intercept of the line.

SOLUTION

$$\frac{I}{[A^*]} = k_R + k_D + k_Q[N]$$

$$\frac{1}{\phi} = \frac{I}{k_R[A^*]} = 1 + \frac{k_D}{k_R} + \frac{k_Q}{k_R}[N]$$

(Solution continued on page 380)

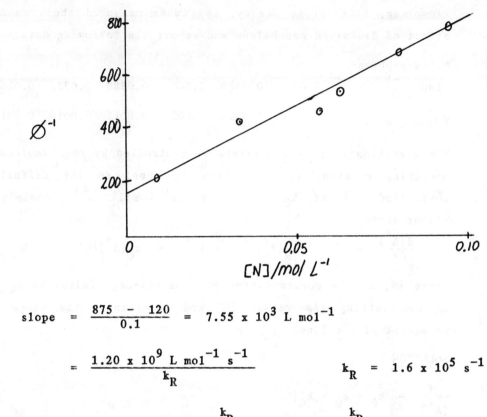

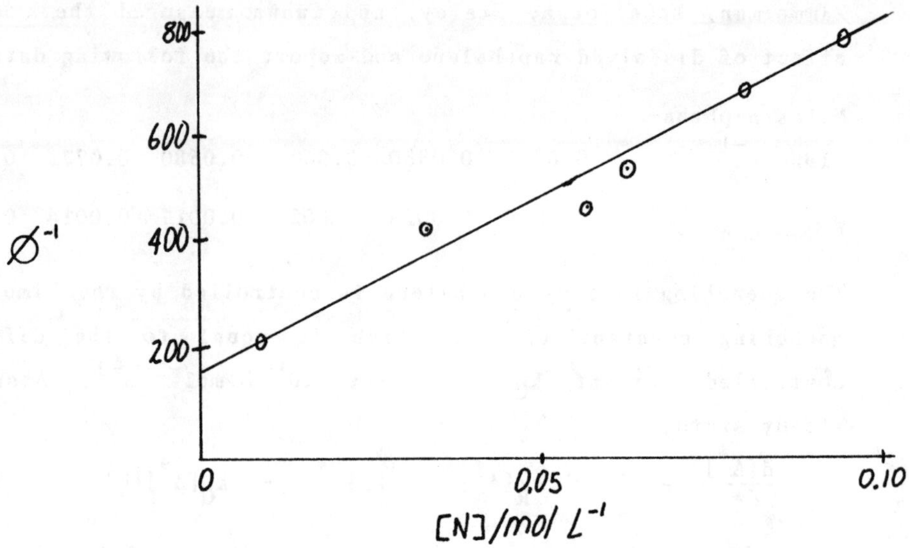

$$\text{slope} = \frac{875 - 120}{0.1} = 7.55 \times 10^{3}\ \text{L mol}^{-1}$$

$$= \frac{1.20 \times 10^{9}\ \text{L mol}^{-1}\ \text{s}^{-1}}{k_R} \qquad k_R = 1.6 \times 10^{5}\ \text{s}^{-1}$$

$$\text{intercept} = 120 = 1 + \frac{k_D}{k_R} = 1 + \frac{k_D}{1.6 \times 10^{5}\ \text{s}^{-1}}$$

$$k_D = 1.9 \times 10^{7}\ \text{s}^{-1}$$

23.13 The following calculations are made on a uranyl oxalate actinometer, on the assumption that the energy of all wavelengths between 254 and 435 nm is completely absorbed. The actinometer contains 20 cm^3 of 0.05 mol L^{-1} oxalic acid, which also is 0.01 mol L^{-1} with respect to uranyl sulfate. After 2 hr of exposure to ultraviolet light, the solution required 34 cm^3 of potassium permanganate, $KMnO_4$, solution to titrate the undecomposed oxalic acid. The same volume, 20 cm^3, of unilluminated solution required 40 cm^3 of the $KMnO_4$ solution. If the average energy of the quanta in this range may be taken as corresponding to a wavelength of 350 nm, how many joules were absorbed per second in this experiment? (ϕ = 0.57)

SOLUTION

Moles of oxalic acid decomposed $= (0.02\ L)(0.05\ mol\ L^{-1})\left(\frac{40 - 34}{40}\right)$

$= 1.5 \times 10^{-4}$ mol

Moles of photons required $= \dfrac{1.5 \times 10^{-4}\ mol}{(2 \times 60 \times 60\ s)(0.57)} = 3.65 \times 10^{-8}\ mol\ s^{-1}$

$$E = \frac{hc}{\lambda} = \frac{(6.626 \times 10^{-34}\ J\ s)(2.998 \times 10^{8}\ m\ s^{-1})}{3.5 \times 10^{-7}\ m}$$

$= 5.68 \times 10^{-19}$ J

Energy flux $= (3.65 \times 10^{-8}\ mol^{-1}\ s^{-1})(6.02 \times 10^{23}\ mol^{-1}) \times (5.68 \times 10^{-19}\ J)$

$= 1.25 \times 10^{-2}\ J\ s^{-1}$

23.14 The quantum yield for the photolysis of acetone $(CH_3)_2CO = C_2H_6 + CO$ at 300 nm is 0.2. How many moles per second of CO are formed if the intensity of the 300-nm radiation absorbed is 10^{-2} J s^{-1}?

SOLUTION

$$E = \frac{hc}{\lambda} = \frac{(6.626 \times 10^{-34}\ J\ s)(3 \times 10^{8}\ m\ s^{-1})}{300 \times 10^{-9}\ m} = 6.63 \times 10^{-19}\ J$$

$$\frac{(10^{-2}\ J\ s\quad)(0.2)}{(6.02 \times 10^{23}\ mol^{-1})(6.63 \times 10^{-19}\ J)} = 5 \times 10^{-9}\ mol\ s^{-1}$$

23.15 A solution of a dye is irradiated with 400 nm light to produce a steady concentration of triplet state molecules. If the triplet state yield is 0.9, and the triplet state lifetime is 20×10^{-6} s, what light intensity, expressed in watts, is required to maintain a steady triplet concentration of 5×10^{-6} mol L^{-1} in a liter of solution. Assume that all of the light is absorbed.

SOLUTION

$$\frac{d[T_1]}{dt} = 0.9\ I - \frac{1}{20 \times 10^{-6}}[T_1] = 0$$

$$I = \frac{(5 \times 10^{-6}\ \text{mol L}^{-1})(1\ \text{L})}{(20 \times 10^{-6}\ \text{s})(0.9)} = 0.278\ \text{mol s}^{-1}$$

$$E = \frac{N_A hc}{\lambda} = \frac{(6.022 \times 10^{23}\ \text{mol}^{-1})(6.62 \times 10^{-34}\ \text{J s})(3 \times 10^{8}\ \text{m s}^{-1})}{400 \times 10^{-9}\ \text{m}}$$

$$= 2.99 \times 10^{5}\ \text{J mol}^{-1}$$

$$I = (0.278\ \text{mol s}^{-1})(2.99 \times 10^{5}\ \text{J mol}^{-1}) = 83\ \text{kW}$$

23.16 The photochemical chlorination of chloroform

$$CHCl_3 + Cl_2 = CCl_4 + HCl$$

is believed to proceed by the following mechanism.

$$Cl_2 + h\nu \xrightarrow{I_a} 2Cl$$

$$Cl + CHCl_3 \xrightarrow{k_1} CCl_3 + HCl$$

$$CCl_3 + Cl_2 \xrightarrow{k_2} CCl_4 + Cl$$

$$2CCl_3 + Cl_2 \xrightarrow{k_3} 2CCl_4$$

Derive the steady state rate law for the production of carbontetrachloride.

SOLUTION

$$\frac{d[CCl_4]}{dt} = k_2[CCl_3][Cl_2] + 2k_3[CCl_3]^2[Cl_2]$$

$$\frac{d[CCl_3]}{dt} = k_1[Cl][CHCl_3] - k_2[CCl_3][Cl_2] - 2k_3[CCl_3]^2[Cl_2] = 0$$

$$\frac{d[Cl]}{dt} = 2I_a - k_1[Cl][CHCl_3] + k_2[CCl_3][Cl_2] = 0$$

Adding the two steady state expressions yields

$$I_a = k_3[CCl_3]^2[Cl_2]$$

This makes it possible to eliminate $[CCl_3]$ from the first equation to obtain

$$\frac{d[CCl_4]}{dt} = k_2 I_a^{1/2}\, [Cl_2]^{1/2} / k_3^{1/2} + 2I_a$$

23.17 Given that solar radiation at noon at a certain place on the earth's surface is 4.2 J cm^{-2} min^{-1}, what is the maximum power output in W m^{-2}?

SOLUTION

$$\frac{(4.2\ \mathrm{J\ cm^{-2}\ min^{-1}})(100\ \mathrm{cm\ m^{-1}})^2}{60\ \mathrm{s\ min^{-1}}} = 700\ \mathrm{W\ m^{-2}}$$

23.18 If a good agricultural crop yields about 2 tons $acre^{-1}$ of dry organic material per year with a heat of combustion of about 16.7 kJ g^{-1}, what fraction of a year's solar energy is stored in an agricultural crop if the solar energy is about 4184 J min^{-1} ft^{-2} and the sun shines about 500 min day^{-1} on the average: 1 acre = 43,560 ft^2 and 1 ton = 907,000 g.

SOLUTION

$$\text{solar energy} = (43{,}560\ \text{ft}^2\ \text{acre}^{-1})(500\ \text{min day}^{-1})(365\ \text{day year}^{-1}) \times (4184\ \text{J ft}^{-2}\ \text{min}^{-1})$$

$$= 3.33 \times 10^{13}\ \text{J acre}^{-1}\ \text{year}^{-1}$$

$$\text{energy stored} = (2\ \text{ton acre}^{-1}\ \text{year}^{-1})(907{,}000\ \text{g ton}^{-1})()16{,}700\ \text{J g}^{-1})$$

$$= 303 \times 10^{10}\ \text{J acre}^{-1}\ \text{year}^{-1}$$

$$\text{fraction stored} = \frac{3.03 \times 10^{10}\ \text{J}}{3.33 \times 10^{13}\ \text{J}} \approx 10^{-3}$$

23.19 Sunlight between 290 and 313 nm can produce sunburn (erythema) in 30 min. The intensity of radiation between these wavelengths in summer and at 45° latitude is about 50 μW cm^{-2}. Assuming that 1 photon produces chemical change in 1 molecule, how many molecules in a square centimeter of human skin must be photochemically affected to produce evidence of sunburn?

SOLUTION

$$E = \frac{hc}{\lambda} = \frac{(6.63 \times 10^{-34})(3 \times 10^{8})}{3 \times 10^{-7}} = 6.63 \times 10^{-19}\ \text{J}$$

$$(50 \times 10^{-6}\ \text{J s}^{-1}\ \text{cm}^{-2})(30 \times 60\ \text{s}) = .09\ \text{J cm}^{-2}$$

$$\frac{.09}{6.63 \times 10^{-19}} = 1.36 \times 10^{17}\ \text{molecules cm}^{-2}$$

23.20 Calculate the longest wavelength of light that can theoretically decompose water at 25 °C in a one-photon electrochemical process to give $H_2(g)$ and $\frac{1}{2}\,O_2(g)$ in their standard states. Given: $\Delta G^{o} = 237.2$ kJ mol^{-1} for $H_2O(l) = H_2(g) + \frac{1}{2}\,O_2(g)$.

SOLUTION

$\Delta G^\circ = N_A hc/\lambda$

$$\lambda = \frac{N_A hc}{\Delta G^\circ} = \frac{(6.022\text{x}10^{23}\ \text{mol}^{-1})(6.626\text{x}10^{-34}\ \text{J s})(2.998\text{x}10^{8}\ \text{m s}^{-1})}{237{,}200\ \text{J mol}^{-1}}$$

$= 504$ nm

23.21 2.51×10^{-6} mol s^{-1}

23.22 (a) 0.1709 J s^{-1} (b) 0.3988 J s^{-1}
(c) 0.1709 watt (d) 0.3988 watt

23.23 2.18×10^{-19} J

23.24 7.84×10^{4} s

23.25 10^{-7} mol L^{-1}

23.26 9×10^{-9} mol L^{-1}

23.27 47.1 s

23.28 0.050 J s^{-1}

23.29 $NO_2 + h\nu \longrightarrow NO_2^*$

$NO_2^* + NO_2 \longrightarrow 2\ NO + O_2$

Decreased concentration of NO_2 after long illumination makes decomposing collisions less likely. Also the reverse reaction becomes more important as the concentrations of the products build up.

23.31 3.94 tons

23.32 (a) 10^{14} (b) 4.53×10^{-9} g day^{-1}

23.33 617 g

23.34 24.4 tons

23.35 504 nm

CHAPTER 24: Irreversible Processes in Solution

24.1 Ten cubic centimeters of water at 25 °C is forced through 20 cm of 2-mm diameter capillary in 4 s. Calculate the pressure required and the Reynolds number.

SOLUTION

$$P = \frac{8Vl\eta}{\pi r^4 t} = \frac{(8)(10 \times 10^{-6}\ m^3)(0.2\ m)(8.95 \times 10^{-4}\ kg\ m^{-1}\ s^{-1})}{\pi(10^{-3}\ m)^4(4\ s)}$$

$$= 1.14 \times 10^3\ N\ m^{-2}$$

$$\text{Reynolds number} = \frac{d\bar{v}\rho}{\eta}$$

$$\bar{v} = \frac{\text{volume of liquid per unit time}}{\text{area of tube}}$$

$$= \frac{(10 \times 10^{-6}\ m^3)/(4\ s)}{\pi(10^{-3}\ m)^2} = 0.796\ m\ s^{-1}$$

Reynolds number

$$= \frac{(2 \times 10^{-3}\ m)(0.796\ m\ s^{-1})(10^3\ kg\ m^{-3})}{(8.95 \times 10^{-4}\ kg\ m^{-1}\ s^{-1})} = 1780$$

24.2 A steel ball (ρ = 7.86 g cm^{-3}) 0.2 cm in diameter falls 10 cm through a viscous liquid (ρ_o = 1.50 g cm^{-3}) in 25 s. What is the viscosity at this temperature?

SOLUTION

$$\eta = \frac{2r^2(\rho - \rho_o)g}{9\frac{dx}{dt}}$$

$$= \frac{2(1 \times 10^{-3}\ m)^2[(7.86 - 1.50) \times 10^3\ kg\ m^{-3}](9.8\ m\ s^{-2})}{9\left(\frac{0.10\ m}{25\ s}\right)}$$

$$= 3.46\ Pa\ s$$

24.3 Estimate the rate of sedimentation of water droplets of 1-μm diameter in air at 20 °C. The viscosity of air at this temperature is 1.808×10^{-5} Pa s.

SOLUTION

$$\frac{dx}{dt} = \frac{2r^2(\rho - \rho_o)g}{9\eta}$$

$$= \frac{2(0.5 \times 10^{-6}\ \text{m})^2(0.998 \times 10^3\ \text{kg m}^{-3})(9.8\ \text{m s}^{-2})}{9(1.808 \times 10^{-5}\ \text{Pa s})}$$

$$= 3.01 \times 10^{-5}\ \text{m s}^{-1}$$

24.4 Using data in Table 24.1 and equation 24.6, estimate the activation energy for water molecules to move into a vacancy at 25 °C.

SOLUTION

$$E_a = \frac{RT_1T_2}{T_2 - T_1} \ln \frac{\eta_1}{\eta_2}$$

$$= \frac{(8.314\ \text{J K}^{-1}\ \text{mol}^{-1})(273.15\ \text{K})(323.15\ \text{K})}{(50\ \text{K})(10^3\ \text{J kJ}^{-1})} \ln \frac{1793}{549}$$

$$= 17.4\ \text{kJ mol}^{-1}$$

24.5 The relative viscosities of a series of solutions of a sample of polystyrene in toluene were determined with an Ostwald viscometer at 25 °C.

$c/10^{-2}$ g cm^{-3}	0.249	0.499	0.999	1.998
η/η_o	1.355	1.782	2.879	6.090

The ratio η_{sp}/c is plotted against c and extrapolated to zero concentration to obtain the intrinsic viscosity. If the constants in equation 24.13 are $K = 3.7 \times 10^{-2}$ and $a = 0.62$ for this polymer, when concentrations are expressed in g/cm^3, calculate the molar mass.

SOLUTION

$c/10^{-2}$ g cm^{-3}	0.249	0.499	0.999	1.998
$\frac{\eta/\eta_o - 1}{c}$	142.6	156.7	188.1	254.8

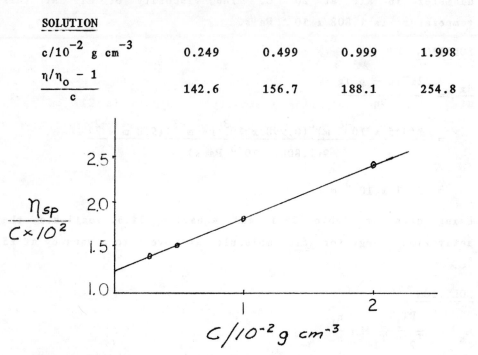

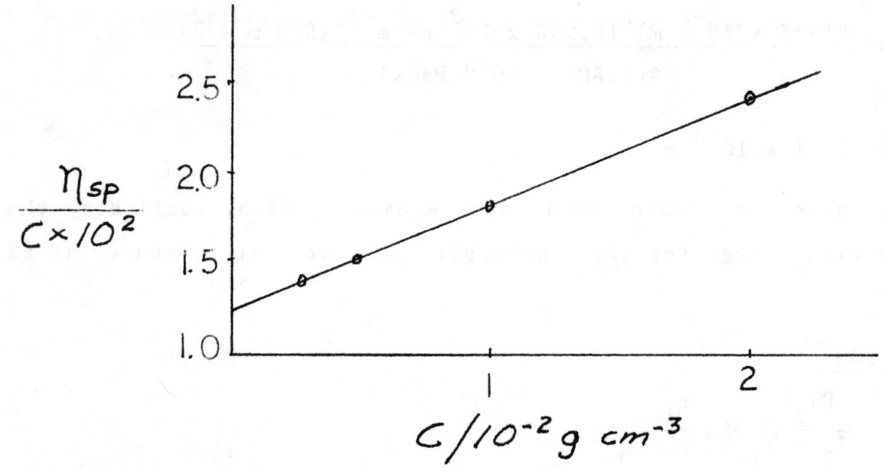

$$\ln \frac{[\eta]}{3.7 \times 10^{-2}} = 0.62 \ln M$$

$$M = \exp\left[\frac{1}{0.62} \ln \frac{126}{3.7 \times 10^{-2}}\right] = 500,000 \text{ g mol}^{-1}$$

24.6 At 34 °C the intrinsic viscosity of a sample of polystyrene in toluene is 84 cm^3 g^{-1}. The empirical relation between the intrinsic viscosity of polystyrene in toluene and molar mass is $[\eta] = 1.15 \times 10^{-2} M^{0.72}$. What is the molar mass of this sample?

SOLUTION

$$\ln \frac{[\eta]}{1.15 \times 10^{-2}} = 0.72 \ln M$$

$$M = \exp\left[\frac{1}{0.72} \ln \frac{84}{1.15 \times 10^{-2}}\right] = 230{,}000 \text{ g mol}^{-1}$$

24.7 Given that the intrinsic viscosity of myosin is 217 $cm^3\ g^{-1}$, approximately what concentration of myosin in water would have a relative viscosity of 1.5?

SOLUTION

$$\frac{\frac{\eta}{\eta_o} - 1}{c} = 217 \text{ cm}^3 \text{ g}^{-1} \qquad \frac{0.5}{c} = 217 \text{ cm}^3 \text{ g}^{-1}$$

$$c = \frac{0.5}{217 \text{ cm}^3 \text{ g}^{-1}} = 2.30 \times 10^{-3} \text{ g cm}^{-3}$$

24.8 A conductance cell was calibrated by filling it with a 0.02 mol L^{-1} solution of potassium chloride ($\kappa = 0.2768\ \Omega^{-1}\ m^{-1}$) and measuring the resistance at 25 °C, which was found to be 457.3 Ω. The cell was then filled with a calcium chloride solution containing 0.555 g of $CaCl_2$ per liter. The measured resistance was 1050 Ω. Calculate (a) the cell constant for the cell, and (b) the conductivity of the $CaCl_2$ solution.

SOLUTION

(a) From equation 24.15

$$K_{cell} = \kappa R = (0.2768\ \Omega^{-1}\ m^{-1})(457.3\ \Omega) = 126.6\ m^{-1}$$

(b) $$\kappa = \frac{K_{cell}}{R} = \frac{126.6\ m^{-1}}{1050\ \Omega} = 0.1206\ \Omega^{-1}\ m^{-1}$$

24.9 It is desired to use a conductance apparatus to measure the concentration of dilute solutions of sodium chloride. If the electrodes in the cell are each 1 cm^2 in area and are 0.2 cm apart, calculate the resistance that will be obtained for 1, 10, and 100 ppm NaCl at 25 °C.

SOLUTION

$$1 \text{ ppm} = 1 \text{ g NaCl in } 10^6 \text{ g } H_2O = \frac{(1 \text{ g})/(58.45 \text{ g mol}^{-1})}{1 \text{ m}^3}$$

$$= 1.71 \times 10^{-2} \text{ mol m}^{-3}$$

Using electric mobilities at infinite dilution,

$$\kappa = Fc(u_{Na} + u_{Cl^-})$$

$$= (96{,}485 \text{ C mol}^{-1})(1.71\times10^{-2} \text{ mol m}^{-3})[(5.192 + 7.913) \times 10^{-8} \text{ m}^2 \text{ V}^{-1} \text{ s}^{-1}]$$

$$= 2.16 \times 10^{-4} \ \Omega^{-1} \text{ m}^{-1}$$

$$R = \frac{1}{\kappa A} = \frac{0.2 \times 10^{-2} \text{ m}}{(2.16 \times 10^{-4} \ \Omega^{-1} \text{ m}^{-1})(0.01 \text{ m})^2} = 9.24 \times 10^4 \ \Omega$$

For 10 ppm $R = 9.25 \times 10^3 \ \Omega$
For 100 ppm $R = 925 \ \Omega$

24.10 A moving boundary experiment is carried out with a 0.1 mol L^{-1} solution of hydrochloric acid at 25 °C ($\kappa = 4.24 \ \Omega^{-1} \text{ m}^{-1}$). Sodium ions are caused to follow the hydrogen ions. Three milliamperes is passed through the tube of 0.3 cm^2 cross-sectional area, and it is oberved that the boundary moves 3.08 cm in 1 hr. Calculate (a) the hydrogen ion mobility, (b) the chloride ion mobility, and (c) the electric field strength.

SOLUTION

$$\text{(a) } E = \frac{I}{A\kappa} = \frac{3 \times 10^{-3} \text{ A}}{(0.3 \times 10^{-4} \text{ m}^2)(4.24 \ \Omega^{-1} \text{ m}^{-1})} = 23.58 \text{ V m}^{-1}$$

$$u = \frac{\Delta x/\Delta t}{E} = \frac{3.08 \times 10^{-2}\ \text{m}}{(60 \times 60\ \text{s})(23.58\ \text{v m}^{-1})}$$

$$= 3.63 \times 10^{-7}\ \text{m}^2\ \text{V}^{-1}\ \text{s}^{-1}$$

(b) $\kappa = Fc(u_{H^+} + u_{Cl^-})$

$$u_{Cl^-} = \frac{\kappa}{Fc} - u_{H^+} = \frac{(4.24\ \Omega^{-1}\ \text{m}^{-1})}{(96{,}485\ \text{C mol}^{-1})(0.1 \times 10^3\ \text{mol}^{-3})} - 3.63 \times 10^{-7}\ \text{m}^2\ \text{V}^{-1}\ \text{s}^{-1}$$

$$= 7.64 \times 10^{-8}\ \text{m}^2\ \text{V}^{-1}\ \text{s}^{-1}$$

(c) See (a)

24.11 Calculate the conductivity of 0.001 mol L^{-1} HCl at 25 °C. The limiting ion mobilities may be used for this problem.

SOLUTION

$$\kappa = Fc(u_{H^+} + u_{Cl^-})$$

$$= (96{,}485\ \text{C mol}^{-1})(1\ \text{mol m}^{-3})[(36.25 + 7.91) \times 10^{-8}\ \text{m}^2\ \text{V}^{-1}\ \text{s}^{-1}]$$

$$= 0.042\ 61\ \Omega^{-1}\ \text{m}^{-1}$$

24.12 One hundred grams of sodium chloride is dissolved in 10,000 L of water at 25 °C, giving a solution that may be regarded in these calculations as infinitely dilute. (a) What is the conductivity of the solution? (b) This dilute solution is placed in a glass tube of 4-cm diameter provided with electrodes filling the tube and placed 20 cm apart. How much current will flow if the potential drop between the electrodes is 80 V?

SOLUTION

(a) $$c = \frac{(100\ \text{g})/(58.5\ \text{g mol}^{-1})}{(10^7\ \text{cm}^3)(10^{-2}\ \text{m cm}^{-1})^3} = 1.71 \times 10^{-1}\ \text{mol m}^{-3}$$

$$\kappa = (96{,}485\ C\ mol^{-1})(1.71 x 10^{-1}\ mol\ m^{-3})(13.105 x 10^{-8}\ m^2\ V^{-1}\ s^{-1})$$

$$= 2.16 x 10^{-3}\ \Omega^{-1}\ m^{-1}$$

$$(b)\ R = \frac{1}{\kappa A} = \frac{0.2\ m}{(2.16 x 10^{-3}\ \Omega^{-1}\ m^{-1})(2 x 10^{-2}\ m)^2 \pi}$$

$$= 7.36 x 10^4\ \Omega$$

$$I = \frac{E}{R} = \frac{80\ V}{7.36 x 10^4\ \Omega} = 1.09 x 10^{-3}\ A$$

24.13 Using Stokes' law, calculate the effective radius of a nitrate ion from its mobility ($74.0 x 10^{-9}\ m^2\ V^{-1}\ s^{-1}$ at 25 °C).

SOLUTION

$$f_i = \frac{|Z_i|e}{u_i} = 6\pi\eta r$$

$$r = \frac{|Z_i|e}{u_i\ 6\pi\eta}$$

$$= \frac{1.602 x 10^{-19}\ C}{(74 x 10^{-9}\ m^2\ V^{-1}\ s^{-1})(6\pi)(8.95 x 10^{-4}\ kg\ m^{-1}\ s^{-1})}$$

$$= 0.128\ nm$$

24.14 It may be shown that the diffusion coefficient at infinite dilution of an electrolyte with two univalent ions is given by

$$D = \frac{2u_1u_2RT}{(u_1 + u_2)F}$$

where u_1 and u_2 are the limiting values of the mobilities of the two ions. What is the diffusion coefficient of potassium chloride in water at 25 °C?

SOLUTION

$$D = \frac{2(7.617 \times 10^{-8}\ \mathrm{m^2\ V^{-1}\ s^{-1}})(7.913 \times 10^{-8}\ \mathrm{m^2\ V^{-1}\ s^{-1}})(R)(298.15\ \mathrm{K})}{(15.53 \times 10^{-8}\ \mathrm{m^2\ V^{-1}\ s^{-1}})(96{,}485\ \mathrm{C\ mol^{-1}})}$$

$$= 1.99 \times 10^{-9}\ \mathrm{m^2\ s^{-1}} \qquad R = 8.314\ \mathrm{J\ K^{-1}\ mol^{-1}}$$

24.15 What is the self-diffusion coefficient of Na^+ in water at 25 °C?

SOLUTION

$$D = \frac{uRT}{|z|F}$$

$$= \frac{(5.192 \times 10^{-8}\ \mathrm{m^2\ V^{-1}\ s^{-1}})(8.314\ \mathrm{J\ K^{-1}\ mol^{-1}})(298.15\ \mathrm{K})}{9.6485 \times 10^{4}\ \mathrm{C\ mol^{-1}}}$$

$$= 1.334 \times 10^{-9}\ \mathrm{m^2\ s^{-1}}$$

24.16 Using a table of probability integral, calculate enough points on a plot of c versus x (like Fig. 24.4c) to draw in the smooth curve for diffusion of 0.1 mol L^{-1} sucrose into water at 25 °C after 4 hr and 29.83 min ($D = 5.23 \times 10^{-10}\ \mathrm{m^2\ s^{-1}}$).

SOLUTION

$$\text{Normal probability function} = \frac{1}{\sqrt{2\pi}}\, e^{-y^2/2} = \text{N.P.F.}$$

Equation 20.38

$$c = \frac{c_o}{2}\left[1 + \frac{2}{\sqrt{\pi}} \int_o^{x/2\sqrt{Dt}} e^{-\beta^2}\, d\beta\right]$$

In order to get this equation in the form of the normal probability function let $\beta^2 = t^2/2$. Thus $\beta = t/\ 2$ and $d\beta = dt/\ 2$.

$$c = \frac{c_o}{2}\left[1 + \frac{2}{\sqrt{2\pi}} \int_o^{x/\sqrt{2Dt}} e^{-t^2/2}\, dt\right]$$

$$= \frac{c_o}{2}\left[1 + \frac{1}{\sqrt{2\pi}}\underbrace{\int_{-x/\sqrt{2Dt}}^{x/\sqrt{2DT}} e^{-t^2/2}\,dt}\right]$$

This area is obtainable from tables of the normal probability integral (N.P.I.)

$$\frac{c_o}{2} = [1 \pm \text{Area of N.P.I.}]$$

The + sign is used when $x > 0$, and the − sign is used when $x < 0$.

At $x = 1$ cm the Gaussian parameter is

$$\frac{x}{\sqrt{2Dt}} = \frac{10^{-2}\ \text{m}}{\sqrt{2(5.23 \times 10^{-10}\ \text{m}^2\ \text{s}^{-1})(269.83 \times 60\ \text{s})}} = 2.43$$

$\lvert x\rvert$ cm	$\frac{x}{\sqrt{2Dt}}$	Area N.P.I.	$c(x < 0)$	$c(x > 0)$
0	0	0	0.05 mol L^{-1}	0.05 mol L^{-1}
0.1	0.243	0.1922	0.040	0.060
0.2	0.486	0.3732	0.031	0.069
0.3	0.729	0.5340	0.023	0.077
0.4	0.972	0.6690	0.016	0.084
0.6	1.458	0.8556	0.007	0.093
0.8	1.944	0.9480	0.003	0.097
1.0	2.43	0.9850	0.002	0.098
1.2	2.916	0.9964	0.001	0.099

(Solution continued on page 395)

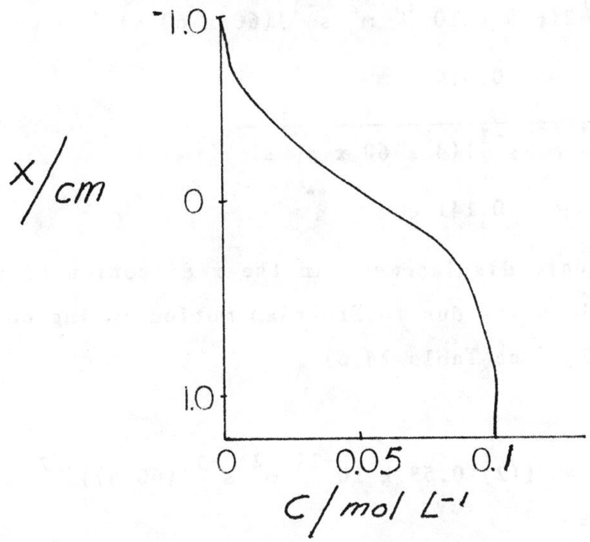

24.17 A sharp boundary is formed between a dilute aqueous solution of sucrose and water at 25 °C. After 5 hr the standard deviation of the concentration gradient is 0.434 cm. (a) What is the diffusion coefficient for sucrose under these conditions? (b) What will be the standard deviation after 10 hr?

SOLUTION

$$\text{(a) } D = \frac{\sigma^2}{2t} = \frac{(4.34 \times 10^{-3}\ \text{m})^2}{2(5 \times 60 \times 60\ \text{s})} = 5.23 \times 10^{-10}\ \text{m}^2\ \text{s}^{-1}$$

$$\text{(b) } \sigma = \sqrt{2Dt} = \sqrt{2(5.23 \times 10^{-10}\ \text{m}^2\ \text{s}^{-1})(10 \times 60 \times 60\ \text{s})}$$

$$= 6.14 \times 10^{-3}\ \text{m}$$

24.18 A sharp boundary is formed between a dilute buffered solution of hemoglobin ($D = 6.9 \times 10^{-11}\ \text{m}^2\ \text{s}^{-1}$) and the buffer at 20 °C. What is the half width of the boundary after 1 hr and 4 hr?

SOLUTION

(a) $\sigma = \sqrt{2Dt} = \sqrt{2(6.9 \times 10^{-11}\ m^2\ s^{-1})(60 \times 60\ s)}$

$= 7.05 \times 10^{-4}\ m = 0.0705\ cm$

(b) $\sigma = \sqrt{2(6.9 \times 10^{-11}\ m^2\ s^{-1})(4 \times 60 \times 60\ s)}$

$= 1.41 \times 10^{-3}\ m = 0.141\ cm$

24.19 What is the root-mean-square displacement in the x direction of a molecule of tobacco mosaic virus due to Brownian motion during one minute in water at 20 °C? (See Table 24.5)

SOLUTION

$\langle(\Delta x)^2\rangle^{1/2} = \sqrt{2Dt} = [(2)(0.53 \times 10^{-11}\ m^2\ s^{-1})(60\ s)]^{1/2}$

$= 0.0252\ nm$

24.20 The diffusion coefficient for serum globulin at 20 °C in a dilute aqueous salt solution is $4.0 \times 10^{-11}\ m^2\ s^{-1}$. If the molecules are assumed to be spherical, calculate their molar mass. Given: $\eta_{H_2O} = 0.001\ 005$ Pa s at 20° C and $v = 0.75\ cm^3\ g^{-1}$ for the protein.

SOLUTION

$$D = \frac{RT}{N_A 6\pi\eta}\left(\frac{4\pi N_A}{3Mv}\right)^{1/3} \qquad M = \frac{4\pi N_A}{3v}\left(\frac{RT}{N_A 6\pi\eta D}\right)^3$$

$$= \frac{4\pi(6.02\times10^{23}\ mol^{-1})}{3(0.75\times10^{-3}\ m^3\ kg^{-1})}\left[\frac{(8.31\ J\ K^{-1}\ mol^{-1})(293\ K)}{(6.02 \times 10^{23}\ mol^{-1})6\pi(1.005 \times 10^{-3}\ J\ m^{-3}\ s)(4.0 \times 10^{-11}\ m^2\ s^{-1})}\right]^3$$

$= 511\ kg\ mol^{-1} = 511{,}000\ g\ mol^{-1}$

24.21 The diffusion coefficient of hemoglobin at 20 °C is 6.9×10^{-11} m^2 s^{-1}. Assuming its molecules are spherical, what is the molar mass? Given: $v = 0.749$ s 10^{-3} m^3 kg^{-1} and $\eta = 0.001\ 005$ J m^{-3} s.

SOLUTION

$$D = \frac{RT}{N_A 6\pi\eta}\left(\frac{4\pi N_A}{3Mv}\right)^{1/3}$$

$$M = \frac{4\pi N_A}{3v}\left(\frac{RT}{DN_A 6\pi\eta}\right)^{3}$$

$$M = \frac{4\pi(6.022\text{x}10^{23}\ \text{mol}^{-1})}{3(0.749\text{x}10^{-3}\ \text{m}^3\ \text{kg}^{-1})}\left[\frac{(8.314\ \text{J K}^{-1}\ \text{mol}^{-1})(293\ \text{K})}{(6.9\text{x}10^{-11}\ \text{m}^2\ \text{s}^{-1})(6.022\text{x}10^{23}\ \text{mol}^{-1})\ 6\pi(0.001\ 005\ \text{J m}^{-3}\ \text{s})}\right]^3$$

$$= 100\ \text{kg mol}^{-1} = 100{,}000\ \text{g mol}^{-1}$$

24.22 Calculate the sedimentation coefficient of tobacco mosaic virus from the fact that the boundary moves with a velocity of 0.454 cm hr^{-1} in an ultracentrifuge at a speed of 10,000 rpm at a distance of 6.5 cm from the axis of the centrifuge rotor.

SOLUTION

$$S = \frac{\frac{dr}{dt}}{\omega^2 r} = \frac{\dfrac{0.454\ \text{cm hr}^{-1}}{3600\ \text{s hr}^{-1}}}{\left(\dfrac{10{,}000 \times 2\pi}{60\ \text{s}}\right)^2 6.5\ \text{cm}}$$

$$= 177 \times 10^{-13}\ \text{s}$$

24.23 The sedimentation coefficient of myoglobin at 20°C is 2.06×10^{-13} s. What molar mass would it have if the molecules were spherical? Given: $v = 0.749 \times 10^{-3}\ m^3\ kg^{-1}$, $\rho = 0.9982 \times 10^3\ kg\ m^{-3}$, and $\eta = 0.001\ 005$ Pa s.

SOLUTION

$$S = \frac{M(1 - v\rho)}{N_A f} = \frac{M(1 - v\rho)}{N_A 6\pi\eta}\left(\frac{4\pi N_A}{3Mv}\right)^{1/3}$$

$$M = \left(\frac{N_A 6\pi\eta S}{1 - v\rho}\right)^{3/2}\left(\frac{3v}{4\pi N_A}\right)^{1/2}$$

$$M = \left[\frac{(6.022 \times 10^{23}\ mol^{-1})(6\pi(0.001\ 005)(2.06 \times 10^{-13})}{1 - (0.749 \times 10^{-3})(0.9982 \times 10^{3})}\right]^{3/2} \times \left(\frac{3 \times 0.749 \times 10^{-3}}{4\pi 6.022 \times 10^{23}}\right)^{1/2}$$

$$= 15.5\ kg\ mol^{-1} = 15{,}500\ g\ mol^{-1}$$

24.24 The sedimentation and diffusion coefficients for hemoglobin corrected to 20 °C in water are 4.41×10^{-13} s and $6.3 \times 10^{-11}\ m^2\ s^{-1}$, respectively. If $v = 0.749\ cm^3\ g^{-1}$ and $\rho_{H_2O} = 0.998\ g\ cm^{-3}$ at this temperature, calculate the molar mass of the protein. If there is 1 mol of iron per 17,000 g of protein, how many atoms of iron are there per hemoglobin molecule?

SOLUTION

$$M = \frac{RTS}{D(1 - v\rho)}$$

$$= \frac{(8.31\ J\ K^{-1}\ mol^{-1})(203\ K)(4.41 \times 10^{-13}\ s)}{(6.3\times10^{-11}\ m^2\ s^{-1})[1 - (0.749\times10^{-3}\ m^3\ kg^{-1})(0.998\times10^3\ kg\ m^{-3})]}$$

$$= 67.6 \text{ kg mol}^{-1} = 67{,}600 \text{ g mol}^{-1}$$

$$\frac{67{,}000 \text{ g mol}^{-1}}{17{,}000 \text{ g g-atom Fe}} = 4 \text{ g-atom Fe mol}^{-1}$$

$$= 4 \text{ Fe per molecule}$$

24.25 Given the diffusion coefficient for sucrose at 20 °C in water ($D = 45.5 \times 10^{-11}$ m^2 s^{-1}), calculate its sedimentation coefficient. The partial specific volume v is 0.630 cm^3 g^{-1}.

SOLUTION

$$s = \frac{MD(1 - v\rho)}{RT}$$

$$= \frac{(342\text{x}10^{-3} \text{ kg mol}^{-1})(45.5\text{x}10^{-11} \text{ m}^2 \text{ s}^{-1})[1 - (0.630 \text{ cm}^3 \text{ g}^{-1})(A)]}{(8.31 \text{ J K}^{-1} \text{ mol}^{-1})(293 \text{ K})}$$

$$= 0.236 \times 10^{-13} \text{ s} \qquad A = 1 \text{ g cm}^{-3}$$

24.26 A sedimentation equilibrium experiment is to be carried out with myoglobin ($M = 16{,}000$ g mol^{-1}) in an ultracentrifuge operating at 15,000 rpm. The bottom of the cell is 6.93 cm from the axis of rotation and the meniscus is 6.67 cm from the axis of rotation. What ratio of concentrations is expected at 20 °C if $v = 0.75 \times 10^{-3}$ m^3 kg^{-1} and $\rho = 1.00 \times 10^3$ kg m^{-3}?

SOLUTION

$$\ln \frac{c_2}{c_1} = \frac{M(1 - v\rho)\,\omega^2 (r_2^{\,2} - r_1^{\,2})}{2RT}$$

$$= \frac{(16 \text{ kg mol}^{-1})(0.25)(2\pi 250 \text{ s}^{-1})\,[(0.0693 \text{ m})^2 - (0.0667 \text{ m})^2]}{2(8.314 \text{ J K}^{-1} \text{ mol}^{-1})(293.15 \text{ K})}$$

$$= 0.716 \qquad \frac{c_2}{c_1} = 2.05$$

24.27 1.13×10^3 Pa m^{-2} The Reynolds number is 447, and so the flow will be laminar.

24.28 95.7 min

24.29 0.66 s

24.30 0.001 41 Pa s

24.31 8.39×10^6 g mol^{-1}

24.32 638,000 g mol^{-1}

24.33 520 nm

24.34 (a) 4.28 Ω^{-1} m^{-1} (b) 0.0768 m (c) 58.4 V m^{-1}

24.35 0.1058 Ω^{-1} m^{-1}

24.36 1.6×10^{-3} Ω^{-1} m^{-1}

24.37 3.16×10^{-8} m^2 V^{-1} s^{-1}

24.38

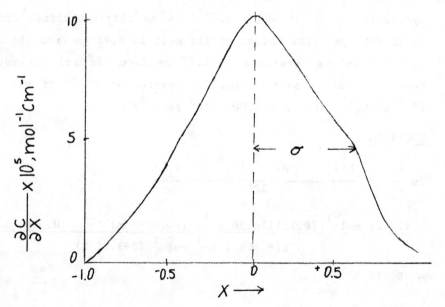

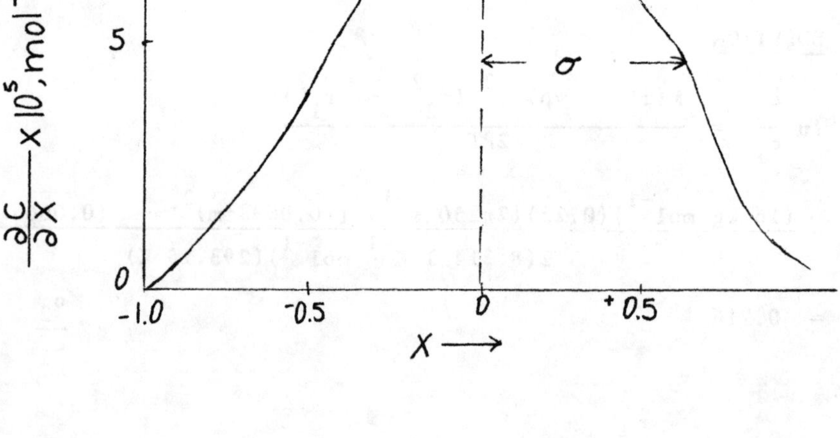

24.39 9.34×10^{-9}, 2.03×10^{-9}, $3.34 \times 10^{-9}\ m^2\ s^{-1}$

24.40 (a) 1.96 hr. (b) 56.4 hr.

24.41 $6.43 \times 10^{-8}\ m^2\ V^{-1}\ s^{-1}$

24.44 (b) $\frac{\partial c}{\partial t} = D\nabla^2 c$

24.45 (a) 104 min (b) 56.5 hr.

24.46 $7.09 \times 10^{-11}\ m^2\ s^{-1}$

24.48 5.41×10^{-4} cm

24.49 27.6×10^6 g mol^{-1}

24.50 0.33 cm

24.51 64,000 g mol^{-1}

PART FOUR

STRUCTURES

CHAPTER 25: Solid-State Chemistry

25.1 What is the equation for the distances between planes 1, 1, 0 for a crystal with mutually perpendicular axes?

SOLUTION

$$\frac{1}{d} = \left(\frac{1}{a^2} + \frac{1}{b^2}\right)^{1/2}$$

$$d = \left(\frac{1}{a^2} + \frac{1}{b^2}\right)^{-1/2} = \left(\frac{b^2 + a^2}{a^2 b^2}\right)^{-1/2} = \frac{ab}{\left(a^2 + b^2\right)^{1/2}}$$

25.2 Calculate the angles at which the first-, second-, and third-order reflections are obtained from planes 500 pm apart, using X rays with a wavelength of 100 pm.

SOLUTION

$$\sin\theta = \frac{\lambda}{2d_{hkl}} = \frac{\lambda}{2(d/n)}$$

$$\sin\theta_1 = \frac{100\ \text{pm}}{2(500\ \text{pm}/1)} \qquad \theta_1 = 5.74^\circ$$

$$\sin\theta_2 = \frac{100\ \text{pm}}{2(500\ \text{pm}/2)} \qquad \theta_2 = 11.54^\circ$$

$$\sin\theta_3 = \frac{100\ \text{pm}}{2(500\ \text{pm}/3)} \qquad \theta_3 = 17.46^\circ$$

25.3 The only metal that crystallizes in a primitive cubic lattice is polonium, which has unit cell side of 334.5 pm. What are the perpendicular distances between planes with indices (110), (111), (210) and (211)?

SOLUTION

$$d_{hkl} = \frac{a}{(h^2 + k^2 + l^2)^{1/2}}$$

$$d_{110} = 334.5 \text{ pm}/2^{1/2} = 236.5 \text{ pm}$$

$$d_{111} = 334.5 \text{ pm}/3^{1/2} = 193.1 \text{ pm}$$

$$d_{210} = 334.5 \text{ pm}/5^{1/2} = 149.6 \text{ pm}$$

$$d_{211} = 334.5 \text{ pm}/6^{1/2} = 136.6 \text{ pm}$$

25.4 Calculate the structure factor for a cubic unit cell of AB in which the B atoms occupy the body centered position. Which reflections will be strong and which weak?

SOLUTION

$$F_{hkl} = f_A + f_B e^{2\pi i(h/2 + k/2 + l/2)}$$

$$= f_A + f_B e^{\pi i(h + k + l)}$$

$$= f_A + f_B(-1)^{h + k + l}$$

If $h + k + l$ is even, $F_{hkl} = f_A + f_B$, strong reflections.

If $h + k + l$ is odd, $F_{hkl} = f_A - f_B$, weak reflections.

25.5 The crystal unit cell of magnesium oxide is a cube 420 pm on an edge. The structure is interpenetrating face centered. What is the density of crystalline MgO?

SOLUTION

$$\frac{4(40.32 \times 10^{-3}\ \text{kg mol}^{-1})}{(6.022 \times 10^{23}\ \text{mol}^{-1})(4.20 \times 10^{-10}\ \text{m})^3} = 3.615 \times 10^3\ \text{kg m}^{-3}$$

$$= 3.615\ \text{g cm}^{-3}$$

25.6 Tungsten forms body-centered cubic crystals. From the fact that the density of tungsten is 19.3 g cm^{-3}, calculate (a) the length of the side of this unit cell, and (b) d_{200}, d_{110}, and d_{222}.

SOLUTION

$$\text{(a)}\ 19.3 \times 10^3\ \text{kg m}^{-3} = \frac{2(183.85 \times 10^{-3}\ \text{kg mol}^{-1})}{(6.022\ 045 \times 10^{23}\ \text{mol}^{-1})a^3}$$

$$a = 316 \times 10^{-12}\ \text{m} = 316\ \text{pm}$$

$$\text{(b)}\ d_{200} = \frac{a}{\sqrt{2^2 + 0 + 0}} = \frac{316\ \text{pm}}{2} = 158\ \text{pm}$$

$$d_{110} = \frac{a}{\sqrt{1 + 1 + 0}} = \frac{316\ \text{pm}}{\sqrt{2}} = 223\ \text{pm}$$

$$d_{222} = \frac{a}{\sqrt{4 + 4 + 4}} = \frac{316\ \text{pm}}{\sqrt{12}} = 91.2\ \text{pm}$$

25.7 Copper forms cubic crystals. When an X-ray powder pattern of crystalline copper is taken using X rays from a copper target (the wavelength of the Kα line is 154.05 pm), reflections are found at $\theta = 21.65°$, $25.21°$, $44.96°$, $47.58°$, and other larger angles. (a) What type of lattice is formed by copper? (b) What is the length of a side of the unit cell at this temperature? (c) What is the density of copper?

SOLUTION

$$d_{hkl} = \frac{\lambda}{2 \sin \theta} = \frac{154.05 \text{ pm}}{2 \sin \theta} = \frac{a}{\sqrt{h^2 + k^2 + l^2}}$$

(a)

θ	d_{hkl}/pm	Ratio d_{hkl} to largest spacing
21.65°	208.77	1.0000
25.21°	180.84	0.8662
37.06°	127.81	0.6122
44.96°	109.00	0.5221
47.58°	104.34	0.4998

Ratios expected for cubic crystals

$$d_{hkl} = \frac{a}{\sqrt{h^2 + k^2 + l^2}}$$

	Primitive	**Body-centered**	**Face-centered**
100	1.0000	——	——
110	0.7071	1.0000	——
111	0.5774	——	1.0000
200	0.5000	0.7071	0.8660
210	0.4472	——	——
211	X	0.5774	——
220	X	0.5000	0.6124
310	X	0.4472	——
311	X	——	0.5222
222	X	X	0.5000

—— reflection absent

X not needed for this problem

Thus copper forms face-centered cubic crystals.

(b) $d_{111} = \dfrac{a}{\sqrt{1 + 1 + 1}} = 208.77 \text{ pm}$

$a = 361.6 \text{ pm}$

(c) $$d = \frac{4(63.546 \times 10^{-3}\ \text{kg mol}^{-1})}{(6.022\ 05 \times 10^{23}\ \text{mol}^{-1})(361.6 \times 10^{-12}\ \text{m}^3)^3}$$

$$= 0.8927 \times 10^3\ \text{kg m}^{-3}$$

25.8 (a) Metallic iron at 20 °C is studied by the Bragg method, in which the crystal is oriented so that a reflection is obtained from the planes parallel to the sides of the cubic crystal, then from planes cutting diagonally through opposite edges, and finally from planes cutting diagonally through opposite corners. Reflections are first obtained at $\theta = 11^\circ\ 36'$, $8^\circ\ 3'$, and $20^\circ\ 26'$, respectively. What type of cubic lattice does iron have at 20 °C? (b) Metallic iron also forms cubic crystals at 1100°C, but the reflections determined as described in (a) occur at $\theta = 9^\circ\ 8'$, $12^\circ\ 57'$, and $7^\circ\ 55'$, respectively. What type of cubic lattice does iron have at 1100 °C? (c) The density of iron at 20 °C is 7.86 g cm^{-3}. What is the length of a side of the unit cell at 20 °C? (d) What is the wavelength of the X rays used? (e) What is the density of iron at 1100 °C?

SOLUTION

(a)(b) The three orientations will give reflections from planes a00, bb0, and ccc, respectively where a, b and c are small integers. The smallest suitable integers are: primitive lattice: a=b=c=1; body-centered lattice: b=1, a=c=2 (100 and 111 reflections are missing); face-centered lattice: c=1, a=b=2.

$$\sin\theta = \frac{\lambda}{2d_{hkl}} = \frac{\lambda\sqrt{h^2 + k^2 + l^2}}{2a}$$

Since $\lambda/2a$ is constant for a particular experiment, the three lattice types can be distinguished simply on the basis of the relative magnitudes of the three θ's. Thus the crystal at 20° C is body-centered since the second angle is smaller than the first and the third is the largest of all (ratio of sin θ is $2 : \sqrt{2} : \sqrt{12}$). At 1100 °C the face-centered form is found (ratio of sin θ is $2 : \sqrt{8} : \sqrt{3}$). Note that a primitive lattice would give ratios of sin θ of $1 : \sqrt{2} : \sqrt{3}$).

(c) $d = 7.86 \times 10^3 \text{ kg m}^{-3} = \dfrac{2(55.847 \times 10^{-3} \text{ kg mol}^{-1})}{(6.022\,045 \times 10^{23} \text{ mol}^{-1})a^3}$

$$a = \left[\frac{(2)(55.847 \times 10^{-3} \text{ kg mol}^{-1})}{(6.022\,045 \times 10^{23} \text{ mol}^{-1})(7.86 \times 10^3 \text{ kg m}^{-3})}\right]^{1/3}$$

$$= 286.8 \times 10^{-12} \text{ m} = 286.8 \text{ pm}$$

(d) $\lambda = 2d_{hkl} \sin\theta = \dfrac{2a \sin\theta}{\sqrt{h^2 + k^2 + l^2}}$

$$\frac{573.6 \text{ pm}}{2} \sin 11^\circ\, 36' = 57.7 \text{ pm}$$

$$\frac{573.6 \text{ pm}}{\sqrt{2}} \sin 8^\circ\, 3' = 56.8 \text{ pm}$$

$$\frac{573.6 \text{ pm}}{\sqrt{12}} \sin 20^\circ\, 26' = 57.8 \text{ pm}$$

Average = 57.4 pm

(e) $d_{hkl} = \dfrac{\lambda}{2 \sin\theta}$

$$d_{200} = \frac{57.4 \text{ pm}}{2 \sin 9^\circ\, 8'} = 180.8 \text{ pm}; \quad a = d_{200}\sqrt{4} = 361.6 \text{ pm}$$

$$d_{220} = \frac{57.4 \text{ pm}}{2 \sin 12^\circ\, 57'} = 128.1 \text{ pm}; \quad a = d_{220}\sqrt{8} = 362.3 \text{ pm}$$

$$d_{111} = \frac{57.4 \text{ pm}}{2 \sin 7^\circ\, 55'} = 208.4 \text{ pm}; \quad a = d_{111}\sqrt{3} = 360.9 \text{ pm}$$

Average = 361.6 pm

$$d = \frac{4(55.847 \times 10^{-3} \text{ kg mol}^{-1})}{(6.022\ 045 \times 10^{23} \text{ mol}^{-1})(361.6 \times 10^{-12} \text{ m})^3}$$

$$= 7.846 \times 10^3 \text{ kg m}^{-3}$$

25.9 Cesium chloride, bromide, and iodide form interpenetrating simple cubic crystals instead of interpenetrating face-centered cubic crystals like the other alkali halides. The length of the side of the unit cell of CsCl is 412.1 pm. (a) What is the density? (b) Calculate the ion radius of Cs^+, assuming that the ions touch along a diagonal through the unit cell and that the ion radius of Cl^- is 181 pm.

<u>SOLUTION</u>

(a) $$d = \frac{(168.36 \times 10^{-3} \text{ kg mol}^{-1})}{(6.022\ 045 \times 10^{23} \text{ mol}^{-1})(412.1 \times 10^{-12} \text{ m})^3}$$

$$= 3.995 \times 10^3 \text{ kg m}^{-3}$$

(b) The diagonal plane through the cubic unit cell is as follows:

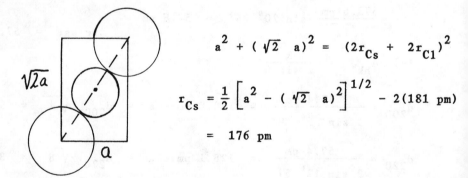

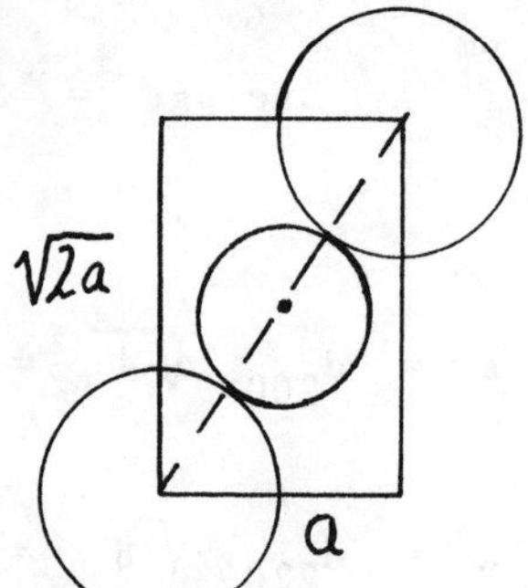

$$a^2 + (\sqrt{2}\ a)^2 = (2r_{Cs} + 2r_{Cl})^2$$

$$r_{Cs} = \frac{1}{2}\left[a^2 - (\sqrt{2}\ a)^2\right]^{1/2} - 2(181 \text{ pm})$$

$$= 176 \text{ pm}$$

25.10 The density of potassium chloride at 18 °C is 1.9893 g cm^{-3}, and the length of a side of the unit cell is 629.082 pm, as determined by X-ray diffraction. Calculate the Avogadro constant using the values of the relative atomic masses given in the front cover.

SOLUTION

$$1.9893 \times 10^{3}\ \text{kg m}^{-3} = \frac{4(74.551 \times 10^{-3}\ \text{kg mol}^{-1})}{N_A(629.082 \times 10^{-12}\ \text{m})^3}$$

$$N_A = 6.0213 \times 10^{23}\ \text{mol}^{-1}$$

25.11 Deslattes, et al., Phys. Rev. Lett., 33, 463 (1974) found the following values for a single crystal of very pure silicon at 25 °C: $\rho = 2.328\ 992\ \text{g cm}^{-3}$, $a = 543.1066$ pm. Silicon has a face-centered cubic lattice like diamond. The atomic mass is 28.085 41 g mol^{-1}. What value of Avogadro's constant is obtained from these values?

SOLUTION

$$\rho = \frac{nM}{N_A a^3}$$

$$N_A = \frac{nM}{\rho a^3} = \frac{(8)(28.085\ 41 \times 10^{-3}\ \text{kg mol}^{-1})}{(2.328\ 992 \times 10^{3}\ \text{kg m}^{-3})(543.106\ 6 \times 10^{-12}\ \text{m})^3}$$

$$= 6.022\ 093 \times 10^{23}\ \text{mol}^{-1}$$

25.12 Insulin forms crystals of the orthohombic type with unit-cell dimensions of 13.0 x 7.48 x 3.09 nm. If the density of the crystal is 1.315 g cm^{-3} and there are six insulin molecules per unit cell, what is the molar mass of the protein insulin?

SOLUTION

$$d = 1.315 \times 10^{3}\ \text{kg m}^{-3} = \frac{6M}{(6.022\ 045 \times 10^{23}\ \text{mol}^{-1})(13.0 \times 7.48 \times 3.09 \times 10^{-27}\ \text{m}^3)}$$

$$M = \frac{1}{6}\,(1.315 \times 10^{3}\ \text{kg m}^{-3})(6.022\ 045 \times 10^{23}\ \text{mol}^{-1})(13.0 \times 7.48 \times 3.09 \times 10^{-27}\ \text{m}^{3})$$

$$= 39.7\ \text{kg mol}^{-1} = 39{,}700\ \text{g mol}^{-1}$$

25.13 Molybdenum forms body-centered cubic crystals and, at 20 °C, the density is 10.3 g cm^{-3}. Calculate the distance between the centers of the nearest molybdenum atoms.

SOLUTION

$$10.3 \times 10^{3}\ \text{kg m}^{3} = \frac{2(95.94 \times 10^{-3}\ \text{kg mol}^{-1})}{(6.022\ 045 \times 10^{23}\ \text{mol}^{-1})a^{3}}$$

$$a = 314 \times 10^{-12}\ \text{m} = 314\ \text{pm}$$

Consider a plane bisecting the cube and cutting diagonally through opposite faces.

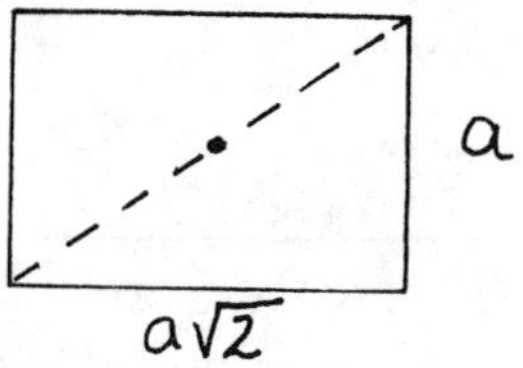

length of diagonal $= \sqrt{a + 2a} = a\sqrt{3}$

distance between atoms $= \dfrac{a\sqrt{3}}{2} = 273$ pm

25.14 Aluminum forms face-centered cubic crystals, and the length of the side of the unit cell is 405 pm at 25 °C. Calculate (a) the density of aluminum at this temperature, and (b) the distances between (200), (220), and (111) planes.

SOLUTION

(a) $$d = \frac{4(26.981\ 54 \times 10^{-3}\ \text{kg mol}^{-1})}{(6.022\ 045 \times 10^{23}\ \text{mol}^{-1})(405.0 \times 10^{-12}\ \text{m})^{3}}$$

$$= 2.698 \times 10^{3}\ \text{kg m}^{-3}$$

(b) $$d_{200} = \frac{a}{\sqrt{h^2 + k^2 + l^2}} = \frac{405.0 \text{ pm}}{2} = 202.5 \text{ pm}$$

$$d_{220} = \frac{a}{\sqrt{8}} = 143.2 \text{ pm}$$

$$d_{111} = \frac{a}{\sqrt{3}} = 233.8 \text{ pm}$$

25.15 Rutile (TiO_2) forms a primitive tetragonal lattice with a = 459.4 pm and c = 296.2 pm. There are two Ti atoms per unit cell, one at (000) and the other at (1/2 1/2 1/2). The four oxygen atoms are located at $\pm$ (uu0) and $\pm$ (1/2 + u, 1/2 − u, 1/2) with u = 0.305. What is the density of the crystal?

SOLUTION

$$V = (459.4 \text{ pm})^2(296.2 \text{ pm}) = 62.51 \times 10^{-30} \text{ m}^3$$

$$d = \frac{2(47.90 + 2 \times 15.99) \times 10^{-3} \text{ kg mol}^{-1}}{(6.022\ 045 \times 10^{23} \text{ mol}^{-1})(62.51 \times 10^{-30} \text{ m}^3)}$$

$$= 4.245 \times 10^3 \text{ kg m}^{-3}$$

25.16 The diamond has a face-centered cubic crystal lattice, and there are eight atoms in a unit cell. Its density is 3.51 g cm^{-3}. Calculate the first six angles at which reflections would be obtained using an X-ray beam of wavelength 71.2 pm.

SOLUTION

$$d = 3.51 \times 10^3 \text{ kg m}^{-3} = \frac{8(12.011 \times 10^{-3} \text{ kg mol}^{-1})}{(6.022\ 045 \times 10^{23} \text{ mol}^{-1})a^3}$$

$$a = 357 \times 10^{-12} \text{ m} = 357 \text{ pm}$$

$$\theta = \sin^{-1} \frac{\lambda}{2a} \sqrt{h^2 + k^2 + l^2} = \sin^{-1} \frac{71.2}{2(357)} \sqrt{h^2 + k^2 + l^2}$$

For face-centered cubic the first six reflections are:

hkl	θ
111	9.95°
200	11.50°
220	16.38°
311	19.31°
222	20.21°
400	23.51°

25.17 Calculate the density of diamond from the fact that it has a face-centered cubic structure with two atoms per lattice point and a unit cell edge of 356.7 pm.

SOLUTION

$$d = \frac{(4)(12)(12.011 \times 10^{-3} \text{ kg mol}^{-1})}{(6.022\ 045 \times 10^{23} \text{ mol}^{-1})(356.7 \times 10^{-12} \text{ m})^3}$$

$$= 3.516 \times 10^3 \text{ kg m}^{-3}$$

25.18 Tungsten has a body-centered cubic structure at room temperature. Since the density is 19.35 g cm^{-3} at 20 °C, what is the atomic radius?

SOLUTION

For a body-centered cubic crystal there are two atoms per unit cell.

$$d = 19.35 \times 10^3 \text{ kg m}^{-3} = \frac{2(183.85 \times 10^{-3} \text{ kg mol}^{-1})}{(6.022\ 045 \times 10^{23} \text{ mol}^{-1})a^3}$$

$$a = 316.0 \text{ pm}$$

$$R = \frac{\sqrt{3}}{4} a = \frac{\sqrt{3}}{4} (316.0 \text{ pm}) = 136.8 \text{ pm}$$

25.19 A close-packed structure of uniform spheres has a cubic unit cell with a side of 800 pm. What is the radius of the spherical molecule?

SOLUTION

Since the cubic close-packed structure is face-centered cubic, a face diagonal is equal to four radii.

$$\text{Length of face diagonal} = \sqrt{128 \times 10^4}$$

$$\text{Radius of sphere:} = \frac{\sqrt{128 \times 10^4}}{4} = 282.8 \text{ pm}$$

25.20 If spherical molecules of 500 pm radius are packed in cubic close packing, and body-centered cubic, what are the lengths of the sides of the cubic unit cells in the two cases?

SOLUTION

Cubic close packing (face-centered cubic)

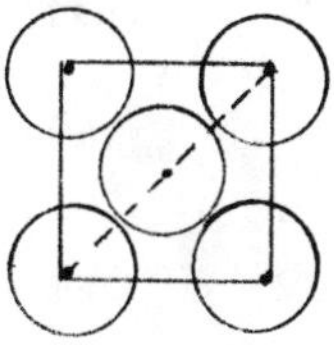

$$\text{Length of diagonal} = 2{,}000 \text{ pm}$$

$$\text{Length of side of unit cell} = \frac{2000}{\sqrt{2}} = 1414.2 \text{ pm}$$

Body-centered cubic

Consider a plane bisecting the cube and cutting diagonally through opposite faces.

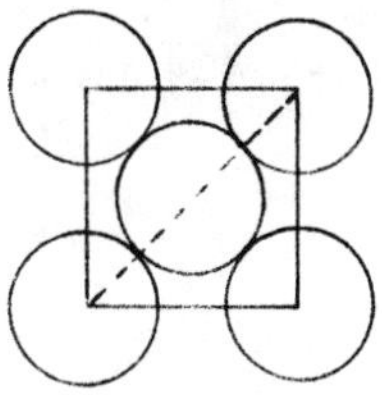

$$\text{Length of diagonal} = 2{,}000 \text{ pm}$$

$$2000^2 = a^2 + (\sqrt{2}\ a)^2 = 3a^2$$

$$a = \frac{2000^2}{\sqrt{3}} = 1154.7 \text{ pm}$$

25.21 Titanium forms hexagonal close-packed crystals. Given the atomic radius of 146 pm, what are the unit cell dimensions and what is the density of the crystal?

SOLUTION

$$a = b = 2(156\ \text{pm}) = 292\ \text{pm} \qquad c = \sqrt{2}\ \ 2\ a/\sqrt{3} = 477\ \text{pm}$$

$$V = a^2\ c(1 - \cos^2\gamma)^{1/2} = a^2\ c\ \sin\gamma$$

$$= (292\ \text{pm})^2(477\ \text{pm})\ \sin 120^\circ = 35.22 \times 10^{-30}\ \text{m}^3$$

$$d = \frac{2(47.90 \times 10^{-3}\ \text{kg mol}^{-1})}{(6.022\ 045 \times 10^{23}\ \text{mol}^{-1})(35.22 \times 10^{-30}\ \text{m}^3)}$$

$$= 4.517 \times 10^3\ \text{kg m}^{-3}$$

25.22 Metallic sodium forms a body-centered cubic unit cell with a = 424 pm. What is the sodium atom radius?

SOLUTION

Three atoms are in contact along the diagonal of the unit cell

$$(4R)^2 = (424\ \text{pm})^2 + (424\ \text{pm})^2 + (424\ \text{pm})^2$$

$$R = \frac{\sqrt{3}}{4}\ 424\ \text{pm} = 184\ \text{pm}$$

Alternatively,

$$1 = \left[a\ (x_2 - x_1)^2 + (y_2 - y_1)^2 + (z_2 - z_1)^2\right]^{1/2}$$

$$= 424\ \text{pm}\left[\frac{1}{4} + \frac{1}{4} + \frac{1}{4}\right]^{1/2} = \frac{\sqrt{3}}{4} = 424\ \text{pm} = 367\ \text{pm}$$

The sodium atom radius is half the distance.

$$\frac{367\ \text{pm}}{2} = 184\ \text{pm}$$

25.23 What neutron energy in electronvolts is required for a wavelength of 100 pm?

SOLUTION

$$\text{Kinetic energy} = \frac{p^2}{2m_n} = \frac{h^2}{\lambda^2 2m_n} \qquad \text{since } \lambda = h/p$$

$$Ee = \frac{h^2}{2\lambda^2 m_n}$$

$$E = \frac{h^2}{2\lambda^2 m_n e}$$

$$= \frac{(6.626 \times 10^{-34}\ \text{J s})^2}{2(10^{-10}\ \text{m})^2(1.675 \times 10^{-27}\ \text{kg})(1.602 \times 10^{-19}\ \text{C})}$$

$$= 0.0818\ \text{V} \quad \text{or} \quad 0.0818\ \text{eV}$$

25.24 A solution of carbon in face-centered cubic iron has a density of 8.105 g cm^{-3} and a unit cell edge of 358.3 pm. Are the carbon atoms interstitial, or do they substitute for iron atoms in the lattice? What is the weight percent carbon?

SOLUTION

$$d = \frac{4(55.847 \times 10^{-3}\ \text{kg mol}^{-1})}{(6.022\ 045 \times 10^{23}\ \text{mol}^{-1})(358.3 \times 10^{-12}\ \text{m})^3}$$

$$= 8.064 \times 10^3\ \text{kg m}^{-3}$$

Since the experimental density is greater, the carbon atoms must be interstitial.

$$\begin{array}{r} 8.105 \times 10^3\ \text{kg m}^{-3} \\ -8.064 \times 10^3 \\ \hline 0.041 \times 10^3 \end{array}$$

$$\frac{0.041 \times 10^3}{8.105 \times 10^3} 100 = 0.51\% \text{ by weight}$$

25.25 At 550 °C the conductivity of solid NaCl is $2 \times 10^{-4}\ \Omega^{-1}\ m^{-1}$. Since the sodium ions are smaller than the chloride ions (see Table 25.1), they are responsible for most of the electric conductivity. What is the ionic mobility of Na^+ under these conditions?

SOLUTION

Since there are 4 Na^+ per unit cell of $a = 564$ pm

$$N = \frac{4}{(564 \times 10^{-12}\ m)^3} = 2.23 \times 10^{28}\ m^{-3}$$

According to equation 25.38

$$u = \kappa/Nq = \frac{2 \times 10^{-4}\ \Omega^{-1}\ m^{-1}}{(2.23 \times 10^{28}\ m^{-3})(1.60 \times 10^{-19}\ C)}$$

$$= 5.61 \times 10^{-14} m\ V^{-1}\ s^{-1}$$

25.26 What fraction of the lattice sites of a crystal are vacant at 300 K if the energy required to move an atom from a lattice site in the crystal to a lattice site on the surface is 1 eV? At 100 K?

SOLUTION

At 300 K $\quad \frac{n}{N} = e^{-E_v/kT}$

$$= e^{-\frac{(1V)(1.602 \times 10^{-19} C)}{(1.38 \times 10^{-23}\ J\ K^{-1})(300\ K)}} = 1.56 \times 10^{-17}$$

At 1000 K $\quad \frac{n}{N} = e^{-\frac{(1V)(1.602 \times 10^{-19}\ C)}{(1.38 \times 10^{-23}\ J\ K^{-1}\ mol^{-1})(1000\ K)}}$

$$= 9.08 \times 10^{-6}$$

25.27 $n = 7$

25.28 (a) 6 (b) 8 (c) 12

25.29 When h + k is odd, the structure factor F(hkl) is zero and the reflection is extinguished.

25.30 (a) body-centered

(b) 314.8

(c) 10.21 $g\ cm^{-3}$

25.31 (a) 628 pm (b) 2.00 $g\ cm^{-3}$

25.32 392.4 pm

25.33 74.69 $g\ mol^{-1}$

25.34 4

25.35 2826 $kg\ m^{-3}$

25.36 (a) 2 (b) 328
(c) 164 (d) 232
(e) 94.7 nm

25.37 287,124 pm

25.38 2.703 $g\ cm^{-3}$

25.39 3.13 $g\ cm^{-3}$

25.40 $8.836 \times 10^3\ kg\ m^{-3}$

25.41 $2.33 \times 10^3\ kg\ m^{-3}$

25.42 12

25.43 152 pm

25.44 346 pm

25.45 (a) $\sqrt{8/3}\ a$ (b) $2\sqrt{2}\ r$ (c) $2r$

25.46	0.524
25.47	0.414
25.48	145.8 pm
25.49	490 $cm^2\ V^{-1}\ s^{-1}$
25.50	3.6×10^{-8} mole fraction

NOTES

NOTES

NOTES

NOTES

NOTES

NOTES

NOTES

NOTES

NOTES

NOTES

NOTES

NOTES